NANOMATERIALS AND NANOTECHNOLOGY FOR COMPOSITES

Design, Simulation, and Applications

NANOMATERIALS AND NANOTECHNOLOGY FOR COMPOSITES

Design, Simulation, and Applications

Edited by

**A. K. Haghi, PhD, Sabu Thomas, PhD,
Ali Pourhashemi, PhD, Abbas Hamrang, PhD, and
Ewa Kłodzińska, PhD**

Apple Academic Press Inc. | Apple Academic Press Inc.
3333 Mistwell Crescent | 9 Spinnaker Way
Oakville, ON L6L 0A2 | Waretown, NJ 08758
Canada | USA

©2015 by Apple Academic Press, Inc.

First issued in paperback 2021

Exclusive worldwide distribution by CRC Press, a member of Taylor & Francis Group
No claim to original U.S. Government works

ISBN 13: 978-1-77463-081-5 (pbk)
ISBN 13: 978-1-77188-065-7 (hbk)

Library of Congress Control Number: 2015935139

Library and Archives Canada Cataloguing in Publication

Nanomaterials and nanotechnology for composites: design, simulation, and applications /
edited by A.K. Haghi, PhD, Sabu Thomas, PhD, Ali Pourhashemi, PhD, Abbas Hamrang, PhD,
and Ewa Kłodzińska, PhD.

Includes bibliographical references and index.
ISBN 978-1-77188-065-7 (bound)
1. Nanocomposites (Materials). 2. Nanoparticles.
3. Nanotechnology. I. Hamrang, Abbas, editor II. Thomas, Sabu, editor III. Haghi, A. K.,
author, editor IV. Pourhashemi, Ali, editor
V. Kłodzińska, Ewa, editor

TA418.9.N35N27 2015   620.1'18   C2015-901600-2

Apple Academic Press also publishes its books in a variety of electronic formats. Some content that appears in print may not be available in electronic format. For information about Apple Academic Press products, visit our website at **www.appleacademicpress.com** and the CRC Press website at **www.crcpress.com**

ABOUT THE EDITORS

A. K. Haghi, PhD

A. K. Haghi, PhD, holds a BSc in urban and environmental engineering from the University of North Carolina (USA); a MSc in mechanical engineering from North Carolina A&T State University (USA); a DEA in applied mechanics, acoustics and materials from Université de Technologie de Compiègne (France); and a PhD in engineering sciences from Université de Franche-Comté (France). He is the author and editor of 165 books and has published 1000 research papers in various journals and conference proceedings. Dr. Haghi has received several grants, consulted for a number of major corporations, and is a frequent speaker to national and international audiences. Since 1983, he served as a professor at several universities. He is currently Editor-in-Chief of the *International Journal of Chemoinformatics and Chemical Engineering* and *Polymers Research Journal* and on the editorial boards of many international journals. He is a member of the Canadian Research and Development Center of Sciences and Cultures (CRDCSC), Montreal, Quebec, Canada.

Sabu Thomas, PhD

Dr. Sabu Thomas is the Director of the School of Chemical Sciences, Mahatma Gandhi University, Kottayam, India. He is also Professor of polymer science and engineering and the Director of the International and Inter University Centre for Nanoscience and Nanotechnology of the same university. He is a fellow of many professional bodies. Professor Thomas has authored or coauthored many papers in international peer-reviewed journals in the area of polymer processing. He has organized several international conferences and has more than 420 publications, 11 books, and two patents to his credit. He has been involved in a number of books both as author and editor. He is a reviewer to many international journals and has received many awards for his excellent work in polymer processing. His h Index is 42. Professor Thomas is listed as the 5th position in the list of Most Productive Researchers in India, in 2008.

Ali Pourhashemi, PhD

Ali Pourhashemi, PhD, is currently Professor of chemical and biochemical engineering at Christian Brothers University (CBU) in Memphis, Tennessee. He was formerly the Department Chair at CBU and also taught at Howard University in Washington, DC, USA. He taught various courses in chemical engineering, and his main area has been teaching the capstone process design as well as supervising industrial internship projects. He is a member of several professional organizations, including the *American Institute of Chemical Engineers*. He is on the international editorial review board of the *International Journal of Chemoinformatics and Chemical Engineering* and is an editorial member of the *International of Journal of Advanced Packaging Technology*. He has published many articles and presented at many professional conferences.

Abbas Hamrang, PhD

Abbas Hamrang, PhD, is Professor of polymer science and technology. He is currently a senior Polymer Consultant and Editor and member of the academic board of various international journals. His research interests include degradation studies of historical objects and archival materials, cellulose-based plastics, thermogravimetric analysis, accelerated aging process, and stabilization of polymers by chemical and nonchemical methods. His previous involvement in academic and industry sectors at international level include deputy vice-chancellor of research & development, senior lecturer, manufacturing consultant, science and technology advisor.

Ewa Kłodzińska, PhD

Ewa Kłodzińska holds a PhD from Nicolaus Copernicus University, Faculty of Chemistry in Torun, Poland. For 10 years, she has been doing research on determination and identification of microorganisms using the electromigration techniques for the purposes of medical diagnosis. Currently she is working at the Institute for Engineering of Polymer Materials and Dyes and investigates surface characteristics of biodegradable polymer material on the basis of zeta potential measurements. She has written several original articles, monographs, and chapters in books for graduate students and scientists. She has made valuable contributions to the theory and practice of electromigration techniques, chromatography, sample preparation, and application of separation science in pharmaceutical and medical analysis. Dr. Ewa Kłodzińska is a member of editorial board of *ISRN Analytical Chemistry* and the *International Journal of Chemoinformatics and Chemical Engineering (IJCCE)*.

CONTENTS

List of Contributors..*ix*

List of Abbreviations..*xiii*

List of Symbols..*xvii*

Preface..*xxi*

1. **Influence of a Strong Electric Field on the Electrical, Transport and Diffusion Properties of Carbon Nanostructures with Point Defects Structure**..1

   S. A. Sudorgin and N. G. Lebedev

2. **Carbon-Polymer Nanocomposites with Polyethylene as a Binder: A Research Note**..15

   Sergei V. Kolesov, Marina V. Bazunova, Elena I. Kulish, Denis R. Valiev, and Gennady E. Zaikov

3. **Application of Polymers Containing Silicon Nanoparticles as Effective UV Protectors**...25

   A. A. Olkhov, M. A. Goldshtrakh, and G. E. Zaikov

4. **Dynamically Vulcanized Thermoelastoplastics Based on Butadiene-Acrylonitrile Rubber and Polypropylene Modified Nanofiller**................39

   S. I. Volfson, G. E. Zaikov, N. A. Okhotina, A. I. Nigmatullina, O. A. Panfilova, and A. A. Nikiforov

5. **Sorption Active Carbon Polymer Nanocomposites: A Research Note**.....47

   Marina Bazunova, Rustam Tukhvatullin, Denis Valiev, Elena Kulish, and Gennadij Zaikov

6. **Modification of Physical and Mechanical Properties of Fiber Reinforced Concrete Using Nanoparticles**..59

   M. Mehdipour

7. **UV-Protective Nanocomposites Films Based on Polyethylene**.................73

   A. A. Olkhov, M. A. Goldshtrakh, G. Nyszko, K. Majewski, J. Pielichowski, and G. E. Zaikov

8. **Nanocomposite Foils Based On Silicate Precursor with Surface Active Compounds: A Research Note**...87

   L. O. Zaskokina, V. V. Osipova, Y. G. Galyametdinov, and A. A. Momzyakov

9. Experimental Investigation on the Effects of Nano Clay on Mechanical
    Properties of Aged Asphalt Mixture..93
    M. Arabani, A. K. Haghi, and R. Tanzadeh

10. Nanostructural Elements and the Molecular Mechanics and
    Dynamics Interactions: A Systematic Study...107
    A. V. Vakhrushev and A. M. Lipanov

11. A Study on Biological Application of Ag and Co/Ag Nanoparticles
    Cytotoxicity and Genotoxicity 139
    Iman E. Gomaa, Samarth Bhatt, Thomas Liehr, Mona Bakr, and Tarek A. El-Tayeb

12. The Transfer Variants of Metal/Carbon Nanocomposites Influence
    on Liquid Media: A Research Note 155
    V. I. Kodolov and V. V. Trineeva

13. Analysis of the Metal/Carbon Nanocomposites Surface Energy:
    A Research Note 163
    V. I. Kodolov and V. V. Trineeva

14. Nanopolymer Fibers: A Very Comprehensive Review in Axial and
    Coaxial Electrospinning Process...171
    Saeedeh Rafiei and A. K. Haghi

15. A Study on Polymer/Organoclay Nanocomposites 351
    K. S. Dibirova, G. V. Kozlov, and G. M. Magomedov

16. A Very Detailed Review on Application of Nanofibers in Energy
    and Environmental 359
    Saeedeh Rafiei, Babak Noroozi, and A. K. Haghi

    Index..419

LIST OF CONTRIBUTORS

M. Arabani
Professor, Faculty of Engineering, University of Guilan, Rasht, Postal code: 3756, I. R. Iran; Tel: +98(131)6690270; Fax: +98 (131) 6690270; E-mail: arabani@guilan.ac.ir

Mona Bakr
The National Institute for Laser Enhanced Sciences, Cairo University, Egypt

Marina V. Bazunova
Scientific Degree: The Candidate of the Chemical Sciences. Post: The Docent of the Department of High-Molecular Connections and General Chemical Technology of the Chemistry Faculty of the Bashkir State University; Official Add: 450076, Ufa, Zaks Validi Street, 32; Tel.: (347) 229-96-86; Mob: 89276388192; E-mail: mbazunova@mail.ru

Samarth Bhatt
Jena University Hospital, Friedrich Schiller University, Institute of Human Genetics, Kollegiengasse 10, D-07743 Jena, Germany

K. S. Dibirova
Dagestan State Pedagogical University, Makhachkala 367003, Yaragskii Street 57, Russian Federation

Tarek A. El-Tayeb
The National Institute for Laser Enhanced Sciences, Cairo University, Egypt

Y. G. Galyametdinov
Doctor of Chemical Sciences, Head of Department of Physical and Colloid Chemistry, KNRTU; E-mail: office@kstu.ru

M. A. Goldshtrakh
Military Institute of Chemistry and Radiometry, 105 Allea of General A. Chrusciela, 00-910 Warsaw, Poland, M. V. Lomonosov State University of Fine Chemical Technology, 119571 Moscow, Vernadskogo prosp 86

Iman E. Gomaa
German University in Cairo, Egypt, Main Entrance of Al-Tagamoa Al-Khames; E-mail: iman.gomaa@guc.edu.eg; Tel: +20-0100 2155053; Fax: +20-2-27590772
A. K. Haghi
Professor, Faculty of Engineering, University of Guilan, Rasht, Postal code: 3756, I. R. Iran; Tel: +98(131)6690270; Fax: +98 (131) 6690270; E-mail: Haghi@guilan.ac.ir

V. I. Kodolov
Basic Research High Educational Centre of Chemical Physics & Mesoscopy, Udmurt Scientific Centre, Ural Division; Russian Academy of Sciences, M. T. Kalashnikov Izhevsk State Technical University

Sergei V. Kolesov
Scientific Degree: The Doctor of the Chemical Sciences. Post: The Professor of the Department of High-Molecular Connections and General Chemical Technology of the Chemistry Faculty of the Bashkir State University; Official add: 450076 Ufa, Zaks Validi Street, 32; Tel.: (347) 229-96-86; E-mail: Kolesovservic@mail.Ru

G. V. Kozlov
Dagestan State Pedagogical University, Makhachkala 367003, Yaragskii Street 57, Russian Federation

Elena I. Kulish
Scientific Degree: The Doctor of the Chemical Sciences. Post: The Professor of the Department of High-Molecular Connections and General Chemical Technology of the Chemistry Faculty of the Bashkir State University. Official add: 450076, Ufa, Zaks Validi Street 32; Tel.: (347) 229-96-86; E-mail: Onlyalena@mail.Ru

N. G. Lebedev
Volgograd State University, Volgograd, Russia; E-mail: lebedev.ng@mail.ru

Thomas Liehr
Jena University Hospital, Friedrich Schiller University, Institute of Human Genetics, Kollegiengasse 10, D-07743 Jena, Germany

A. M. Lipanov
Institute of Mechanics, Ural Branch of the Russian Academy of Sciences, T. Baramsinoy 34, Izhevsk, Russia; E-mail: postmaster@ntm.udm.ru

G. M. Magomedov
Dagestan State Pedagogical University, Makhachkala 367003, Yaragskii Street 57, Russian Federation

K. Majewski
Military Institute of Chemistry and Radiometry, 105 Allea of General A. Chrusciela, 00-910 Warsaw, Poland; E-mail: K.Majewski@wichir.waw.pl

M. Mehdipour
Textile Engineering Department, Guilan University, Rasht, Iran

A. I. Nigmatullina
Chemistry and Processing Technology of Elastomers Department, Kazan National Research Technological University, 68 K Marks street, Kazan, Russia; E-mail: Chembio@sky.chph.ras.ru

A. A. Nikiforov
Chemistry and Processing Technology of Elastomers Department, Kazan National Research Technological University, 68K Marks street Kazan, Russia; E-mail: Chembio@sky.chph.ras.ru

Babak Noroozi
Department of Textile Engineering, University of Guilan, P. O. Box 41635-3756, Rasht, Iran

G. Nyszko
Military Institute of Chemistry and Radiometry, 105 Allea of General A. Chrusciela, 00-910 Warsaw, Poland, E-mail: Grzegorz.Nyszko@wichir.waw.pl

N. A. Okhotina
Chemistry and Processing Technology of Elastomers Department, Kazan National Research Technological University, 68 K Marks street Kazan, Russia; E-mail: Chembio@sky.chph.ras.ru

A. A. Olkhov
N. N. Semenov Institute of Chemical physics Russian Academy of Sciences, 119991 Moscow, street Kosygina, 4; E-mail: aolkhov72@yandex.ru

V. V. Osipova
Ph. D. Department of Physical and Colloid Chemistry, KNRTU

O. A. Panfilova

Chemistry and Processing Technology of Elastomers Department, Kazan National Research Technological University, 68K Marks street, Kazan, Russia, E-mail: Chembio@sky.chph.ras.ru

J. Pielichowski
Cracow University of Technology, Department of Polymer Science and Technology, Warszawska street, 31–155 Krakow, Poland, E-mail: Pielich@pk.edu.pl

Saeedeh Rafiei
Department of Textile Engineering, University of Guilan, P.O. Box 41635-3756, Rasht, Iran

S. A. Sudorgin
Volgograd State University, Volgograd, Russia, Volgograd State Technical University, Volgograd, Russia; E-mail: sergsud@mail.ru

R. Tanzadeh
Department of Civil Engineering University of Guilan, Rasht, Iran. Tel: +98(131)3229883; Fax: +98 (131) 3231116; E-mail: rashidtanzadeh@yahoo.com

V. V. Trineeva
Basic Research High Educational Centre of Chemical Physics & Mesoscopy, Udmurt Scientific Centre, Ural Division; Russian Academy of Sciences, Institute of Mechanics, Ural Division, Russian Academy of Sciences

Rustam Tukhvatullin
Bashkir State University, 32 Zaki Validi Street, 450076 Ufa, Republic of Bashkortostan, Russia

A. V. Vakhrushev
Institute of Mechanics, Ural Branch of the Russian Academy of Sciences, T. Baramsinoy 34, Izhevsk, Russia; E-mail: postmaster@ntm.udm.ru

Denis R. Valiev
Scientific Degree:-Post: The Student of the Department of High-Molecular Connections and General Chemical Technology of the Chemistry Faculty of the Bashkir State University Official add: 450076, Ufa, Zaks Validi Street, 32; Tel. Official: (347) 229-96-86; E-mail: valief@mail.ru

S. I. Volfson
Chemistry and Processing Technology of Elastomers Department, Kazan National Research Technological University, 68K Marks street, Kazan, Russia; E-mail: Chembio@sky.chph.ras.ru

Gennady E. Zaikov
Institute of Biochemical Physics named N. M. Emanuel of Russian Academy of Sciences Scientific Degree: The Doctor of the Chemical Sciences; Official add: 4 Kosygina Street, 119334, Moscow, Russia; E-mail: chembio@sky.chph.ras.ru

L. O. Zaskokina
Master of Department of Physical and Colloid Chemistry, KNRTU

LIST OF ABBREVIATIONS

AC	Activated Carbon
ACF	Activated Carbon Fibers
ACHF	Activated Carbon Hollow Fibers
ACNF	Activated Carbon Nanofiber
AFM	Atomic Force Microscope
AN	Acrylonitrile
BEM	Boundary Element Method
BET	Brunner-Emmett1-Teller
BGK	Bhatnagar-Gross-Krook
BJH	Barrett Joiner Halenda
BSA	Bovine Serum Albumin
CF	Carbon Fiber
CFD	Computational Fluid Dynamics
CH	Calcium Hydroxide
CNT	Carbon Nanotubes
CS	Cellulose
CVD	Chemical Vapor Deposition
DAAD	Deutsche Akademische Austausch Dienst
DDT	Dichlorodiphenyltrichloroethane
DMF	Dimethylformamide
DSB	Double Strand Breaks
EDLCs	Electrochemical Double-Layer Capacitors
EDXS	Energy Dispersive X-Ray Spectrometry
EELS	Electron Energy Loss Spectroscopy
EMEM	Eagle's Minimal Essential Medium
ERKs	Extracellular Signal-Regulated Kinases
FEM	Finite Element Methods
FRC	Fiber-Reinforced Concrete
FTIR	Fourier Transform Infrared
GUC	German University in Cairo
HDPE	High Density Polyethylene
HTS	High Temperature Shearing
HTT	Heat-Treatment Temperature
ISCN	International System for human Cytogenetic Nomenclature
ITZ	Interfacial Transition Zone

LBM	Lattice Boltzmann methods
LDPE	Low Density Polyethylene
LED	Light Emitting Diode
LERT	Large Electrical Relaxation Time Limit
LIBs	Lithium-Ion Batteries
MA	Maleine Anhydride
MD	Molecular Dynamics
MFCs	Microbial Fuel Cells
MMA	Methyl Methacrylate
MWNTs	Multiwall Carbon Nanotubes
NDT	Nottingham Device Test
NEMs	Nano-Electro-Mechanical Systems
NF	Nanofiltration
NILES	National Institute of Laser Enhanced Sciences
NPs	Nanoparticles
NS	Nano-SiO2
NT	Nano-TiO2
ODE's	Ordinary Differential Equations
PAA	Poly (Acrylic Acid)
PAN	Polyacrylonitrile
PBS	Phosphate Buffer Saline
PCN	Polymer Clay Nanocomposites
PDEs	Partial Differential Equations
PEM	Proton Exchange Mat
PGA	Poly(Glycolic Acid)
PHEV	Plug-In Hybrid Electric Vehicles
PLLA	Poly (L-Lactic Acid)
PMMA	Poly(Methyl Methacrylate)
POM	Polarization Optical Microscopy
PP	Polypropylene
PPX	Poly (P-Xylylene)
PTT	Photo Thermal Therapy
PVDC	Polyvinylidene Chloride
PZT	Plumbum Zirconate Titanate
RF	Radio-Frequency
RNP	Responsive Nanoparticle
RSM	Response Surface Methodology
RVE	Representative Volume Element
SAN	Styrene-Co-Acrylonitrile
SBS	Styrene-Butadiene-Styrene
SEI	Solid Electrolyte Interphase
SEM	Scanning Electron Microscope

SERT	Small Electrical Relaxation Time Limit
SMPE	Sulfophenyl Methallyl Ether
SSB	Single Strand Breaks
SSS	Sodium P-Styrene Sulfonate
SWNTs	Single Wall Carbon Nanotubes
TEM	Transmission Electron Microscopy
TFOT	Thin Film Oven Test
TG	Thermogravimetric Method
THF	Tetrahydrofuran
TLBM	Thermal Lattice Boltzmann methods
VA	Vinyl Acetate
VOCs	Volatile Organic Compounds

LIST OF SYMBOLS

A_m	amplitude of nanocomposite vibration
A_{ms}	coefficients of Fourier expansion
b	width of the Specimen
c	velocity of light
d	density of the adsorbate, g/cm^3
$<d>_V$	average arithmetic size
d_f	fractal dimension
F_{max}	maximal force
$\vec{F}_c$	principal vector of forces
$f_s(\mathbf{p,r})$	Fermi distribution function
$\left\|\vec{F}_{bi}\right\|$	force magnitude of the nanoparticle interaction
$\vec{F}_i(t)$	random set of forces at a given temperature
L	span length
$\bar{l}$	mean segment length
M	mass of the dried sample
$\vec{M}_c$	principal moment
Me (d)	median of distribution defining the size d
Mo (d)	position of maximum (a distribution mode)
m_i	mass of the ith atom
m	mass of adsorbed benzene
$m_{absorbed\ water}$	mass of the water
m_{av}	average mass of medium
m_{cl}	mass of metal containing phase
m_{sample}	mass of the sample
N	number of carbon atoms in lattice
N_{imp}	number of adsorbed hydrogen atoms
N_k	number of atoms forming each nanoparticle
n	number of interatomic interaction types
P	maximum indicated load
p_x	parallel component of the graphene sheet
q	charge of ion
R	flexural strength
S'	modulus of elasticity
S''	loss modulus
s	quasimomentum of the electrons in grapheme

$T(r)$ constant in the linear approximation in magnitude
U   constant of Coulomb repulsion of impurity
U_{trans} electromagnetic wave transmission velocity
V   hybridization potential
$\vec{V}_{i0}, \vec{V}_i$ initial and current velocities
V_{lj} matrix element of hybridization

GREEK CHARACTERS

α_{am} amorphous phase relative fraction
α_i friction coefficient in atomic structure
$c_{j\sigma}$ Fermi annihilation
$c_{j\sigma}^+$ creation operators of electrons
ε_{ab} absolute dielectric constant for medium
ε_{prog} nanocomposite surface energy portion
ε_s^{NC} surface energy of nanocomposite
$\varepsilon_{l\sigma}$ energy of electron by impurity
Ω_k area occupied by nanoelement
Φ_{cb} chemical bonds
Φ_{es} electrostatics
Φ_{pg} flat groups
Φ_{hb} hydrogen bonds
$\Phi(\vec{\rho}_{ij})$ potential depending on mutual positions of all atoms
$\vec{\rho}_{ij}$ radius vector determining the position of the ith atom relative to the jth atom
Φ_{ta} torsion angles
Φ_{va} valence angles
Φ_{vv} Vander Waals contacts
λ wave length
$n_{l\sigma}^d$ number of electrons on impurities
ρ medium density
$\vec{\rho}_{cj}$ vector connecting points c and j
g axial viscous normal stress
χ relative fraction of elastically deformed polymer
$y(j,l)$ length distribution
t_Δ electron hopping integral
U_ζ velocity vector of the interface
μ_{ab} absolute magnetic penetrability of medium
φ_m maximum potential of electromagnetic radiation of nanocomposite
$\vec{x}_{i0}, \vec{x}_i$ original and current coordinates

σ_Y	yield stress
v	Poisson's ratio
n_F	Flory exponent

PREFACE

Engineered nanopolymer and nanoparticles, with their extraordinary mechanical and unique electronic properties, have garnered much attention in the recent years. With a broad range of potential applications, including nanoelectronics, composites, chemical sensors, biosensors, microscopy, nanoelectromechanical systems, and many more, the scientific community is more motivated than ever to move beyond basic properties and explore the real issues associated with carbon nanotube-based applications.

Engineered nanopolymer and nanoparticles are exceptionally interesting from a fundamental research point of view. They open up new perspectives for various applications, such as nano-transistors in circuits, field-emission displays, artificial muscles, or added reinforcements in alloys. This text is an introduction to the physical concepts needed for investigating carbon nanotubes and other one-dimensional solid-state systems. Written for a wide scientific readership, each chapter consists of an instructive approach to the topic and sustainable ideas for solutions.

A large part of the research currently being conducted in the fields of materials science and engineering mechanics is devoted to Engineered nanopolymer and nanoparticles and their applications. In this process, modeling is a very attractive investigation tool due to the difficulties in manufacturing and testing of nanomaterials. Continuum modeling offers significant advantages over atomistic modeling. Furthermore, the lack of accuracy in continuum methods can be overtaken by incorporating input data either from experiments or atomistic methods. This book reviews the recent progress in application of Engineered nanopolymer and nanoparticles and their composites. The advantages and disadvantages of different methods are discussed. The ability of continuum methods to bridge different scales is emphasized. Recommendations for future research are given by focusing on what each method has to learn from the nano-scale. The scope of the book is to provide current knowledge aiming to support researchers entering the scientific area of carbon nanotubes to choose the appropriate modeling tool for accomplishing their study and place their efforts to further improve continuum methods.

CHAPTER 1

INFLUENCE OF A STRONG ELECTRIC FIELD ON THE ELECTRICAL, TRANSPORT AND DIFFUSION PROPERTIES OF CARBON NANOSTRUCTURES WITH POINT DEFECTS STRUCTURE

S. A. SUDORGIN[1,2] and N. G. LEBEDEV[1]

[1]Volgograd State University, Volgograd, Russia
[2]Volgograd State Technical University, Volgograd, Russia
E-mail: sergsud@mail.ru, lebedev.ng@mail.ru

CONTENTS

Abstract 2
1.1   Introduction 2
1.2   Model and Basic Relations 3
1.3   Results and Discussion 6
1.4   Conclusion 12
Keywords 12
References 12

ABSTRACT

Examines the influence of defects on the electrical properties of carbon nanostructures in an external electric field. Defects are the hydrogen atoms, which adsorbed on the surface of carbon nanostructures. Carbon nanostructures are considered the single-walled "zigzag" carbon nanotubes Atomic adsorption model of hydrogen on the surface of single-walled "zigzag" carbon nanotubes based on the single-impurity Anderson periodic model. Theoretical calculation of the electron diffusion coefficient and the conductivity of "zigzag" carbon nanotubes alloy hydrogen atoms carried out in the relaxation time approximation. Revealed a decrease in the electrical conductivity and the electron diffusion coefficient with increasing concentration of adsorbed hydrogen atoms. The nonlinearity of the electrical conductivity and the diffusion coefficient of the amplitude of a constant strong electric field at the constant concentration of hydrogen adatoms shown at the figures.

This work was supported by the Russian Foundation for Basic Research (grant № 13–03–97108, grant № 14–02–31801), and the Volgograd State University grant (project № 82–2013-a/VolGU).

1.1   INTRODUCTION

Despite the already long history of the discovery of carbon nanotubes (CNT) [1], the interest in the problem of obtaining carbon nanostructures with desired characteristics unabated, constantly improving their synthesis. Unique physical and chemical properties of CNTs can be applied in various fields of modern technology, electronics, materials science, chemistry and medicine [2]. One of the most important from the point of view of practical applications is the transport property of CNTs.

Under normal conditions, any solid surfaces coated with films of atoms or molecules adsorbed from the environment, or left on the surface in the diffusion process [3]. The most of elements adsorption on metals forms a chemical bond. The high reactivity of the surface of carbon nanotubes makes them an exception. Therefore, current interest is the study of the influence of the adsorption of atoms and various chemical elements and molecules on the electrical properties of carbon nanostructures.

In the theory of adsorption, in addition to the methods of quantum chemistry, widely used the method of model Hamiltonians [3]. In the study of the adsorption of atoms and molecules on metals used primarily molecular orbital approach – self-consistent field, as this takes into account the delocalization of electrons in the metal. Under this approach, the most commonly used model Hamiltonian Anderson [4, 5], originally proposed for the description of the electronic states of impurity atoms in the metal alloys. The model has been successfully applied to study the adsorption of atoms on the surface of metals and semiconductors [6], the adsorption of hydrogen on the surface of graphene [7] and carbon nanotubes [8, 9].

In this chapter, we consider the influence of the adsorption of atomic hydrogen on the conducting and diffusion properties of single-walled "zigzag" CNTs. Interaction of hydrogen atoms adsorbed to the surface of carbon nanotubes is described in terms of the periodic Anderson model. Since the geometry of the CNT determines their conductive properties, then to describe the adsorption on the surface of CNTs using this model is justified. Transport coefficients (conductivity and diffusion coefficient) CNT electron calculated by solving the Boltzmann equation [10] in the relaxation time approximation.

This technique was successfully applied by authors to calculate the ideal transport characteristics of carbon nanotubes [11], graphene bilayer graphene [12] and graphene nanoribbons [13].

1.2   MODEL AND BASIC RELATIONS

However, with the discovery of new forms of carbon model can be successfully applied to study of the statistical properties of CNTs and graphene. Carbon atom in the nanotube forms three chemical connection σ-type. Lodging with nearest neighbor atoms with three-sp^2 hybridization of atomic orbitals. The fourth p-orbital involved in chemical bonding π-type which creates π-shell nanotube describing state of itinerant electrons, that define the basic properties of CNTs and graphene. This allows us to consider the state of π-electron system in the framework of the Anderson model. The model takes into account the kinetic energy of electrons and their Coulomb interaction at one site and neglected energy inner-shell electrons of atoms and electrons involved in the formation of chemical bonds σ-type, as well as the vibrational energy of the atoms of the crystal lattice.

In general, the periodic Anderson model [5] considers two groups of electrons: itinerant s-electrons and localized d-electrons. Itinerant particles are considered free and localized – interact by Coulomb repulsion on a single node. With the discovery of new forms of carbon model can be successfully applied to study the statistical properties of carbon structures are the CNT and the graphene. Carbon atom in the graphene layer has three forms chemical bonds σ-type with its immediate neighbors. The fourth orbital p-type forms a chemical bond π-type, describing the state of itinerant electrons. States localized electrons created by the valence orbitals (in this case, the p-type) impurity atoms. This allows us to consider the state of π-electrons in the framework of the Anderson model. The model takes into account the kinetic energy of the electrons in the crystal and impurity electrons interacting through a potential hybridization, and neglects the energy of the electrons of the inner shells of atoms and electrons involved in the formation of chemical bonds σ-type, as well as the vibrational energy of the atoms of the crystal lattice [5].

In the periodic Anderson model state of the electrons of the crystal containing impurities in the π-electron approximation and the nearest neighbor approximation is described by the effective Hamiltonian, having the following standard form [5]:

$$H = \sum_{j,\Delta,\sigma} t_\Delta \left(c_{j\sigma}^+ c_{j+\Delta\sigma} + c_{j+\Delta\sigma}^+ c_{j\sigma} \right) + \sum_{l,\sigma} \varepsilon_{l\sigma} n_{l\sigma}^d + \sum_l U n_{l\uparrow}^d n_{l\downarrow}^d +$$
$$+ \sum_{l,j,\sigma} \left(V_{lj} c_{j\sigma}^+ d_{l\sigma} + V_{lj}^* d_{l\sigma}^+ c_{j\sigma} \right) \tag{1}$$

where t_Δ is the electron hopping integral between the neighboring lattice sites of the crystal; U is the constant of the Coulomb repulsion of the impurity; $c_{j\sigma}$ and $c_{j\sigma}^+$ are the Fermi annihilation and creation operators of electrons in the crystal node j with spin σ; $d_{j\sigma}$ and $d_{j\sigma}^+$ are the Fermi annihilation and creation operators of electrons on the impurities l with spin σ; $n_{l\sigma}^d$ is the operator of the number of electrons on impurities l with spin σ; $\varepsilon_{l\sigma}$ is the energy of the electron by the impurity l with spin σ; V_{lj} is the matrix element of hybridization of impurity electron l and atom j of the crystal.

After the transition to k-space by varying the crystal by Fourier transformation of creation and annihilation of electrons and crystal use the Green function method, the band structure of single-walled CNTs with impurities adsorbed hydrogen atoms takes the form [8, 9]:

$$E(\mathbf{k}) = \frac{1}{2} \left[\varepsilon_k + \varepsilon_{l\sigma} \pm \left(\left(\varepsilon_k - \varepsilon_{l\sigma} \right)^2 + 4 \frac{N_{imp}}{N} |V|^2 \right)^{\frac{1}{2}} \right], \tag{2}$$

where N – number of carbon atoms in the lattice, determines the size of the crystal, N_{imp} – the number of adsorbed hydrogen atoms, V – hybridization potential, $\varepsilon_{l\sigma} = -5.72$ eV – electron energy impurities – the band structure of an ideal single-walled nanotubes, for tubes, for example, "zigzag" type dispersion relation is defined as follows [1]:

$$E(\mathbf{p}) = \pm\gamma \sqrt{1 + 4\cos\left(ap_x\right)\cos\left(\pi s/n\right) + 4\cos^2\left(\pi s/n\right)} \tag{3}$$

where $a = 3d/2\hbar$, $d = 0.142$ nm is the distance between adjacent carbon atoms in graphene, $\mathbf{p} = (p_x, s)$ is the quasimomentum of the electrons in graphene, p_x is the parallel component of the graphene sheet of the quasimomentum and $s = 1, 2, ..., n$ are the quantization numbers of the momentum components depending on the width of the graphene ribbon. Different signs are related to the conductivity band and to the valence band accordingly.

Used in the calculation of the Hamiltonian parameters: the value of the hopping integral $t_0 = 2.7$ eV, hybridization potential $V = -1.43$ eV estimated from quantum chemical calculations of the electronic structure of CNTs within the semiempirical MNDO [14]. Electron energy impurity $\varepsilon_{l\sigma} = -5.72$ eV was assessed using the method described in Refs. [6, 7].

Consider the effect of the adsorption of atomic hydrogen on the response of single-walled "zigzag" CNTs to an external electric field applied along the x-axis is directed along the axis of the CNT (Fig. 1.1).

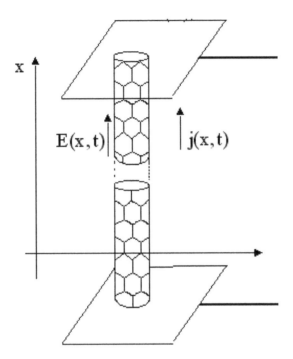

FIGURE 1.1 Geometry configuration. CNT type "zigzag" is in an external electric field. Field strength vector **E** is directed along the axis of the CNT.

Method of calculating the transport coefficients of electrons in carbon nanotubes described in detail in Refs. [11–13]. Evolution of the electronic system was simulated in the semiclassical approximation of the relaxation time. Electron distribution function in the state with momentum $p = (p_x, s)$ is of the t – approximation using Boltzmann equation [10]:

$$\frac{\partial f_s(\mathbf{p,r})}{\partial t} + \mathbf{F}\frac{\partial f_s(\mathbf{p,r})}{\partial \mathbf{p}} = \frac{f_s(\mathbf{p,r}) - f_{0s}(\mathbf{p,r})}{\tau}, \qquad (4)$$

where $f_s(\mathbf{p,r})$ – the Fermi distribution function $\mathbf{F} = e\mathbf{E}$ – acting on the particle constant electrostatic force.

To determine the dependence of the diffusion and conductive characteristics of CNTs on the external electric field using the procedure outlined in Ref. [15]. The longitudinal component of the current density $j = j_x$ has the following form:

$$j(x) = \sigma(\mathbf{E})\mathbf{E} + D(\mathbf{E})\frac{\nabla_x n}{n} \tag{5}$$

For the case of a homogeneous temperature distribution $T(r) = const$ in the linear approximation in magnitude [11], expressions for the transport coefficients of single-walled nanotubes: conductivity and diffusivity of electrons. Electrical conductivity of CNT type "zigzag" given following expression [11]:

$$\sigma(E) = \sum_s \int_{-\pi}^{\pi} \sum_m A_{ms} m f_{0s}(p_x, x) \frac{E}{E^2 m^2 + 1} [\sin(mp_x) + Em\cos(mp_x)] dp_x \tag{6}$$

Expression for the diffusion coefficient of electrons in CNT type "zigzag" has the form [11]:

$$D(E) = \sum_s \int_{-\pi}^{\pi} f_{0s}(p_x, x) \sum_m A_{ms} m \sum_{m'} A_{m's} m' \left\{ \frac{[E^2(m^2 + m'^2) + 1][EmR + M]}{K} + \right.$$

$$\left. + \frac{[E^3(m'^3 - 2m^2 m') + Em']T}{K} \right\} dp_x + \sum_s \int_{-\pi}^{\pi} f_{0s}(p_x, x) \sum_m A_{ms} m \sum_{m'} A_{m's} m' \frac{F}{P} dp_x, \tag{7}$$

where the following notation:

$K = [E^4(m^4 + m'^4 - 2m^2 m'^2) + 2E^2(m^2 + m'^2) + 1][E^2 m^2 + 1]$

$P = [E^2 m^2 + 1]^2 [E^2 m'^2 + 1]$

$R = \cos(mp_x)\sin(m'p_x) + \cos(mp_x)\cos(m'p_x) - \sin(mp_x)\sin(m'p_x)$

$M = \sin(mp_x)\sin(m'p_x) + \sin(mp_x)\cos(m'p_x) + \cos(mp_x)\sin(m'p_x)$

$T = [\cos(mp_x)\cos(m'p_x) - Em\sin(mp_x)\cos(m'p_x)]$

$F = [\sin(m'p_x) + Em\cos(m'p_x)][\sin(mp_x) + 2Em\cos(mp_x) - E^2 m^2 \sin(mp_x)]$

$A_{ms}, A_{m's}$ are the coefficients of the Fourier expansion of the dispersion relation of electrons in CNT, m and m' order Fourier series. For the convenience of visualization and qualitative analysis performed procedure and select the following dimensionless relative unit of measurement of the electric field E0 = 4.7×106 V/m.

1.3   RESULTS AND DISCUSSION

To investigate the influence of an external constant electric field on the transport properties of single-walled CNT type "zigzag" with adsorbed hydrogen atoms selected the following system parameters: temperature T ≈ 300 K, the relaxation time is $\tau \approx 10^{-12}$ s in accordance with the data [16]. For numerical analysis considered type semiconducting CNT (10,0).

It should be noted that a wide range of external field behavior of the specific conductivity σ(E) for nanotubes with hydrogen adatoms has the same qualitative nonlinear dependence as for the ideal case of nanoparticles, which was discussed in detail in Ref. [11]. In general, the dependence of conductivity on the electric field has a characteristic for semiconductors form tends to saturate and decreases monotonically with increasing intensity. This phenomenon is associated with an increase in electrons fill all possible states of the conduction band. Behavior of electrical conductivity under the influence of an external electric field is typical for semiconductor structures with periodic and limited dispersion law [17].

Figure 1.2 shows the dependence of conductivity σ(E) on the intensity of the external electric field E for ideal CNT (10,0) and CNT (10,0) with adsorbed hydrogen at relatively low fields. The graphs show that the addition of single adsorbed atom (adatom) hydrogen reduces the conductivity by a small amount (about 2×10^{-3} S/m). Lowering the conductivity of the hydrogen atom in the adsorption takes place due to the fact that one of the localized electron crystallite forms a chemical bond with the impurity atom and no longer participates in the charge transport by CNT.

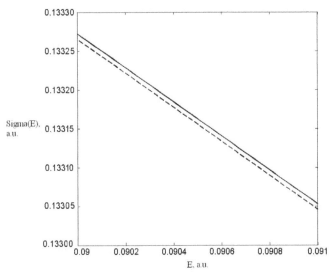

FIGURE 1.2 Dependence of the conductivity σ(E) on the magnitude of tension external electric field E: for ideal CNT (10,0) – solid line and the CNT (10,0) with hydrogen adatom – dashed line. x-axis is a dimensionless quantity of the external electric field E (unit corresponds to 4.7×10^6 V/m), the y-axis is dimensionless conductivity σ(E) (unit corresponds to 1.9×10^3 S/m).

Also analyzed the dependence of the conductivity $\sigma(E)$ on the intensity of the external electric E for CNT (10,0) type, containing different concentrations of hydrogen adatoms (Fig. 1.3). The increasing of the number of adsorbed atoms reduces the conductivity of "zigzag" CNT proportional to the number of localized adsorption bonds formed. When you add one hydrogen adatom conductivity of CNT type (10,0) is reduced by 0.06%, adding 100 adatoms by 0.55%, adding 300 adatoms by 1.66%, adding 500 adatoms by 2.62%.

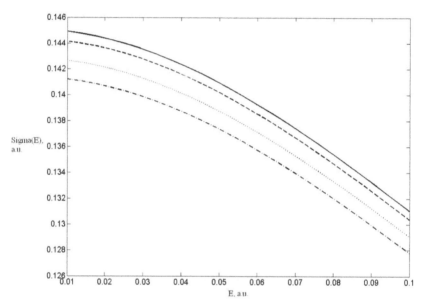

FIGURE 1.3 Dependence of the conductivity $\sigma(E)$ on the magnitude of tension E external electric impurity for CNT (10,0) one hydrogen adatom – solid line; 100 adatoms – dashed line; 300 adatoms – dotted line; 500 adatoms – dash-dot line. x-axis is a dimensionless quantity of the external electric field E (unit corresponds to 4.7×10^6 V/m), the y-axis is dimensionless conductivity $\sigma(E)$ (unit corresponds to 1.9×10^3 S/m).

Figure 1.4 shows that this behavior is typical for semiconductor conductivity of CNTs with different diameters. With the increasing diameter of the nanotubes have high electrical conductivity, since they contain a larger amount of electrons, which may participate in the transfer of electrical charge. The graphs in Fig. 1.4 shows for the (5,0), (10,0) and (20,0) CNT with the addition of 100 hydrogen adatoms.

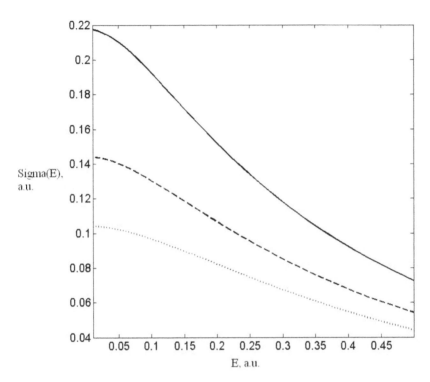

FIGURE 1.4 Dependence of the conductivity σ(E) on the magnitude of tension external electric E for different types of CNTs with the addition of hydrogen adatoms 100 (20,0) – solid line, (10,0) – dashed line; (5,0) – the dotted line. x-axis is a dimensionless quantity of the external electric field E (unit corresponds to 4.7×10^6 V/m), the y-axis is dimensionless conductivity σ(E) (unit corresponds to 1.9×10^3 S/m).

The electron diffusion coefficient $D(E)$ from the electric field in the single-walled "zigzag" CNT with adsorbed hydrogen atoms has a pronounced nonlinear character (Fig. 1.5). Increase of the field leads to an increase in first rate, and then to his descending to a stationary value. This phenomenon is observed for all systems with intermittent and limited electron dispersion law [17]. Electron diffusion coefficient can be considered constant in the order field amplitudes $E \approx 5 \times 10^6$ V/m. The maximum value of the diffusion coefficient for semiconductor CNTs observed at field strengths of the order of $E \approx 4.8 \times 10^5$ V/m.

When adding the adsorbed hydrogen atoms the electron diffusion coefficient, as well as the conductivity is reduced by 0.05% (Fig. 1.5). This behavior of the diffusion coefficient in an external electric field is observed for different concentrations of hydrogen adatoms (Fig. 1.6) and semiconductor CNTs with different diameters by adding 100 adatoms (Fig. 1.7).

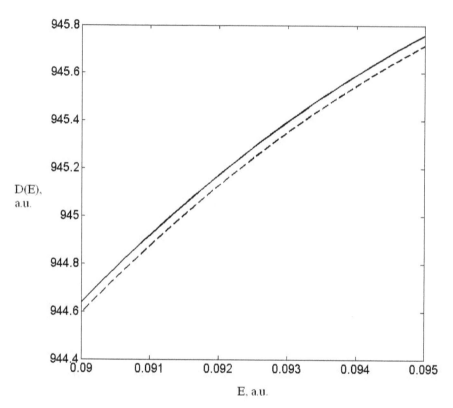

FIGURE 1.5   Dependence of the electron diffusion coefficient $D(E)$ on the intensity of the external electric field E: for CNT (10,0) ideal – solid line and hydrogen adatom – dashed line. x-axis is a dimensionless quantity of the external electric field E (unit corresponds to 4.7×10^6 V/m), the y-axis is a dimensionless diffusion coefficient $D(E)$ (unit corresponds to 3.5×10^2 A/m).

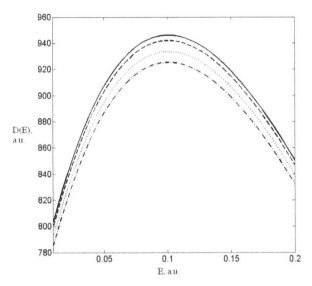

FIGURE 1.6   Dependence of the electron diffusion coefficient $D(E)$ on the intensity of the external electric E for impurity CNT (10,0) one hydrogen adatom – solid line; 100 adatoms – dashed line; 300 adatoms – dotted line; 500 adatoms – dash-dot line. x-axis is a dimensionless quantity of the external electric field E (unit corresponds to 4.7×10^6 V/m), the y-axis is a dimensionless diffusion coefficient $D(E)$ (unit corresponds to 3.5×10^2A/m).

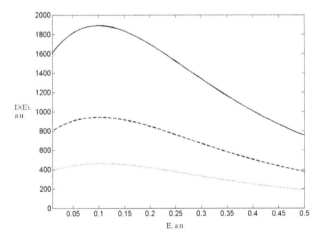

FIGURE 1.7   Dependence of the electron diffusion coefficient $D(E)$ on the intensity of the external electric E for different types of CNTs with the addition of hydrogen adatoms 100 (20,0) – solid line, (10,0) – dashed line; (5,0) – the dotted line. x-axis is a dimensionless quantity of the external electric field E (unit corresponds to 4.7×10^6 V/m), the y-axis is a dimensionless diffusion coefficient $D(E)$ (unit corresponds to 3.5×10^2A/m).

The presented results can be used for the preparation of carbon nanotubes with desired transport characteristics and to develop the microelectronic devices, which based on carbon nanoparticles.

1.4 CONCLUSION

We formulate the main results in the conclusion.
1. The method for theoretical calculation of the semiconducting "zigzag" CNT transport properties with adsorbed hydrogen atoms developed. Analytical expressions for the conductivity and the electron diffusion coefficient in "zigzag" CNT with hydrogen adatoms in the presence of an electric field.
2. Numerical calculations showed nonlinear dependence of the transport coefficients on the electric field. For strong fields coefficients tend to saturate.
3. Atomic hydrogen adsorption of the semiconducting "zigzag" CNT reduces their conductivity by several percent. The electron diffusion coefficient also decreases with increasing concentration of adsorbed hydrogen atoms, and a decrease of the diffusion coefficient is more pronounced than the decrease of electrical conductivity for each of the above types of semiconducting CNTs at a larger number of adatoms.
4. Transport properties of nanotubes with adatoms increases with the diameter. A physical explanation for the observed effect.

KEYWORDS

- **adsorption**
- **carbon nanostructures**
- **conductivity**
- **diffusion coefficient**
- **"zigzag" nanotubes**

REFERENCES

1. Diachkov, P. N. (2010). Electronic Properties and Applications of Nanotubes M. Bin, the Laboratory of Knowledge, 488p.
2. Roco, M. C., Williams, R. S., & Alivisatos, A. P. (2002). Nanotechnology in the Next Decade Forecast the Direction of Research Springer-Verlag, 292p.
3. Bolshov, L. A., Napartovich, A. P., Naumovets, A. G., & Fedorus, A. G. (1977). UFN T *122(1)*, S. 125.
4. Anderson, P. W. (1961). Phys Rev. *124*, 41.

5. Izyumov, A., Chashchin, I. I., & Alekseev, D. S. (2006). The Theory of Strongly Correlated Systems Generating functional method Moscow-Izhevsk: NITs "Regular and Chaotic Dynamics" 384.

6. Davydov, S. U., & Troshin, S. V. (2007). Phys. T. *49 (8),* S. 1508.

7. Davydov, S. Y., & Sabirov, G. I. (2010). Letters ZHTF T. 36*(24),* S. 77.

8. Pak, A. V., & Lebedev, N. G. (2012). Chemical Physics, *31(3),* 82–87.

9. Pak, A. V., & Lebedev, N. G. (2013). ZhFKh T. *87,* S 994.

10. Landau, L. D., & Lifshitz, E. M. (1979). Physical Kinetics M. Sci. Lit 528 c.

11. Belonenko, M. B., Lebedev, N. G., & Sudorgin, S. A. (2011). Phys. T. *53,* S. 1841.

12. Belonenko, M. B., Lebedev, N. G., & Sudorgin, S. A. (2012). Technical Physics, *82(7),* 129–133.

13. Sudorgin, S. A., Belonenko, M. B., & Lebedev, N. G. (2013). Physical Script, *87(1),* 015602(1–4).

14. Stepanov, N. F. (2001). Quantum Mechanics and Quantum Chemistry Springer, Verlag, 519 sec.

15. Buligin, A. S., Shmeliov, G. M., & Maglevannaya, I. I. (1999). Phys. T. *41,* S. 1314.

16. Maksimenko, S. A., & Slepyan, G. Ya. (2004). Nano Electromagnetic of Low-Dimensional Structure in Handbook of Nanotechnology Nanometer Structure: Theory, Modeling, and Simulation. Bellingham: SPIE Press, 145p.

17. Dykman, I. M., & Tomchuk, P. M. (1981). Transport Phenomena and Fluctuations in Semiconductors. Kiev. Science, Dumka, 320c.

CHAPTER 2

CARBON-POLYMER NANOCOMPOSITES WITH POLYETHYLENE AS A BINDER: A RESEARCH NOTE

SERGEI V. KOLESOV, MARINA V. BAZUNOVA, ELENA I. KULISH, DENIS R. VALIEV, and GENNADY E. ZAIKOV

CONTENTS

Abstract 16
2.1 Introduction..16
2.2 Experimental Part..17
2.3 Results and Discussion 18
2.4 Conclusions..22
Keywords 22
References...22

ABSTRACT

A new approach of obtaining of the molded composites on the basis of the mixtures of the powders of nano-dispersed polyethylene, cellulose and the ultra-dispersed carbonic materials is developed. These materials possess the assigned sorption properties and the physic-mechanical characteristics. They are suitable for the usage at the process of cleaning and separation of gas mixture.

2.1   INTRODUCTION

In solving problems of environmental protection, medicine, cleaning and drying processes of hydrocarbon gases are indispensable effective sorbents, including polymer nanocomposites derived from readily available raw materials.

The nature of the binder and the active components, and molding conditions are especially important at the process of sorption-active composites creating. These factors ultimately exert influence on the development of the porous structure of the sorbent particles and its performance. In this regard, it is promising to use powders of various functional materials having nanoscale particle sizes at the process of such composites creating. First, high degree of homogenization of the components facilitates their treatment process. Secondly, the high dispersibility of the particles allows them to provide a regular distribution in the matrix, whereby it is possible to achieve improved physical and mechanical properties. Third, it is possible to create the composites with necessary sorption, magnetic, dielectric and other special properties combining volumetric content of components [1].

Powders of low density polyethylene (LDPE) prepared by high temperature shearing (HTS) used as one of prospective components of the developing functional composite materials [2, 3].

Development of the preparation process and study of physicochemical and mechanical properties of sorbents based on powder mixtures of LDPE, cellulose (CS) and carbon materials are conducted. As the basic sorbent material new – ultrafine nanocarbon (NC) obtained by the oxidative condensation of methane at a treatment time of 50 min (NC1) and 40 min (NC2) having a specific surface area of 200 m^2/g and a particle size of 30–50 nm is selected [4]. Ultrafine form of NC may give rise to technological difficulties, for example, during regeneration of NC after using in gaseous environments, as well as during effective separation of the filtrate from the carbon dust particles. This imposes restrictions on the using of NC as an independent sorbent. In this connection, it should be included in a material that has a high porosity. LDPE and CS powders have great interest for the production of such material. It is known that a mixture of LDPE and CS powders have certain absorption properties, particularly, they were tested as sorbents for purification of water surface from petroleum and other hydrocarbons [5].

Thus, the choice of developing sorbents components is explained by the following reasons:

1. LDPE has a low softening point, allowing to conduct blanks molding at low temperatures. The very small size of the LDPE particles (60 to 150 nm) ensures regular distribution of the binder in the matrix. It is also important that the presence of binder in the composition is necessary for maintaining of the material's shape, size, and mechanical strength.

2. Usage of cellulose in the composite material is determined by features of its chemical structure and properties. CS has developed capillary-porous structure, that's why it has well-known sorption properties [5] towards polar liquids, gases and vapors.

3. Ultrafine carbon components [nanocarbon, activated carbon (AC)] are used as functionalizing addends due to their high specific surface area.

2.2   EXPERIMENTAL PART

Ultrafine powders of LDPE, CS and a mixture of LDPE/CS are obtained by high temperature shearing under simultaneous impact of high pressure and shear deformation in an extrusion type apparatus with a screw diameter of 32 mm [3].

Initial press-powders obtained by two ways. The first method is based on the mechanical mixing of ready LDPE, CS and carbon materials' powders. The second method is based on a preliminary high-shear joint grinding of LDPE pellets and sawdust in a specific ratio and mixing the resulting powder with the powdered activated carbon (БАУ-А mark) and the nanocarbon after it.

Composites molding held by thermobaric compression at the pressure of 127 kPa. Measuring of the tablets strength was carried out on the automatic catalysts strength measurer ПК-1.

The adsorption capacity (A) of the samples under static conditions for condensed water vapor, benzene, n-heptane determined by method of complete saturation of the sorbent by adsorbate vapor in standard conditions at 20°C [6] and calculated by the formula: $A = m/(M \cdot d)$, where m is the mass of the adsorbed benzene (acetone, n-heptane), g; M is the mass of the dried sample, g; d is the density of the adsorbate, g/cm^3.

Water absorption coefficient of polymeric carbon sorbents is defined by the formula: $K = \dfrac{m_{absorbed..water}}{m_{sample}} \times 100\%$, where $m_{absorbed water}$ is the mass of the water, retained by the sorbent sample, m_{sample} is the mass of the sample.

Experimental error does not exceed 5% in all weight methods at $P = 0.95$ and the number of repeated experiments $n = 3$.

2.3   RESULTS AND DISCUSSION

Powder components are used as raw materials for functional composite molding (including the binder LDPE), because molding of melt polymer mixtures with the active components has significant disadvantages. For example, the melt at high degrees of filling loses its fluidity, at low degrees of filling flow rate is maintained, but it is impossible to achieve the required material functionalization.

It is known that amorphous-crystalline polymers, which are typical heterogeneous systems, well exposed to high-temperature shear grinding process. For example, the process of HTS of LDPE almost always achieves a significant results [3]. Disperse composition is the most important feature of powders, obtained as result of high-temperature shear milling. Previously, on the basis of the conventional microscopic measurement, it was believed that sizes of LDPE powder particles obtained by HTS are within 6–30 micrometers. Electron microscopy gives the sizes of 60 to 150 nm. The active powder has a fairly high specific surface area (up to 2.2 m^2/g).

The results of measurement of the water absorption coefficient and of the static capacitance of LDPE powder by n-heptane vapor are equal to 12% and 0.26 cm^3/g, respectively. Therefore, the surface properties of LDPE powder more developed than the other polyethylene materials'.

Selection of molding conditions of sorbents based on mixtures of LDPE, CS and ultrafine carbon materials' powders.

Initial press-powders are obtained by two ways. The first method is based on the mechanical mixing of ready LDPE, CS and carbon materials' powders and the second method is based on a preliminary high-shear joint grinding of LDPE pellets and sawdust in a specific ratio and mixing the resulting powder with the powdered activated carbon and the nanocarbon after it. The method of molding – thermobaric pressing at a pressure of 127 kPa.

The mixture of LDPE/CS compacted into cylindrical pellets at a temperature of 115–145°C was used as a model mixture for selection of composites molding conditions. Pressing temperature should be such that the LDPE softens but not melts, and at the same time forms a matrix to prevent loss of specific surface area in the ready molded sorbent due to fusion of pores with the binder. The composites molded at a higher temperature, have a lower coefficient of water absorption than the tablets produced at a lower temperature, that's why the lowest pressing temperature (120°C) is selected. At a higher content of LDPE the water absorption coefficient markedly decreases with temperature.

Cellulose has a high degree of swelling in water (450%) [5], this may lead to the destruction of the pellets. Its contents in samples of composites, as it has been observed by the sorption of water, should not exceed 30 wt.%. There is a slight change of geometric dimensions of the pellets in aqueous medium at an optimal value of the water absorption coefficient when the LDPE content is 20 wt.%.

Samples of LDPE/CS with AC, which sorption properties are well studied, are tested for selecting of optimal content of ultrafine carbon. The samples containing more than 50 wt.% of AC have less water absorption coefficient values. Therefore, the total content of ultrafine carbon materials in all samples must be equal to 50 wt.%.

Static capacitance measurement of samples, obtained from mechanical mixtures of powders of PE, CS and AC, conducted on vapors of n-heptane and benzene, to determine the effect of the polymer matrix on the sorption properties of functionalizing additives. With a decrease of the content of AC in the samples with a fixed (20 wt.%) amount of the binder, reduction of vapor sorption occurs. It indicates that the AC does not lose its adsorption activity in the composition of investigated sorbents.

Strength of samples of sorbents (Fig. 2.1) is in the range of 620–750 N. The value of strength is achieved in the following molding conditions: t = 120°C and a pressure of 127 kPa.

Thus, optimal weight composition of the matrix of LDPE/CS composition – 20/30 wt.% with 50 wt.% containing of carbon materials.

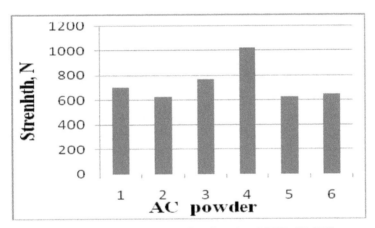

FIGURE 2.1   Comparison of strength of pellets, based on LDPE, CS (different species of wood) and AC powders [1 – sorbent of LDPE/AC/CS = 20/50/30 wt.% based on the powders of jointly dispersed pellets of LDPE and softwood sawdust with subsequently addition of AC; 2 – sorbent of LDPE/AC/CS = 20/50/30 wt.% based on the powders of jointly dispersed pellets of LDPE and hardwood sawdust with subsequently addition of AC; 3 – sorbent of LDPE/AC/CS = 20/50/30 wt.% based on the mechanical mixtures of the individual powders of LDPE, CS from softwood and AC; 4 – AC tablet; 5 – sorbent of LDPE/CS = 20/80 wt.%; 6 – sorbent of LDPE/AC = 20/80 wt.%].

Sorption properties of carbon – polymer composites by condensed vapors of volatile liquids

For a number of samples of sorbents static capacitance values by benzene vapor is identified (Fig. 2.2). They indicate that the molded mechanical mixture of 20/25/25/30 wt.% LDPE/AC/NC1/CS has a maximum adsorption capacity that greatly exceeds the capacity of activated carbon. High sorption capacity values by benzene vapor appears to be determined by weak specific interaction of π-electron system of the aromatic ring with carbocyclic carbon skeleton of the nanocarbon [7].

Static capacitance of obtained sorbents by heptane vapors significantly inferiors to capacity of activated carbon (Fig. 2.3), probably it is determined by the low polarizability of the molecules of low-molecular alkanes. Consequently, the investigated composites selectively absorb benzene and can be used for separation and purification of mixtures of hydrocarbons.

Molded composite based on a mechanical mixture of LDPE/AC/NC1/CS = 20/25/25/30 wt.% has a sorption capacity by acetone vapor comparable with the capacity of activated carbon (0.36 cm³/g) (Fig. 2.4).

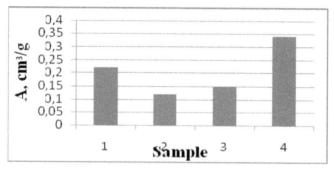

FIGURE 2.2   Static capacitance of sorbents, A (cm³/g) by benzene vapor (20°C) [1 –molded mechanical mixture of LDPE/AC/NC1/CS= 20/25/25/30wt.%; 2 – molded mechanical mixture of LDPE/AC/NC2/CS = 20/25/25/30 wt.%; 3 – molded mechanical mixture of LDPE/AC/CS=20/50/30 wt.%; 4 – AC medical tablet (controlling)].

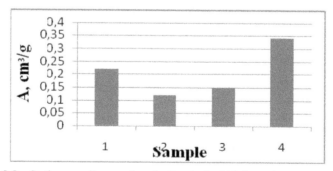

FIGURE 2.3   Static capacitance of sorbents, A (cm³/g) by n-heptane vapor (20°C). 1 – molded mechanical mixture of LDPE/AC/NC1/CS= 20/25/25/30wt.%; 2 – molded mechanical mixture of LDPE/AC/NC2/CS = 20/25/25/30 wt.%; 3 – molded mechanical mixture of PE/AC/CS=20/50/30 wt.%; 4 – AC medical tablet (controlling).

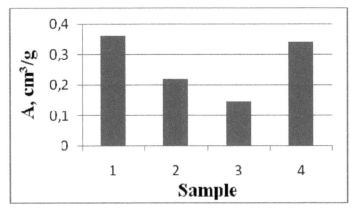

FIGURE 2.4 Static capacitance of sorbents, A (cm³/g) acetone vapor (20°C). 1 – molded mechanical mixture of LDPE/AC/NC1/CS= 20/25/25/30wt.%; 2 – molded mechanical mixture of LDPE/AC/NC2/CS = 20/25/25/30 wt.%; 3 – molded mechanical mixture of LDPE/AC/CS=20/50/30 wt.%; 4 – AC medical tablet (controlling).

Sorbents' samples containing NC2 have low values of static capacity by benzene, heptanes and acetone vapor. It can be probably associated with partial occlusion of carbon material pores by remnants of resinous substances – by products of oxidative condensation of methane, and insufficiently formed porous structure.

The residual benzene content measuring data (Table 2.1) shows that the minimal residual benzene content after its desorption from the pores at t = 70°C for 120 min observes in case of sorbent LDPE/AC/NC1/CS composition = 20/25/25/30 wt.%. It allows to conclude that developed sorbents have better ability to regenerate under these conditions in comparison with activated carbon.

TABLE 2.1 Sorbents' Characteristics: Total Pore Volume $V_{tot.}$; Static Capacitance (A) by Benzene Vapors at the Sorption Time of 2 days; Residual Weight of the Absorbed Benzene After Drying at t = 70°C for 120 min

LDPE/AC/NC/CS sorbent composition, wt.%	$V_{tot.}$, cm³/g	A, cm³/g	Residual benzene content as a result of desorption, %
20/25/25/30	1.54	0.5914	2.9
20/50/ – /30	1.21	0.1921	10.3
– /100/ –/–	1.60	0.3523	32.0

2.4   CONCLUSIONS

Thus, the usage of nanosized LDPE as a binder gives a possibility to get the molded composite materials with acceptable absorption properties. Optimal conditions for molding of sorbents on the basis of mixtures of powdered LDPE, cellulose and ultrafine carbon materials were determined: temperature 120°C and pressure of 127 kPa, content of the binder (polyethylene) is 20 wt.%.

Varying the ratio of the components of the compositions on the basis of ternary and quaternary mixtures of powdered LDPE, cellulose and ultrafine carbon materials it is possible to achieve the selectivity of sorption properties by vapors of certain volatile liquids. Established that molded mechanical mixture of LDPE/AC/NC1/CS 20/25/25/30wt.% has a static capacity by condensed vapors of benzene and acetone 0.6 cm³/g and 0.36 cm³/g, respectively, what exceeds the capacity of activated carbon. The static capacitance of the compositions by the n-heptane vapors is 0.21 cm³/g, therefore, the proposed composites are useful for separation and purification of gaseous and steam mixtures of different nature.

Developed production method of molded sorption-active composites based on ternary and quaternary mixtures of powdered LDPE, cellulose and ultrafine carbon materials can be easily designed by equipment and can be used for industrial production without significant changes.

KEYWORDS

- cellulose
- high-temperature shift crushing
- nano-carbon
- polyethylene
- sorbents

REFERENCES

1. Akbasheva, E. F., & Bazunova, M. V. (2010). Tableted sorbents based on cellulose powder mixtures, polyethylene and ultra-dispersed carbon. Materials Open School Conference of the CIS "Ultrafine and Nanostructured Materials" (11–15 October 2010), Ufa: Bashkir State University, 106.
2. Enikolopyan, N. S., Fridman, M. L., & Karmilov, A. Yu. (1987). Elastic Deformation Grinding of Thermo Plastic Polymers, Reports as USSR, *296(1)*, 134–138.
3. Akhmetkhanov, R. M., Minsker, K. S., & Zaikov, G. E. (2006). On the Mechanism of Fine Dispersion of Polymer Products at Elastic Deformation Effects. Plasticheskie Massi, *(8)*, 6–9.

4. Aleskovskiy, V. B., & Galitseisky, K. B. (20.11.2006). Russian Patent: "Method of Ultrafine Carbon", 2287543.
5. Raspopov, L. N., Russiyan, L. N., & Zlobinsky, Y. I. (2007). Waterproof Composites Comprising Wood and Polyethylene di Revkevich U. P. spersion Russian Polymer Science Journal, *50(3),* 547–552.
6. Keltsev, N. V. (1984). Fundamentals of Adsorption Technology Moscow: Chemistry, 595p.
7. Valinurova, E. R., Kadyrov, A. D., & Kudasheva, F. H. (2008). Adsorption Properties of Carbon Rayon Vestn Bashkirs Universal, *13(4)*, 907–910.

CHAPTER 3

APPLICATION OF POLYMERS CONTAINING SILICON NANOPARTICLES AS EFFECTIVE UV PROTECTORS

A. A. OLKHOV[1], M. A. GOLDSHTRAKH[1], and G. E. ZAIKOV[2]

[1]Moscow M. V. Lomonosov State University of Fine Chemical Technology, 119571 Moscow, Vernadskogo prosp. 86. E-mail: aolkhov72@yandex.ru

[2]N. M. Emanuel Institute of Biochemical physics Russian Academy of Sciences 119991 Moscow, Street Kosygina, 4

CONTENTS

Abstract 26
3.1 Introduction 26
3.2 Objects of Research 28
3.3 Experimental Methods 29
3.4 Results and Discussion 30
Keywords 37
References 37

ABSTRACT

High-strength polyethylene films containing 0.5–1.0 wt.% of nanocrystalline silicon (nc-Si) were synthesized. Samples of nc-Si with an average core diameter of 7–10 nm were produced by plasmochemical method and by laser-induced decomposition of monosilane. Spectral studies revealed almost complete (up to ~95%) absorption of UV radiation in 200–400 nm spectral region by 85 micron thick film if the nc-Si content approaches to 1.0 wt.%. The density function of particle size in the starting powders and polymer films containing immobilized silicon nanocrystallites were obtained using the modeling a complete profile of X-ray diffraction patterns, assuming spherical grains and the lognormal distribution. The results of X-ray analysis shown that the crystallite size distribution function remains almost unchanged and the crystallinity of the original polymer increases to about 10% with the implantation of the initial nc-Si samples in the polymer matrix.

3.1   INTRODUCTION

In recent years, considerable efforts have been devoted for search new functional nanocomposite materials with unique properties that are lacking in their traditional analogs. Control of these properties is an important fundamental problem. The use of nanocrystals as one of the elements of a polymer composite opens up new possibilities for targeted modification of its optical properties because of a strong dependence of the electronic structure of nanocrystals on their sizes and geometric shapes. An increase in the number of nanocrystals in the bulk of composites is expected to enhance long-range correlation effects on their properties. Among the known nanocrystals, nanocrystalline silicon (nc-Si) attracts high attention due to its extraordinary optoelectronic properties and manifestation of quantum size effects. Therefore, it is widely used for designing new generation functional materials for nanoelectronics and information technologies. The use of nc-Si in polymer composites calls for a knowledge of the processes of its interaction with polymeric media. Solid nanoparticles can be combined into aggregates (clusters), and, when the percolation threshold is achieved, a continuous cluster is formed.

An orderly arrangement of interacting nanocrystals in a long-range potential minimum leads to formation of periodic structures. Because of the well-developed interface, an important role in such systems belongs to adsorption processes, which are determined by the structure of the nanocrystal surface. In a polymer medium, nanocrystals are surrounded by an adsorption layer consisting of polymer, which may change the electronic properties of the nanocrystals. The structure of the adsorption layer has an effect on the processes of self-organization of solid-phase particles, as well as on the size, shape, and optical properties of resulting aggregates. According to data obtained for metallic [1] and semiconducting [2] clusters, aggregation and adsorption in three-phase systems with nanocrystals have an effect on the optical properties of the whole system. In this context, it is important to reveal the structural features of systems containing nanocrystals, characterizing aggregation and adsorption processes in these systems, which will make it possible to establish a correlation between the structural and the optical properties of functional nano-composite systems.

Silicon nanoclusters embedded in various transparent media are a new, inter-esting object for physicochemical investigation. For example, for particles smaller than 4 nm in size, quantum size effects become significant. It makes possible to control the luminescence and absorption characteristics of materials based on such particles using of these effects [3, 4]. For nanoparticles about 10 nm in size or larg-er (containing $\sim 10^4$ Si atoms), the absorption characteristics in the UV and visible ranges are determined in many respects by properties typical of massive crystalline or amorphous silicon samples. These characteristics depend on a number of fac-tors: the presence of structural defects and impurities, the phase state, etc. [5, 6]. For effective practical application and creation on a basis nc-Si the new polymeric materials possessing useful properties: sun-protection films [7] and the coverings [8] photoluminescent and electroluminescent composites [9, 10], stable to light dyes [11], embedding of these nanosized particles in polymeric matrixes becomes an important synthetic problem.

The method of manufacture of silicon nanoparticles in the form of a powder by plasma chemical deposition, which was used in this study, makes possible to vary the chemical composition of their surface layers. As a result, another possibility of controlling their spectral characteristics arises, which is absent in conventional methods of manufacture of nanocrystalline silicon in solid matrices (e.g., in α-SiO$_2$) by implantation of charged silicon particles [5] or radio frequency deposition of silicon [2]. Polymer composites based on silicon nanopowder are a new object for comprehensive spectral investigation. At the same time, detailed spectral analysis has been performed for silicon nanopowder prepared by laser-induced decompo-sition of gaseous SiH$_4$ (see, for example, [6, 12]). It is of interest to consider the possibility of designing new effective UV protectors based on polymer containing silicon nanoparticles [13]. An advantage of this nanocomposite in comparison with other known UV protectors is its environmental safety, that is, ability to hinder the formation of biologically harmful compounds during UV-induced degradation of components of commercial materials. In addition, changing the size distribution of nanoparticles and their concentration in a polymer and correspondingly modifying the state of their surface, one can deliberately change the spectral characteristics of nanocomposite as a whole. In this case, it is necessary to minimize the transmission in the wavelength range below 400 nm (which determines the properties of UV-protectors [13]) by changing the characteristics of the silicon powder.

3.2   OBJECTS OF RESEARCH

In this study, the possibilities of using polymers containing silicon nanoparticles as effective UV protectors are considered. First, the structure of nc-Si obtained under different conditions and its aggregates, their adsorption and optical properties was studied in order to find ways of control the UV spectral characteristics of multiphase polymer composites containing nanocrystalline silicon. Also, the purpose of this work was to investigate the effect of the concentration of silicon nanoparticles embedded in polymer matrix and the methods of preparation of these nanoparticles on the spectral characteristics of such nanocomposites. On the basis of the data obtained, recommendations for designing UV protectors based on these nanocomposites were formulated.

nc-Si consists of core–shell nanoparticles in which the core is crystalline silicon coated with a shell formed in the course of passivation of nc-Si with oxygen and/or nitrogen. nc-Si samples were synthesized by an original procedure in an argon plasma in a closed gas loop. To do this, we used a plasma vaporizer/condenser operating in a low-frequency arc discharge. A special consideration was given to the formation of a nanocrystalline core of specified size. The initial reagent was a silicon powder, which was fed into a reactor with a gas flow from a dosing pump. In the reactor, the powder vaporized at 7000–10,000°C. At the outlet of the high-temperature plasma zone, the resulting gas–vapor mixture was sharply cooled by gas jets, which resulted in condensation of silicon vapor to form an aerosol. The synthesis of nc-Si in a low-frequency arc discharge was described in detail in Ref. [3].

The microstructure of nc-Si was studied by transmission electron microscopy (TEM) on a Philips NED microscope. X-ray powder diffraction analysis was carried out on a Shimadzu Lab XRD-6000 diffractometer. The degree of crystallinity of nc-Si was calculated from the integrated intensity of the most characteristic peak at $2\theta = 28°$. Low-temperature adsorption isotherms at 77.3 K were measured with a Gravimat-4303 automated vacuum adsorption apparatus. FTIR spectra were recorded on in the region of 400–5000 cm^{-1} with resolution of about 1 cm^{-1}.

Three samples of nc-Si powders with specific surfaces of 55, 60, and 110 m^2/g were studied. The D values for these samples calculated by Eq. (2) are 1.71, 1.85, and 1.95, respectively; that is, they are lower than the limiting values for rough objects. The corresponding D values calculated by Eq. (3) are 2.57, 2.62, and 2.65, respectively. Hence, the adsorption of nitrogen on nc-Si at 77.3 K is determined by capillary forces acting at the liquid–gas interface. Thus, in argon plasma with addition of oxygen or nitrogen, ultra disperse silicon particles are formed, which consist of a crystalline core coated with a silicon oxide or oxynitride shell. This shell prevents the degradation or uncontrollable transformation of the electronic properties of nc-Si upon its integration into polymer media. Solid structural elements (threads or nanowires) are structurally similar, which stimulates self-organization leading to fractal clusters. The surface fractal dimension of the clusters determined from the nitrogen adsorption isotherm at 77.3 K is a structurally sensitive parameter, which characterizes both the structure of clusters and the morphology of particles and aggregates of nanocrystalline silicon.

As the origin materials for preparation film nanocomposites served polyethylene of low density (LDPE) marks 10803-020 and ultra disperse crystal silicon. Silicon powders have been received by a method plazmochemical recondensation of coarse-crystalline silicon in nanocrystalline powder. Synthesis nc-Si was carried out in argon plasma in the closed gas cycle in the plasma evaporator the condenser working in the arc low-frequency category. After particle synthesis nc-Si were exposed microcapsulating at which on their surfaces the protective cover from SiO_2, protecting a powder from atmospheric influence and doing it steady was created at storage. In the given work powders of silicon from two parties were used: nc-Si-36 with a specific surface of particles ~36 m^2/g and nc-Si-97 with a specific surface ~97 m^2/g.

Preliminary mixture of polyethylene with a powder nc-Si firms "Brabender" (Germany) carried out by means of closed hummer chambers at temperature 135±5°C, within 10 min and speed of rotation of a rotor of 100 min^{-1}. Two compositions LDPE + nc-Si have been prepared: (1) composition PE + 0.5% ncSi-97 on a basis ncSi-97, containing 0.5 weights silicon %; (2) composition PE + 1% ncSi-36 on a basis ncSi-36, containing 1.0 weights silicon %.

Formation of films by thickness 85±5 micron was spent on semiindustrial extrusion unit ARP-20–150 (Russia) for producing the sleeve film. The temperature was 120–190°C on zones extruder and extrusion die. The speed of auger was 120 min^{-1}. Technological parameters of the nanocomposites choose, proceeding from conditions of thermostability and the characteristic viscosity recommended for processing polymer melting.

3.3   EXPERIMENTAL METHODS

Mechanical properties and an optical transparency of polymer films, their phase structure and crystallinity, and also communication of mechanical and optical properties with a microstructure of polyethylene and granulometric structure of modifying powders nc-Si were observed.

Physicomechanical properties of films at a stretching (extrusion) measured in a direction by means of universal tensile machine EZ-40 (Germany) in accordance with Russian State Standard GOST-14236-71. Tests are spent on rectangular samples in width of 10 mm, and a working site of 50 mm. The speed of movement of a clip was 240 mm/min. The 5 parallel samples were tested.

Optical transparency of films was estimated on absorption spectra. Spectra of absorption of the obtained films were measured on spectrophotometer SF-104 (Russia) in a range of wavelengths 200–800 nanometers. Samples of films of polyethylene and composite films PE + 0.5% ncSi-36 and PE + 1% ncSi-36 in the size 3×3 cm were investigated. The special holder was used for maintenance uniform a film tension.

X-ray diffraction analysis by wide-angle scattering of monochromatic X-rays data was applied for research phase structure of materials, degree of crystallinity of a polymeric matrix, the size of single-crystal blocks in powders nc-Si and in a polymeric matrix, and also functions of density of distribution of the size crystalline particles in initial powders nc-Si

X-ray diffraction measurements were observed on Guinier diffractometer: chamber G670 Huber [14] with bent Ge (111) monochromator of a primary beam which are cutting out line $K\alpha_1$ (length of wave $l = 1.5405981$ Å) characteristic radiation of X-ray tube with the copper anode. The diffraction picture in a range of corners 2q from 3° to 100° was registered by the plate with optical memory (IP-detector) of the camera bent on a circle. Measurements were spent on original powders nc-Si-36 and nc-Si-97, on the pure film LDPE further marked as PE, and on composite films PE + 0.5% ncSi-97 and PE + 1.0% ncSi-36. For elimination of tool distortions effect diffractogram standard SRM660a NIST from the crystal powder LaB_6 certificated for these purposes by Institute of standards of the USA was measured. Further it was used as diffractometer tool function.

Samples of initial powders ncSi-36 and ncSi-97 for X-ray diffraction measurements were prepared by drawing of a thin layer of a powder on a substrate from a special film in the thickness 6 microns (MYLAR, Chemplex Industries Inc., Cat. No: 250, Lot No: 011671). Film samples LDPE and its composites were established in the diffractometer holder without any substrate, but for minimization of structure effect two layers of a film focused by directions extrusion perpendicular each other were used.

Phase analysis and granulometric analysis was spent by interpretation of the X-ray diffraction data. For these purposes the two different full-crest analysis methods [15, 16] were applied: (1) method of approximation of a profile diffractogram using analytical functions, polynoms and splines with diffractogram decomposition on making parts; (2) method of diffractogram modeling on the basis of physical principles of scattering of X-rays. The package of computer programs WinXPOW was applied to approximation and profile decomposition diffractogram ver. 2.02 (Stoe, Germany) [17], and diffractogram modeling at the analysis of distribution of particles in the sizes was spent by means of program PM2K (version 2009) [18].

3.4   RESULTS AND DISCUSSION

Results of mechanical tests of the prepared materials are presented to Table 3.1 from which it is visible that additives of particles nc-Si have improved mechanical characteristics of polyethylene.

TABLE 3.1   Mechanical characteristics of nanocomposite films based of LDPE and nc-Si.

Sample	Tensile strength, kg/cm²	Relative elongation-at-break, %
PE	100 ± 12	200–450
PE + 1% ncSi-36	122 ± 12	250–390
PE + 0.5% ncSi-97	118 ± 12	380–500

The results presented in the table show that additives of powders of silicon raise mechanical characteristics of films, and the effect of improvement of mechanical properties is more expressed in case of composite PE + 0.5% ncSi-97 at which in comparison with pure polyethylene relative elongation-at-break has essentially grown.

Transmittance spectra of the investigated films are shown on Fig. 3.1.

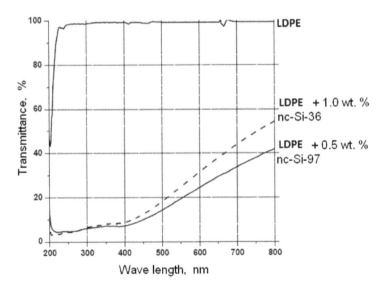

FIGURE 3.1   Transmittance spectra of the investigated films LDPE and nanocomposite films PE + 0.5% ncSi-97 and PE + 1.0% ncSi-36.

It is visible that additives of powders nc-Si reduce a transparency of films in all investigated range of wavelengths, but especially strong decrease transmittance (almost in 20 times) is observed in a range of lengths of waves of 220–400 nanometers, that is, in UV areas.

The wide-angle scattering of X-rays data were used for the observing phase structure of materials and their component. Measured X-ray diffractograms of initial powders ncSi-36 and ncSi-97 on intensity and Bragg peaks position completely corresponded to a phase of pure crystal silicon (a cubic elementary cell of type of diamond – spatial group $Fd\overline{3}m$, cell parameter $a_{\text{Si}} = 0.5435$ nanometers).

For the present research granulometric structure of initial powders nc-Si is of interest. Density function of particle size in a powder was restored on X-ray diffractogram a powder by means of computer program PM2K [18] in which the method [19] modeling's of a full profile diffractogram based on the theory of physical processes of diffraction of X-rays is realized. Modeling was spent in the assumption of the spherical form of crystalline particles and logarithmically normal distributions of their sizes. Deformation effects from flat and linear defects of a crystal lattice were considered. Received function of density of distribution of the size crystalline particles for initial powders nc-Si are represented graphically on Fig. 3.2, in the signature to which statistical parameters of the found distributions are resulted. These distributions are characterized by such important parameters, as $Mo(d)$ – position of maximum (a distribution mode); $<d>_V$ – average size of crystalline particles based on volume of the sample (the average arithmetic size) and $Me(d)$ – the median of distribution defining the size d, specifying that particles with diameters less than this size make half of volume of a powder.

The results represented on Fig. 3.2 show that initial powders nc-Si in the structure have particles with the sizes less than 10 nanometers which especially effectively absorb UV radiation. The both powders modes of density function of particle size are very close, but median of density function of particle size of a powder ncSi-36 it is essential more than at a powder ncSi-97. It suggests that the number of crystalline particles with diameters is less 10 nanometers in unit of volume of a powder ncSi-36 much less, than in unit of volume of a powder ncSi-97. As a part of a powder ncSi-36 it is a lot of particles with a diameter more than 100 nanometers and even there are particles more largely 300 nanometers whereas the sizes of particles in a powder ncSi-97 don't exceed 150 nanometers and the basic part of crystalline particles has diameter less than 100 nanometers.

The phase structure of the obtained films was estimated on wide-angle scattering diffractogram only qualitatively. Complexity of diffraction pictures of scattering and structure don't poses the quantitative phase analysis of polymeric films [20]. At the phase analysis of polymers often it is necessary to be content with the comparative qualitative analysis, which allows watching evolution of structure depending on certain parameters of technology of production. Measured wide-angle X-rays scattering diffractograms of investigated films are shown on Fig. 3.3. Diffractograms have a typical form for polymers. As a rule, polymers are the two-phase systems consisting of an amorphous phase and areas with distant order, conditionally named crystals. Their diffractograms represent [20] superposition of intensity of scattering by the amorphous phase, which is looking like wide halo on the small-angle area (in this case in area 2q between 10° and 30°), and intensity Bragg peaks scattering by a crystal phase.

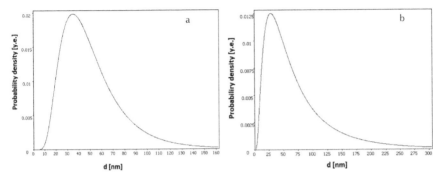

FIGURE 3.2 Density function of particle size in powders ncSi, received from X-ray diffractogram by means of program PM2K:

(a) – ncSi-97   Mo(d) = 35 nm   Me(d) = 45 nm $<d>_V$ = 51 nm;
(б) – ncSi-36   Mo(d) = 30 nm   Me(d) = 54 nm $<d>_V$ = 76 nm.

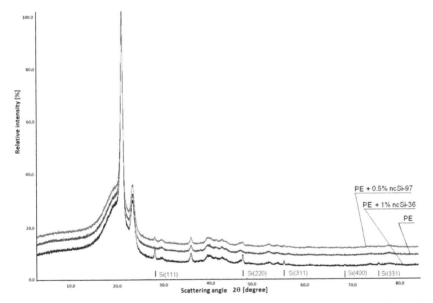

FIGURE 3.3 Diffractograms of the investigated composite films in comparison with diffractogram of pure polyethylene. Below vertical strokes specify reference positions of diffraction lines of silicon with their interference indexes (hkl).

Data on Fig. 3.3 is presented in a scale of relative intensities (intensity of the highest peak is accepted equal 100%). For convenience of consideration curves are represented with displacement on an axis of ordinates. The scattering plots without displacement represented completely overlapping of diffractogram profiles of composite films with diffractogram of a pure LDPE film, except peaks of crystal silicon, which weren't present on PE diffractogram. It testifies that additives of powders nc-Si practically haven't changed crystal structure of polymer.

The peaks of crystal silicon are well distinguishable on diffractograms of films with silicon (the reference positions with Miller's corresponding indexes are pointed below). Heights of the peaks of silicon with the same name (i.e., peaks with identical indexes) on diffractograms of the composite films PE + 0.5% ncSi-97 and PE + 1.0% ncSi-36 differ approximately twice that corresponds to a parity of mass concentration Si set at their manufacturing.

Degree of crystallinity of polymer films (a volume fraction of the crystal ordered areas in a material) in this research was defined by diffractograms Fig. 3.3 for a series of samples only semiquantitative (more/less). The essence of the method of crystallinity definition consists in analytical division of a diffractogram profile on the Bragg peaks from crystal areas and diffusion peak of an amorphous phase [20], as is shown in Fig. 3.4.

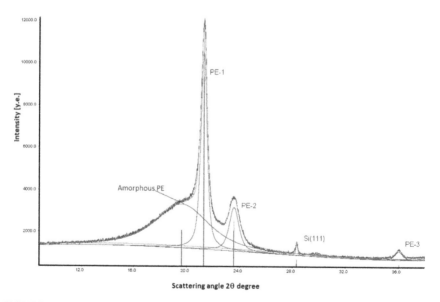

FIGURE 3.4 Diffractogram decomposition on separate peaks and a background by means of approximation of a full profile by analytical functions on an example of the data from sample PE+1%ncSi-36 (Fig. 3.3). PE-n designate Bragg peaks of crystal polyethylene with serial numbers n from left to right. Si (111) – Bragg silicon peak ncSi-36. Vertical strokes specify positions of maxima of peaks.

Peaks profiles of including peak of an amorphous phase, were approximated by function pseudo-Foigt, a background 4 order polynoms of Chebysheva. The nonlinear method of the least squares minimized a difference between intensity of points experimental and approximating curves. The width and height of approximating functions, positions of their maxima and the integrated areas, and also background parameters were thus specified. The relation of integrated intensity of a scattering profile by an amorphous phase to full integrated intensity of scattering by all phases except for particles of crystal silicon gives a share of amorphy of the sample, and crystallinity degree turns out as a difference between unit and an amorphy fraction.

It was supposed that one technology of film obtaining allowed an identical structure. It proved to be true by coincidence relative intensities of all peaks on diffractograms Fig. 3.3, and samples consist only crystal and amorphous phases of the same chemical compound. Therefore, received values of degree of crystallinity should reflect correctly a tendency of its change at modification polyethylene by powders nc-Si though because of a structure of films they can quantitatively differ considerably from the valid concentration of crystal areas in the given material. The found values of degree of crystallinity are represented in Table 3.2.

TABLE 3.2 Characteristics of the Ordered (Crystal) Areas in Polyethylene and Its çcomposites with nc-Si

PE			PE + 1% ncSi-36			PE + 0.5% ncSi-97		
Crystallinity	46%		47.5%			48%		
2q [°]	d [E]	ε	2q [°]	d [E]	ε	2q [°]	d [E]	ε
21.274	276	8.9	21.285	229	7.7	21.282	220	7.9
23.566	151	12.8	23.582	128	11.2	23.567	123	11.6
36.038	191	6.8	36.035	165	5.8	36.038	162	5.8
Average values	206	9.5×10^{-3}		174	8.2×10^{-3}		168	8.4×10^{-3}

One more important characteristic of crystallinity of polymer is the size d of the ordered areas in it. For definition of the size of crystalline particles and their maximum deformation e in X-ray diffraction analysis [21] Bragg peaks width on half of maximum intensity (Bragg lines half-width) is often used. In the given research the sizes of crystalline particles in a polyethylene matrix calculated on three well-expressed diffractogram peaks Fig. 3.3. The peaks of polyethylene located at corners 2q approximately equal 21.28 °, 23.57 ° and 36.03 ° (peaks PE-1, PE-2 and PE-3 on Fig. 3.4 see) were used. The ordered areas size d and the maximum relative deformation e of their lattice were calculated by the joint decision of the equations of Sherrera and Wilson [21] with use of half-width of the peaks defined as a result of approximation by analytical functions, and taking into account experimentally measured diffractometer tool function. Calculations were spent by means of program $WinX^{POW}$ size/strain. Received d and e, and also their average values for investigated films are presented in Table 3.2. The updated positions of maxima of diffraction peaks used at calculations are specified in the table.

The offered technology allowed the obtaining of films LDPE and composite films LDPE + 1% ncSi-36 and LDPE + 0.5% ncSi-97 an identical thickness (85 microns). Thus concentration of modifying additives nc-Si in composite films corresponded to the set structure that is confirmed by the X-ray phase analysis.

By direct measurements it is established that additives of powders ncSi have reduced a polyethylene transparency in all investigated range of lengths of waves, but especially strong transmittance decrease (almost in 20 times) is observed in a range of lengths of waves of 220–400 nanometers, that is, in UV areas. Especially strongly effect of suppression UV of radiation is expressed in LDPE film + 0.5% ncSi-97 though concentration of an additive of silicon in this material is less. It is possible to explain this fact to that according to experimentally received function of density of distribution of the size the quantity of particles with the sizes is less 10 nanometers on volume/weight unit in a powder ncSi-97 more than in a powder ncSi-36.

Direct measurements define mechanical characteristics of the received films – durability at a stretching and relative lengthening at disrupture (Table 3.1). The received results show that additives of powders of silicon raise durability of films approximately on 20% in comparison with pure polyethylene. Composite films in comparison with pure polyethylene also have higher lengthening at disrupture, especially this improvement is expressed in case of composite PE + 0.5% ncSi-97. Observable improvement of mechanical properties correlates with degree of crystallinity of films and the average sizes of crystal blocks in them (Table 3.2). By results of the X-ray analysis the highest crystallinity at LDPE film + 0.5% ncSi-97, and at it the smallest size the crystal ordered areas that should promote durability and plasticity increase.

This work is supported by grants RFBR № 10-02-92000 and RFBR № 11-02-00868 also by grants FCP "Scientific and scientific and pedagogical shots of innovative Russia," contract № 2353 from 17.11.09 and contract № 2352 from 13.11.09.

KEYWORDS

- **nanocrystalline silicon**
- **polyethylene**
- **polymer nanocomposites**
- **spectroscopy**
- **UV-protective film**
- **X-ray diffraction analysis**

REFERENCES

1. Karpov, S. V., & Slabko, V. V. (2003). Optical and Photo Physical Properties of Fractally Structured Metal Sols, Novosibirsk: Sib. Otd. Ross. Akad Nauk.
2. Varfolomeev, A. E., Volkov, A. V., Godovskii, D. Yu. et al. (1995). Pis'ma Zh. Eksp, Teor Fiz, *62*, 344.
3. Delerue, C., Allan, G., & Lannoo, M. (1999). J. Lumin, *80, (65)*.
4. Soni, R. K., Fonseca, L. F., Resto, O. et al. (1999). J. Lumin *83–84*, 187.
5. Altman, I. S., Lee, D., Chung, J. D. et al. (2001). Phys. Rev. B: Condens. Matter Material Physics *63*, 161402.
6. Knief, S., & Von Niessen, W. (1999). Phys. Rev. B: Condensation. Matter Material Phys *59*, 12940.
7. Olkhov, A. A., Goldschtrakh, M. A., & Ischenko, A. A. (2009). RU Patent №: 2009145013.
8. Bagratashvili, V. N., Tutorskii, I. A., Belogorokhov, A. I. et al (2005). Reports of Academy of Science, *Physical Chemistry, 405*, 360.
9. Kumar, V. et al. (2008). Nanosilicon, Elsevier Ltd. XIII 368p.
10. Carl, C., & Koch, N. Y., Eds. (2009). Nanostructured Materials. Processing, Properties, and Applications, William Andrew Publishing, 752.
11. Ischenko, A. A., Dorofeev, S. G., Kononov, N. N. et al. (2009). RU Patent №: 2009146715.
12. Kuzmin, G. P., Karasev, M. E., Khokhlov, E. M. et al. (2000). *Laser Physics, 10*, 939.
13. Beckman, J., & Ischenko, A. A. (2003). RU Patent No: 2 227 015.
14. Stehl, K. (2000). The Huber G670 Imaging-Plate Guinier Camera Tested on Beam line I711 at the MAX II synchrotron. *J. Appl. Crystal. 33*, 394–396.
15. Fetisov, G. V. (2010). The X-ray Phase Analysis Chapter 11, 153–184. Analytical Chemistry and Physical and Chemical Methods of the Analysis T 2 Red Ischenko, A. A., M: ITc Academy, 416p.
16. Scardi, P., & Leoni, M. (2006). Line Profile Analysis: Pattern Modeling Versus Profile Fitting. J. Appl. Cryst *39*, 24–31.
17. WINX^POW Version 1.06 STOE & CIE GmbH Darmstadt Germany 1999.
18. Leoni, M., Confente, T., & Scardi, P. (2006). PM2K: A Flexible Program implementing Whole Powder Pattern Modeling Z. Kristallogr Suppl. *23*, 249–254.
19. Scardi, P. (2008). Recent Advancements in Whole Powder Pattern Modeling Z. Kristallogr Suppl. *27*, 101–111.
20. Strbeck, N. (2007). X-Ray scattering of Soft Matter, Springer-Verlag Berlin Heidelberg, 238p.
21. Iveronova, V. I. & Revkevich U. P. (1978). The Theory of Scattering of X-rays. M. MGU 278 p.

CHAPTER 4

DYNAMICALLY VULCANIZED THERMOELASTOPLASTICS BASED ON BUTADIENE-ACRYLONITRILE RUBBER AND POLYPROPYLENE MODIFIED NANOFILLER

S. I. VOLFSON, G. E. ZAIKOV, N. A. OKHOTINA,
A. I. NIGMATULLINA, O. A. PANFILOVA, and A. A. NIKIFOROV

Chemistry And Processing Technology Of Elastomers Department, Kazan National Research Technological University, 68 K. Marks str., Kazan, Russia;
E-mail: Chembio@sky.chph.ras.ru

CONTENTS

Abstract 40
4.1   Introduction 40
4.2   Experimental Part 41
4.3   Results and Discussion 41
4.4. Conclusions 45
Keywords 46
References 46

ABSTRACT

Thermoplastic vulcanizates based on polypropylene and nitrile-butadiene rubber, containing modified organoclay were developed. It was shown that composites containing 1 to 5 pbw of Cloisite 15A montmorillonite added to rubber show improved physical-mechanical characteristics. Their swelling degree in AI-92 and motor oil was determined. The swelling degree of composites in petrol and motor oil decreases substantially, by 20–63%, due to the introduction of Cloisite 15A montmorillonite. Modification of thermoplastic vulcanizates using layered silicates raised the degradation onset temperature and decreases weight loss upon high temperature heating.

4.1   INTRODUCTION

Polymer nanocomposites are the most effective advanced materials for different areas of application any researches of polymer based on reinforced nanocomposites, which are very perspective direction of modern polymer science. It is well known [1, 2], that the polymers filled with small quantities (about 5–7 pbw) nanoparticles demonstrate great improvement of thermomechanical and barrier properties. The most part of published works is devoted to polyolefines filled with nanoparticles of laminated silicates, mainly montmorillonite (MMT).

The production and application of dynamically vulcanized thermoelastoplastics (DTEP, TPV) are intensively developed in the last years. That is connected with the fact, that TPV combined the properties of the vulcanized elastomers during utilization and thermoplastics in the processing. By their characteristics, the method of manufacturing, as well as processing TPV are in principle significantly differ from both plastics and elastomers. Due to their high level of properties the production of DTEP increases approximately by 10–15% per year in the world. Dynamic thermoelastoplastic elastomers are complicated composition materials consisting of continuous phase of thermoplastic and discrete phase of cured elastomers [3–9].

Dynamically vulcanized thermoelastoplastics based on butadiene-acrylonitrile elastomer (NBR) and polypropylene (PP) possess the special interest for production of car parts and petroleum industry. At the same time low compatibility of components during mixing of polar elastomer phase and nonpolar thermoplastic phase in pair NBR – PP is observed. That leads to decreased elastic-strength and thermal properties of composition material of this type. The effective way of improvement TPV properties is the use of natural nanosize filler – organoclay, particularly, montmorillonite. Montmorillonite could create organomineral complexes due to lability of laminated structure of MMT swelling during intercalation of organic substances. So, it was very interesting to investigate the influence of small quantity of modified montmorillonite clay on DTEP properties.

4.2 EXPERIMENTAL PART

Butadiene-acrylonitrile elastomer with 18% content acrylonitrile (NBR-18) and polypropylene Balen grade 01030 were used for manufacturing of DTEP in mixing camera of the plasticorder Brabender at 180°C during 10 min [10–12]. Ratio rubber-polypropylene was 70:30 and 50:50 in the TPV, the sulfur-based vulcanizing system was used. As a filler was used MMT Cloisite 15A brand of USA Rockwood company with variable capacity 125 ekv/100 g, modified by quaternary ammonium salts $[(RH)_2(CH_3)_2N]^+Cl^-$, where R is a residue of hydrogenated fatty acids C_{16}-C_{18}. The content of MMT varied from 1 to 3 pbw at 100 pbw of polymer phase MMT was preliminarily added into elastomer or into polypropylene.

The structural characteristics of MMT were determined by powder X-ray method at diffractometer D8ADVANCE Brucker company in Breg-Brentano in the mode of the step scanning. Monochromating radiation was performed by anticathode with wave length $\lambda = 0.154178$ nm. The X-ray tube action mode was 40 kV and 30 mA. The size of the slit was $1 \times V\ 20$.

The elastic-strength characteristics by device Inspect mini of Monsanto company at 50 mm/min speed were determined. The elastic-hysteresis properties of modified DTEP were estimated by dynamic rheometer RPA 2000 Alfa technologic company (USA).

During the test cycle shift deformation of the sample was performed. The torsional factor G^* by use of Fourier transformation was divided in two components: true G' and virtual G", which characterize elastic and plastic properties of the material consequently. The modulus of elasticity S' and loss modulus S", the same as the ratio of plastic and elastic components (tg δ) were also determined by RPA 2000 device.

The tests of modified dynamic thermoplastic elastomers by RPA 2000 were performed in two modes: (i) at constant temperature 100°C, constant frequently of deformation 1 Hz and variable of deformation from 1 to 10%; (ii) at constant temperature 100°C, constant of deformation 1% and variable deformation frequently from 0.1 to 10 Hz.

The thermal behavior of composites was investigated by simultaneous thermo-analizing device STA 409 PC NETZSCH company by thermogravimetric method (TG).

The swelling degree was estimated in liquid aggressive media (such as AI-92 petrol or motor oil) during 72 h at various temperatures (depending on medium type): 23, 70, and 125°C;

4.3 RESULTS AND DISCUSSION

The investigation of the DTEP structural characteristics demonstrated the whole series reflexes with d~3.3 (the strongest reflex) 1.6–1.1 nm, which is typical for MMT

Cloisite 15A on the diffractogram of DTEP sample with 3 pbw MMT. At the same time the reflex in small angle area (1.5–3° 2q) with interplanar distance 4.2–4.8 nm for DTEP with 1 pbw MMT was observed.

This reflex is the first basal diffraction of organomontmorillonite from the second basal diffraction. It is characterized by slight increase of the background and third reflex is absent in this case. It is connected with exfoliation MMT in polymer matrix. And we can forecast the significant improvement of physical-mechanical and thermal properties of DTEP in this case.

Physical-mechanical characteristics of composites are presented in Table 4.1.

TABLE 4.1   Elastic-Strength Properties of MMT-Modified DTEP

Properties	Content MMT, pbw		
	0	1	3
Mixing MMT with PP			
DTEP 70:30			
Conditional tensile strength, MPa	3.72	4.24	4.14
Relative elongation, %	140	170	160
Modulus of elasticity, MPa	93	118	100
DTEP 50:50			
Conditional tensile strength, MPa	8.2	8.8	8.4
Relative elongation, %	119	168	153
Modulus of elasticity, MPa	170	285	200
Mixing MMT with rubber			
DTEP 70:30			
Conditional tensile strength, MPa	3.72	4.32	4.27
Modulus of elasticity, MPa	93	143	131
DTEP 50:50			
Conditional tensile strength, MPa	8.2	9.0	8.5
Relative elongation, %	119	172	163
Modulus of elasticity, MPa	170	329	292

As seen from Table 4.1, composites containing 1 to 5 pbw of Cloisite 15A mont-morillonite added to rubber show improved physical-mechanical characteristics. In comparison with unfilled TPV, the modulus of elasticity increases by factor of $1.94 \div 2.11$, tensile strength a increases by factor of $1.09 \div 1.11$, and relative elongation increases by factor of $1.37 \div 1.47$.

It is known that increase of modulus of elasticity S' and decrease of loss modulus S" as well as tangent of mechanical losses $tg\ \delta$ should lead to improvement of physical-mechanical properties of polymer materials. The data, received by RPA 2000 method, at the first (Fig. 4.1) and second (Fig. 4.2) modes of testing of modified by MMT DTEP on Figs. 4.1 and 4.2 are presented.

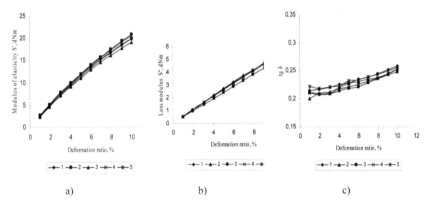

a)   b)   c)

FIGURE 4.1   Dependence of modulus of elasticity (a), loss modulus (b) and tangent of mechanical losses (c) on deformation ratio for DTEP with ratio rubber-polypropylene 70:30 with different content of MMT: 1–0 pbw MMT; 2–1 pbw MMT in the rubber; 3–1 pbw MMT in the polypropylene; 4–3 pbw MMT in the rubber; 5–3 pbw MMT in the polypropylene.

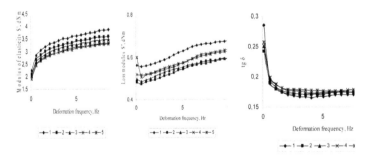

FIGURE 4.2   Dependence of modulus of elasticity (a), loss modulus (b) and tangent of mechanical losses (c) on deformation frequency for DTEP with ratio rubber: polypropylene 70:30 with different content of MMT: 1–0 pbw MMT; 2–1 pbw MMT in the rubber; 3–1 pbw MMT in the polypropylene; 4–3 pbw MMT in the rubber; 5–3 pbw MMT in the polypropylene.

According to results of these tests, we can propose the increase of elastic-hysteresis properties in the raw DVTEP + 1 pbw MMT in the rubber > DTEP + 1 pbw MMT in the polypropylene > DVTEP + 3 pbw MMT in the rubber > DTEP + 3 pbw MMT in the polypropylene.

The mixing of MMT with PP the same as with butadiene-acrylonitrile rubber leads to improvement of elastic-strength properties of DTEP especially modulus of elasticity, which increased by 27–54%.

The best results are achieved at preliminary mixing of MMT with NBR-18. It is important to underline that the less quantity of MMT is used the greater of the possibility of its exfoliation as separate particles in polymer matrix consequently, the improvement of elastic-strength properties was achieved. With increase of filler content agglomerates of greater size are formed. They play the role of structure defects and induce the decrease of conditional strength. The best elastic-strength properties of DTEP, modified by 1 pbw MMT are explained by better distribution of filler in the rubber phase.

Since the materials based on the formulated DTEP are intended to be exposed to higher temperatures, the effect of modified layer silicates as additives on TPV thermal stability was studied. Thermal test results are shown in Table 4.2.

TABLE 4.2   Thermal Behavior of DTEP (70:30 NBR to PP)

Properties	Content MMT, pbw		
	0	**1**	**3**
Mixing MMT with PP			
Degradation onset temperature, °C	269.2	327	314
Mixing MMT with rubber			
Degradation onset temperature, °C	269.2	350.4	343.4

As shown in Table 4.2, modification of DTEP using layered silicates increases the degradation onset temperature and decreases the weight loss upon high temperature heating. The unfilled TPV has the degradation onset temperature of 269°C. TPV compositions containing 1 and 3 pbw of Cloisite 15A MMT added to rubber show the best thermal stability (350 and 343°C, respectively). Introduction of layered silicate into DTEP at equal proportions of NBR and PP also results in the degradation onset temperature rising by 30°C.

Due to DTEP application in automobile and petroleum industries it is important to exam their resistance to aggressive media, such are petrol Euro 3 and motor oil. The examined composites exposed in aggressive media for 72 h at different temperatures and this data are represented in Table 4.3.

TABLE 4.3   Swelling Degree, %, of DTEP (70:30 NBR to PP), Modified by Montmorillonite

Conditions	Content MMT, pbw		
	0	**1**	**3**
Mixing MMT with PP			
23°C, petrol	15	5.8	5.6
23°C, motor oil	4.5	3.8	3.4
70°C, motor oil	11.2	4.3	7.1
125°C, motor oil	15.5	6.1	8.6
Mixing MMT with rubber			
23°C, petrol	15	8	7.6
23°C, motor oil	4.5	3.1	3.1
70°C, motor oil	11.2	4.4	5
125°C, motor oil	15.5	6.2	6.5

As can be seen from table 3, the filler significantly decrease the swelling degree and, consequently, increase the resistance to aggressive media.

When MMT is mixed with PP as well as with NBR the best resistance to aggressive media is achieved by use of 1 pbw of additive. The materials are more stable in motor oil, than in petrol. The swelling degree of DTEP is increase with increasing of temperature exposition in motor oil independently on composition of materials.

4.4. CONCLUSIONS

Thermoplastic vulcanizates have been formulated which are based on polypropylene and nitrile-butadiene rubber with a modified organoclay as a filler. It was established that the best physical-mechanical and thermal properties by preliminary mixing of small quantities nanofiller with elastomer are achieved. The good correlation of results of dynamic analysis of polymers by RPA 2000 and structural, physical-mechanical, thermal and barrier characteristics of DTEP, modified by MMT was achieved.

KEYWORDS

- **montmorillonite**
- **nanofiller**
- **nitrile-butadiene rubber**
- **organoclay**
- **physical-mechanical characteristics**
- **polypropylene**
- **thermoplastic**
- **vulcanization**

REFERENCES

1. Volfson, S. I. (2004). Dynamically Vulcanized Thermoplastics: Production, Processing, Properties Nauka Moscow.
2. Pinnavaia, T. J., & Beall G. W. (Ed.) (2001). John Willey &Sons Polymer Clay Nanocomposites: New York, 349p.
3. Polymer Nano composites: Synthesis, Characterization, and Modeling Krishnamoorti, R., Vaia, R. A. (Eds.) (2001): American Chemical Society: Washington, 242p.
4. Utrachki L. E. (2004). Clay Containing Polymeric Nanocomposites Monograph to be published by Rapra in 600p.
5. Alexandre, M., & Dubois, Ph. (2000). Polymer-Layered Silicate Nanocomposites: Preparation, Properties, and Uses of a New Class of Materials. *Material Science Engineering, 28,* 1–63.
6. Mikitaev, A. C., Kaladzhyan, A. A., Lednev, O. B., & Mikitaev, M. A. (2004). Nanocomposite Polymer Materials Based on Organoclay, Plastic Mass, *12,* 45–50.
7. Sheptalin, R. A., Koverzyanova, E. V., Lomakin, S. M., & Osipchik, V. S. (2004). The Features of Inflammability and Thermal Destruction of Nanocomposites of Elastic Polyurethane Foam Based on Organic Modified Layered Alumino silicate, *Plastic Mass, 4,* 20–26.
8. Lomakin, S. M., & Zaikov, G. E. (2005). Polymer Nanocomposites with Decreased Inflammability Based on Layered Silicates. High Molecular Substance, Series B, *47(1),* 104–120.
9. Mikitaev, A. C., Kaladzhyan, A. A., Lednev, O. B. et al. (2005). Nanocomposite Polymer Materials Based on Organoclay with Increased Fire Resistance, *Plastic Mass, 4,* 36–43.
10. Vol'fson, S. I., Nigmatullina, A. I., Sabirov, R. K., Lyzina, T. Z., Naumkina, N. I., & Gubaidullina, A. M. (2010). Effect of Organoclay on Properties of Dynamic Thermo Elastoplastics Russian Journal of Applied Chemistry, *83(1),* 123–126.
11. Nigmatullina, A. I., Volfson, S. I., Okhotina, N. A., & Shaldibina, M. S. (2010). Properties of Dynamically Vulcanized Thermoplastics containing modified Polypropylene and a Layered Filler Vestnik of Kazan Technological University, *9,* 329–333.
12. Volfson, S., Okhotina, I. N., Nigmatullina, A. A., Sabirov, I. R. K., Kuznetsova, O. A., Akhmerova, L. Z. (2012). Elastic and Hysteretic Properties of Dynamically Cured Thermo Elastoplastics Modified by Nanofiller. Plastic Mass, *4,* 42–45.

CHAPTER 5

SORPTION-ACTIVE CARBON-POLYMER NANOCOMPOSITES: A RESEARCH NOTE

MARINA BAZUNOVA[1], RUSTAM TUKHVATULLIN[1], DENIS VALIEV[1], ELENA KULISH[1], and GENNADIJ ZAIKOV[2]

[1]Bashkir State University, 32 Zaki Validi Street, 450076 Ufa, Republic of Bashkortostan, Russia

[2]Institute of Biochemical Physics named N.M. Emanuel of Russian Academy of Sciences, 4 Kosygina Street, 119334, Moscow, Russia

CONTENTS

Abstract 48
5.1   Introduction..48
5.2   Experimental Part..49
5.3   Results and Discussion 50
5.4   Conclusions...56
Keywords 56
References...57

ABSTRACT

The method of obtaining of the molded composites on the basis of the mixtures of the powders of nano-dispersed polyethylene, cellulose and the ultra-dispersed car-bonic materials is developed. These materials possess the assigned sorption proper-ties and the physic-mechanical characteristics. They are suitable for the usage at the process of cleaning and separation of gas mixture.

5.1   INTRODUCTION

Currently, the creation of materials with high adsorption activity to a range of defi-nite substances by controlling their surface structure has significant interest. Par-ticularly, selective sorbents for separation processes, dividing or concentration of the components of different nature mixtures are developed on the basis of such composites.

The nature of the binder and the active components, and molding conditions are especially important at the process of sorption-active composites creating. These factors ultimately exert influence on the development of the porous structure of the sorbent particles and its performance. In this regard, it is promising to use powders of various functional materials having nanoscale particle sizes at the process of such composites creating. First, the high dispersibility of the particles allows them to pro-vide a regular distribution in the matrix, whereby it is possible to achieve improved physical and mechanical properties. Secondly, high degree of homogenization of the components facilitates their treatment process. Third, it is possible to create the composites with necessary magnetic, sorption, dielectric and other special proper-ties combining volumetric content of components [1].

Powders of low density polyethylene (LDPE) prepared by high temperature shearing (HTS) used as one of prospective components of the developing functional composite materials [2, 3].

Development of the preparation process and study of physicochemical and me-chanical properties of sorbents based on powder mixtures of LDPE, cellulose (CS) and carbon materials are conducted. As the basic sorbent material new – ultrafine nanocarbon (NC) obtained by the oxidative condensation of methane at a treatment time of 50 min (NC1) and 40 min (NC2) having a specific surface area of 200 m^2/g and a particle size of 30–50 nm is selected [4]. Highly dispersed form of NC may give rise to technological difficulties, for example, during regeneration of NC after using in gaseous environments, as well as during effective separation of the filtrate from the carbon dust particles. This imposes restrictions on the using of NC as an independent sorbent. In this connection, it should be included in a material that has a high porosity. LDPE and CS powders have great interest for the production of such material. It is known that a mixture of LDPE and CS powders have certain absorp-tion properties, particularly, they were tested as sorbents for purification of water surface from petroleum and other hydrocarbons [5].

Thus, the choice of developing sorbents components is explained by the following reasons:

1. LDPE has a low softening point, allowing to conduct blanks molding at low temperatures. The small size of the LDPE particles (60 to 150 nm) ensures regular distribution of the binder in the matrix. It is also important that the presence of binder in the composition is necessary for maintaining of the material's shape, size, and mechanical strength.

2. Usage of cellulose in the composite material is determined by features of its chemical structure and properties. CS has developed capillary-porous structure, that's why it has well-known sorption properties [5] towards polar liquids, gases and vapors.

3. Ultrafine carbon components [nanocarbon, activated carbon (AC)] are used as functionalizing addends due to their high specific surface area.

5.2   EXPERIMENTAL PART

Fine powders of LDPE, CS and a mixture of LDPE/CS are obtained by high temperature shearing under simultaneous impact of high pressure and shear deformation in an extrusion type apparatus with a screw diameter of 32 mm [3].

Initial press-powders obtained by two ways. The first method is based on the mechanical mixing of ready LDPE, CS and carbon materials' powders. The second method is based on a preliminary high-shear joint grinding of LDPE pellets and sawdust in a specific ratio and mixing the resulting powder with the powdered activated carbon (БАУ-А mark) and the nanocarbon after it.

Composites molding held by thermobaric compression at the pressure of 127 kPa.

Water absorption coefficient of polymeric carbon sorbents is defined by the formula: $K = \frac{m_{\text{поглощённ. воды}}}{m_{\text{образца}}} \times 100\%$, wherein $m_{\text{absorbed water}}$ is mass of the water, retained by the sorbent sample, m_{sample} is mass of the sample.

The adsorption capacity (A) of the samples under static conditions for condensed water vapor, benzene, n-heptane determined by method of complete saturation of the sorbent by adsorbate vapor in standard conditions at 20°C [6] and calculated by the formula: A=m/(M·d), wherein m is the mass of the adsorbed benzene (acetone, n-heptane), g; M is the mass of the dried sample, g; d is the density of the adsorbateб, g/cm³.

Measuring of the tablets strength was carried out on the automatic catalysts strength measurer ПК-1.

Experimental error does not exceed 5% in all weight methods at P = 0.95 and the number of repeated experiments n = 3.

5.3   RESULTS AND DISCUSSION

LDPE (obtained by the method of HTS) powder particles' size, dispersity and surface properties study

Powder components are used as raw materials for functional composite molding (including the binder LDPE), because molding of melt polymer mixtures with the active components has significant disadvantages. For example, the melt at high degrees of filling loses its fluidity, at low degrees of filling flow rate is maintained, but it is impossible to achieve the required material functionalization.

It is known that amorphous-crystalline polymers, which are typical heterogeneous systems, well exposed to high-temperature shear grinding process. For example, the process of HTS of LDPE almost always achieves a significant results [3]. Disperse composition is the most important feature of powders, obtained as result of high-temperature shear milling. Previously, on the basis of the conventional microscopic measurement, it was believed that sizes of LDPE powder particles obtained by HTS are within 6–30 micrometers. Electron microscopy (Fig. 5.1) gives the sizes of 60 to 150 nm. The active powder has a fairly high specific surface area (up to 2.2 m^2/g).

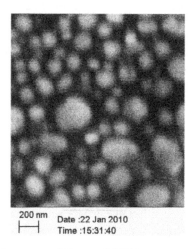

FIGURE 5.1   Electron microscopy of dispersed LDPE.

The results of measurement of the water absorption coefficient and of the static capacitance of LDPE powder by n-heptane vapor are equal to 12% and 0.26 cm^3/g, respectively. Therefore, the surface properties of LDPE powder more developed than the other polyethylene materials'.

Selection of molding conditions of sorbents based on mixtures of LDPE, CS and ultrafine carbon materials' powders

Initial press-powders obtained by two ways. The first method is based on the mechanical mixing of ready LDPE, CS and carbon materials' powders. The second method is based on a preliminary high-shear joint grinding of LDPE pellets and sawdust in a specific ratio and mixing the resulting powder with the powdered activated carbon and the nanocarbon after it. The method of molding – thermobaric pressing at a pressure of 127 kPa.

The mixture of LDPE/CS compacted into tablets at a temperature of 115–145°C was used as a model mixture for selection of composites molding conditions. Pressing temperature should be such that the LDPE softens but not melts, and at the same time forms a matrix to prevent loss of specific surface area in the ready molded sorbent due to fusion of pores with the binder. Data on the dependence of the coefficient of water absorption from temperature of the mixture of LDPE/CS (20/80 wt.%) pressing is presented in Table 5.1. It makes possible to determine the LDPE softening-melting temperature limits. The composites molded at a higher temperature, have a lower coefficient of water absorption than the tablets produced at a lower temperature, that's why the lowest pressing temperature (120°C) is selected. At a higher content of LDPE the water absorption coefficient markedly decreases with temperature.

TABLE 5.1 Dependence of Water Absorption Coefficient (K) of the Composites Based on a Mixture of LDPE/CS Composition (20/80 wt.%) from the Pressing Temperature

t, °C	115	119	125	128	133	136	138	142	145
K, %	180	180	182	177	168	169	165	150	149

Cellulose has a high degree of swelling in water (450%) [5], this may lead to the destruction of the tablets. Its contents in samples of composites, as it has been observed by the sorption of water, should not exceed 30 wt.%. There is a slight change of geometric dimensions of the tablets in aqueous medium at an optimal value of the water absorption coefficient when the LDPE content is 20 wt.%.

Samples of LDPE/CS with AC, which sorption properties are well studied, are tested for selecting of optimal content of ultrafine carbon. As follows from Table 5.2, the samples containing more than 50 wt.% of AC have less water absorption coefficient values. Therefore, the total content of ultrafine carbon materials in all samples must be equal to 50 wt.%.

TABLE 5.2   Water Absorption Coefficient of PE/AC/CS Composites Samples

Composition of PE/AC/CS sorbent, wt.%	20/5/75	20/15/65	20/25/55	20/35/45	20/45/35	20/50/30	20/65/15
K, %	136.3	136.5	135.9	138.1	135.6	133.7	120.7

Static capacitance measurement of samples, obtained from mechanical mixtures of powders of PE, CS and AC, conducted on vapors of n-heptane and benzene, to determine the effect of the polymer matrix on the sorption properties of functionalizing additives. As can be seen (Table 5.3), with a decrease of the content of AC in the samples with a fixed (20 wt.%) amount of the binder, reduction of vapor sorption occurs. It indicates that the AC does not lose its adsorption activity in the composition of investigated sorbents.

TABLE 5.3   Static Capacitance of Sorbents, A (cm^3/g), by Benzene and n-heptane Vapors (20 °C)

Composition of PE/AC/CS sorbent, wt.%		20/0/75	20/15/65	20/35/55	20/45/35	20/60/20	20/75/15	20/80/0
A cm^3/g	by n-heptane vapors	0.08	0.11	0.16	0.19	0.23	0.25	0.25
	by benzene vapors	0.06	0.08	0.17	0.18	0.23	0.25	0.26

Strength of samples of sorbents (Fig. 5.2) is in the range of 620–750 N. The value of strength is achieved in the following molding conditions: t = 120 °C and a pressure of 127 kPa.

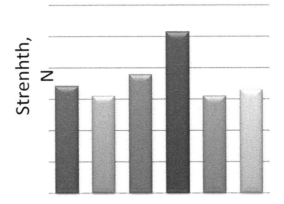

FIGURE 5.2   Comparison of strength of tablets, based on LDPE, CS (different species of wood) and AC powders. 1 – sorbent of LDPE/AC/CS = 20/50/30 wt.% based on the powders of jointly dispersed pellets of LDPE and softwood sawdust with subsequently addition of AC; 2 – sorbent of LDPE/AC/CS = 20/50/30 wt.% based on the powders of jointly dispersed pellets of LDPE and hardwood sawdust with subsequently addition of AC; 3 – sorbent of LDPE/AC/CS = 20/50/30 wt.% based on the mechanical mixtures of the individual powders of LDPE, CS from softwood and AC; 4 – AC tablet; 5 – sorbent of LDPE/CS = 20/80 wt.%; 6 – sorbent of LDPE/AC = 20/80 wt.%.

Thus, optimal weight composition of the matrix of LDPE/CS composition – 20/30 wt.% with 50 wt.% containing of carbon materials.

Sorption properties of carbon – polymer composites by condensed vapors of volatile liquids.

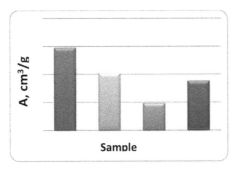

FIGURE 5.3   Static capacitance of sorbents, A (cm³/g) by benzene vapor (20°C). 1 –molded mechanical mixture of LDPE/AC/NC1/CS= 20/25/25/30wt.%; 2 – molded mechanical mixture of LDPE/AC/NC2/CS = 20/25/25/30 wt.%; 3 – molded mechanical mixture of LDPE/AC/CS=20/50/30 wt.%; 4 – AC medical tablet (controlling).

For a number of samples of sorbents static capacitance values by benzene vapor is identified (Fig. 5.3). They indicate that the molded mechanical mixture of 20/25/25/30 wt.% LDPE/AC/NC1/CS has a maximum adsorption capacity that greatly exceeds the capacity of activated carbon. High sorption capacity values by benzene vapor appears to be determined by weak specific interaction of π-electron system of the aromatic ring with carbocyclic carbon skeleton of the nanocarbon [7].

Static capacitance of obtained sorbents by heptane vapors significantly inferiors to capacity of activated carbon (Fig. 5.4), probably it is determined by the low polarizability of the molecules of low-molecular alkanes. Consequently, the investigated composites selectively absorb benzene and can be used for separation and purification of mixtures of hydrocarbons.

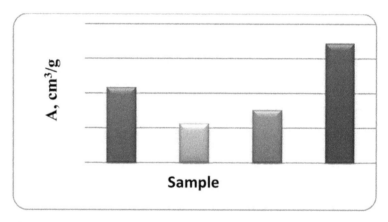

FIGURE 5.4   Static capacitance of sorbents, A (cm³/g) by n-heptane vapor (20°C). – molded mechanical mixture of LDPE/AC/NC1/CS= 20/25/25/30wt.%; 2 – molded mechanical mixture of LDPE/AC/NC2/CS = 20/25/25/30 wt.%; 3 – molded mechanical mixture of PE/AC/CS=20/50/30 wt.%; 4 – AC medical tablet (controlling).

Molded composite based on a mechanical mixture of PENP/AC/NC1/CS = 20/25/25/30 wt.% has a sorption capacity by acetone vapor comparable with the capacity of activated carbon (0.36 cm³/g) (Fig. 5.5).

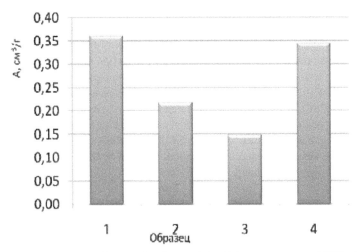

FIGURE 5.5 Static capacitance of sorbents, A (cm³/g) acetone vapor (20°C). 1 – molded mechanical mixture of LDPE/AC/NC1/CS= 20/25/25/30wt.%; 2 – molded mechanical mixture of LDPE/AC/NC2/CS = 20/25/25/30 wt.%; 3 – molded mechanical mixture of LDPE/AC/CS=20/50/30 wt.%; 4 – AC medical tablet (controlling).

Sorbents' samples containing NC2 have a low values of static capacity by benzene, heptanes and acetone vapor. It can be probably associated with partial occlusion of carbon material pores by remnants of resinous substances – by products of oxidative condensation of methane, and insufficiently formed porous structure.

The residual benzene content measuring data (Table 5.4) shows that the minimal residual benzene content after its desorption from the pores at t = 70°C for 120 min observes in case of sorbent LDPE/AC/NC1/CS composition = 20/25/25/30 wt.%. It allows to conclude that developed sorbents have better ability to regenerate under these conditions in comparison with activated carbon.

TABLE 5.4 Sorbents' Characteristics: Total Pore Volume V_{tot}; Static Capacitance (A) by Benzene Vapors at the Sorption Time of 2 days; Residual wt. Fraction of the Absorbed Benzene After Drying at t = 70 °C for 120 min

LDPE/AC/NC/CS sorbent composition, wt.%	V_{tot}, cm³/g	A, cm³/g	Residual benzene content as a result of desorption, %
20/25/25/30	1.54	0.5914	2.9
20/50/ – /30	1.21	0.1921	10.3
–/100/ –/–	1.6	0.3523	32.0

Thus, the usage of nanosized LDPE as a binder gives a possibility to get the molded composite materials with acceptable absorption properties. Varying the ratio of the components of the compositions on the basis of ternary and quaternary mixtures of powdered LDPE, cellulose and ultrafine carbon materials it is possible to achieve the selectivity of sorption properties by vapors of certain volatile liquids. These facts allow to suggest that the proposed composites are expedient to use for separation and purification of gaseous and steam mixtures of different nature.

Developed production method of molded sorption-active composites based on ternary and quaternary mixtures of powdered LDPE, cellulose and ultrafine carbon materials can be easily designed by equipment and can be used for industrial production without significant changes.

5.4 CONCLUSIONS

1. It is revealed by electron microscopy that the dispersed LDPE particle sizes are 60–150 nm. It allows to obtain functional composite materials with a regular distribution of components and with the necessary physicochemical properties, depending on their volume content, on its base.

2. Optimal conditions for molding of sorbents on the basis of three-and four-component mixtures of powdered LDPE, cellulose and ultrafine carbon materials were determined: temperature 120°C and pressure of 127 kPa, content of the binder (LDPE) = 20 wt.%.

3. Established that molded mechanical mixture of LDPE/AC/NC1/CS (20/25/25/30wt.%) has a static capacity (by condensed vapors of benzene and acetone) = 0.6 cm^3/g and 0.36 cm^3/g, respectively, what exceeds the capacity of activated carbon. The static capacitance of the compositions by the n-heptane vapors is 0.21 cm^3/g, therefore, the proposed composites are useful for separation and purification of gaseous and steam mixtures of different nature.

KEYWORDS

- cellulose
- high-temperature shift crushing
- nano-carbon
- polyethylene
- sorbents

REFERENCES

1. Akbasheva, E. F., & Bazunova, M. V. (2010). Tableted Sorbents Based on Cellulose Powder Mixtures, Polyethylene and Ultra Dispersed Carbon. Materials Open School Conference of the CIS "Ultrafine and Nanostructured Materials" 11–15 October 2010, Ufa: Bashkir State University, 106.
2. Enikolopyan, N. S., Fridman, M. L., & Karmilov, A. Yu. (1987). Elastic Deformation Grinding of Thermoplastic Polymers, Reports as USSR, *296(1)*, 134–138.
3. Akhmetkhanov, R. M., Minsker, K. S., & Zaikov, G. E. (2006). On the Mechanism of Fine Dispersion of Polymer Products at Elastic Deformation Affects Plasticheskie Massi. *8*, 6–9.
4. Aleskovskiy, V. B., & Galitseisky, K. B. (20.11.2006). Russian Patent: "Method of Ultrafine Carbon." 2287543.
5. Raspopov, L. N., Russiyan, L. N., & Zlobinsky, Y. I. (2007). Waterproof composites Comprising Wood and Polyethylene dispersion Russian Polymer Science Journal, *50(3)*, 547–552.
6. Keltsev, N. V. (1984). Fundamentals of Adsorption Technology. Moscow: Chemistry, 595p.
7. Valinurova, E. R., Kadyrov, A. D., & Kudasheva, F. H. (2008). Adsorption Properties of Carbon Rayon Vestn Bashkirs Universal, *13(4)*, 907–910.

CHAPTER 6

MODIFICATION OF PHYSICAL AND MECHANICAL PROPERTIES OF FIBER REINFORCED CONCRETE USING NANOPARTICLES

M. MEHDIPOUR

Textile Engineering Department, Guilan University, Rasht, Iran

CONTENTS

Abstract 60
6.1   Introduction..60
6.2   Materials and Methods..62
6.3   Results and Discussions...66
6.4   Conclusion 71
Keywords 71
References...71

ABSTRACT

This chapter reports the effects of simultaneous application of polypropylene (PP) fibers and SiO_2 and TiO_2 nano-particles on some physical and mechanical properties of concrete. Fibers and nano-particles are used in different proportions in concrete mixture according the results of Design of Experiment analysis. Important hardened concrete properties such as 7-days and 28-days compressive strength, flexural strength, and water absorption were evaluated. Optimization of the mixture design was conducted in two stages. In first stage, SiO_2 nano-particles and PP fibers with different proportions were added to the concrete mixture and 7-days and 28-days compressive strength, tests were carried out. The results were compared and mixture of sample with optimum strengths was selected for the second stage. In the second stage, TiO_2 nano-particles were added to the optimum mixture and physical and mechanical properties of samples were evaluated. The results indicated that adding two weight percent's of SiO_2 nano-particles and PP fibers simultaneously, can effectively improve the cracking level and reduce the content of $Ca(OH)_2$ and produce more Calcium-Silicate-Hydrate (C-S-H) resulting in a denser, more homogeneous Interfacial Transition Zone microstructure, respectively.

6.1   INTRODUCTION

Concrete is widely used in structural engineering due to its high compressive strength, low-cost and abundant raw material. But common concrete has some shortcomings, for example, shrinkage and cracking, low tensile and flexural strength, poor toughness, high brittleness, low shock resistance and so on, that restrict its applications. It is well documented some mechanical properties of concrete such as compressive and flexural strength and abrasion resistance can be significantly improved by addition fibers like polypropylene (PP), glass, carbon, and Nano materials like Al_2O_3, TiO_2, SiO_2, and carbon nanotube (CNT). It is believed that application of these nano particles in cement matrix up to 3 wt% could accelerate formation crystalline silicate regions in reaction between nano SiO_2 and cement matrix (C-S-H) as a result of increased crystalline $Ca(OH)_2$ amount at the early age of hydration and hence increase compressive strength of concrete [1–4]. Vital role of PP fiber creates some connecting bridge between cracks in the concrete microstructure. These cracks produce for some reasons such as alkaline hydrated reaction of cement in concrete paste, thermal contraction, drying shrinkage, and autogenously shrinkage. Fiber-reinforced concrete (FRC) has been successfully used in construction with its excellent flexural-tensile strength, resistance to splitting, impact resistance and excellent permeability. It is also an effective way to increase toughness, shock resistance, and resistance to plastic shrinkage cracking of the mortar [2, 3]. The uniform dispersion of the fibers and other additives is an important view and without this property, addition of fiber and any other additives have inverse effect on mechanical properties of concrete [5, 6].

The performance of PP fibers as reinforcement in cement-based materials for different fiber length and proportion has been investigated by Bagherzadeh et al. [3, 4]. It was concluded that PP fibers, in spite of their hydrophobic nature and weak wetting by cement paste, show better bonding than other fibers. Also, it was concluded that with an increase in fiber length or decrease in fiber diameter, crack width decreases significantly. It is also reported [6, 9] that adding PP fibers in concrete materials decrease internal pressure of concrete and can enhance the fire resistance of concrete structures, since it can be helpful to avoid explosive spalling.

Nano-scale materials due to their special property such as high surface to volume ratios have high ability to improve the physical and mechanical properties of cementitious material [3, 4]. Nano-SiO_2 with range diameter less than 100 nm is one of the most common nano-particles in concrete industry. Addition of nano-SiO_2 to mixture of concrete cause activation of SiO_2 which can react with $Ca(OH)_2$ of cement rapidly and produce C-S-H that decrease porosity of cement matrix. Formation of C-S-H in the cement matrix improves mechanical properties of concrete [13]. Also nano-SiO_2 creates a net in interfacial transition zone (ITZ), a small region beside the particles of coarse aggregate and cement, can make it more homogeneous which can improve concrete microstructure and increase its workability [5, 14]. Nano-particles dispersed in the cement paste uniformly, could accelerate cement hydration due to their high-activity [8]. However, when nano-particles cannot be well dispersed, their aggregation will create weak zones in the cement matrix and consequently, the homogeneous hydrated microstructure in ITZ could not be formed and low mechanical properties will be expected [2–8]. Only few in-depth studies related to nano modifications of cement based materials and their enhanced performance on compressive and tensile strength properties are reported in literature. However, this field requires bottom-up approach to tailor, modify, replace, or include nano materials in the already available matrices such as cement or cement based materials. Another nano-particle material which is believed that can improve the mechanical properties of cementitious materials is nano-TiO_2. Addition of the nano-TiO_2 with its tetragonal crystalline structure can improve some mechanical properties such as abrasion resistance of concrete. Due to its ability to produce OH and O_2 radicals with photo catalytic reactions, it could be useful to achieve self-cleaning effect in surface of cement-based composites [7]. It is also reported that flexural strength of the concrete with 1% weight of nano-TiO_2 is much better than the concrete with only pp fibers or nano-SiO_2 [12].

There are well-documented reports in the area of adding fibers and Nano-particles in concrete; however, there is still a lack of studies on the best volume fractions of fibers and particles in the concrete mixture. Also there is a lack of study in investigation of simultaneously effect of fibers and nano-particles in mixture design of cementitious materials. This study presents comprehensive experimental data regarding the effects of adding PP fibers with different fiber lengths and diameter, and different volume fractions, along with different nano particle types and portions on the physical and mechanical properties of FRC.

6.2   MATERIALS AND METHODS

6.2.1   MATERIALS

PP fibers used in this study were provided by the Shimifaraiand Company. Specifications of the fibers employed in this study are presented in Table 6.1. Ordinary Portland cement conforming to ISO 12269 was used for the concrete mixtures. River sand with a specific gravity of 2.65 (g/cm^3) and fineness modulus of 2.64 was used as the fine aggregate. A commercially water reduction agent (SP), Glenium – 110p (because w/c in mixture design should be stayed constant), was purchased from BASF company. Naphthalene- based super plasticizer from LanYa Concrete Admixtures Company purchased too. Nano particles including TiO_2 and SiO_2 were acquired from the Degussa Co. (Germany). Specifications of the nano particles employed in this study are presented in Table 6.2.

TABLE 6.1   Properties of Polypropylene Fiber and Nano-SiO_2 and Nano-TiO_2 Particles

	Density (g/cm^3)	Tensile strength (MPa)	Elastic modulus (GPa)	Length (mm)	denier
PP fiber	0.9	350	3.4–3.6	12	5
	Diameter (nm)	Specific surface area (m^2/g)	Density (g/cm^3)	Purity (%)	
SiO_2	12 3	200 5	0.6	99.5	
TiO_2	15 5	50 5	0.75	99.4	

6.2.2   MIXING AND CURING

Trial mixtures were prepared to obtain target strength of 40 MPa at 28 days, along with a good workability. The final mixture design used in this study was composed of 350 kg/m^3 cement, 184 kg/m^3 water, 0.50 water/cement percentage (W/C), and 1.35% naphthalene- based super plasticizer. The total dosage of fibers was maintained at 0–3%, primarily from the point of view of providing good workability and no balling of fibers during mixing.

The sand, cement, and other additives were first mixed dry in a pan mixer with a capacity of 100 kg for a period of 2 min. The naphthalene-based super plasticizer and nano particles then mixed thoroughly with the mixing water and added to the mixer in order to avoid the agglomeration of nano particles. First, fibers were dispersed by hand and then were mixed in the mixture for a total time of 4 min to achieve a uniform distribution throughout the concrete. Fresh concrete was cast in steel molds and compacted on a vibrating table. For the curing function, the specimens were kept covered in their molds for 24 h. The specimens had less bleeding than the control concrete specimen. After demolding, concrete specimens were placed in 20±2°C water for 28 days for curing. They were removed from water and placed in the laboratory environment for 2 h before carry out the tests. All tests were performed according to relevant standards.

6.2.3   PHYSICAL AND MECHANICAL TESTS

6.2.3.1   COMPRESSIVE STRENGTH

A universal testing machine with a capacity of 100 tones was used for testing the compressive strengths of all 100×100×100 mm cube Specimens. These specimens at 7-days and 28-days from casting were tested at a loading rate of 14 N/mm²/min according to BS 1881–116 standard. The compressive strength was interpreted as the stress generated from the result of compression load per area of specimen surface. The results for each specimen are based on an average value of three replicate Specimens.

6.2.3.2   FLEXURAL STRENGTH

Flexural strength at 28-days of curing test was conducted according to the requirements of ASTM C 293 using three 100×100×500 mm beams under mid-point loading on a simply supported span of 300 mm. According to ASTM standards, the results of the flexural strength test are interpreted by calculating flexural stress as follows:

$$R= \frac{PL}{bd^2} \tag{1}$$

where R is the flexural strength (modulus of rupture), P is the maximum indicated load, L is the span length, b is the width of the Specimen, and d is the depth of the Specimen.

6.2.3.3   WATER ABSORPTION

Water penetration at 28-days of curing tests of two 100×100×100 mm concrete mixture ‹optimum and plain sample was conducted according to the requirements of ASTM C1585, in due to difference between wet and dry sample of concrete.

6.2.4   DESIGN OF EXPERIMENTS

The use of statistical methods has helped the rapid development of nanotechnology in terms of data collection, hypothesis testing, and quality control. Response Surface Methodology (RSM) is an empirical modeling technique used to evaluate the relation between a set of controllable experimental factors and observed results [11]. At first stage of this study, concrete mixture samples designed by D-optimal method. D-optimal designs are based on a computer-aided exchange procedure, which creates the optimal set of experiments. Dx7 (statistic's software, that selections optimized samples and conditions due to design of experiments) selected 11 mixture designs, and two cubic specimens from each mixture design were made to measure compressive strength both at 7 days and 28 days after casting (Fig. 6.1).

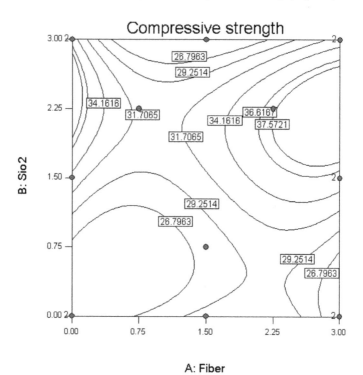

FIGURE 6.1   RSM plots for compare effects of Polypropylene fiber and nano-SiO$_2$ on 7-days compression strength of samples (MPa).

In second stage, nano-TiO$_2$ added to three mixture samples simultaneously with nano-SiO$_2$ and Polypropylene fiber. Eight prepared samples were tested to evaluate the 28-days compressive and 28-days flexural strengths (Fig. 6.2).

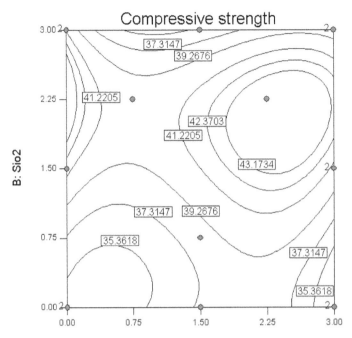

FIGURE 6.2   RSM plots for compared effects of Polypropylene fiber and nano-SiO$_2$ on 28-days compression strength of samples (MPa).

6.2.5   SCANNING ELECTRON MICROSCOPE (SEM)

The concrete specimen cured after 90 days was broken into small pieces. Those pieces with dimensions around 7 mm by 7 mm by 4 mm and containing aggregates and tri-calcium silicate (3CaO·SiO$_2$) were selected as samples for the SEM experiment because expected that this products showed optimum results. These samples were washed using 1% HCL for 20 s, and then dried at 60–70°C. Consequently, the absolute ethyl alcohol was employed to stop hydration of concrete samples, which were then coated with golden conductive films in the vacuum coating machine (DMX – 220A) using the vaporization method.

The SEM (JSM – 840) was employed to observe the microstructures of samples. In this experiment, the interface between the aggregate and the cement paste was positioned at the center of the SEM in order to observe the interfacial transition zone. Figure 6.3 shows the microstructures of the plain and Nano-particles and fibers reinforced concretes (NF) and its microstructure's of the aggregate-cement interfacial transition zone with magnification of 5000x.

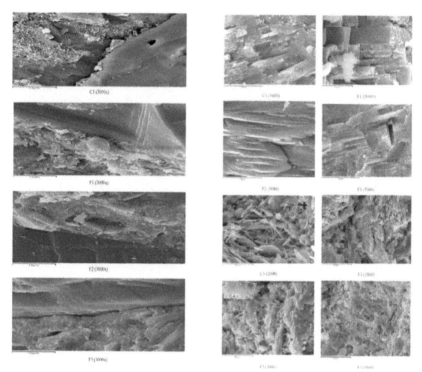

FIGURE 6.3   Microstructures of the plain and NF reinforced concretes with magnification of 5000x.

6.3   RESULTS AND DISCUSSIONS

6.3.1   MECHANICAL AND PHYSICAL PROPERTIES

Concrete mixture proportions and the measured mechanical and physical properties including compressive strength, flexural strength and water penetration of both the plain concrete and NF (Nano-particles and fibers reinforced concretes) mixtures are presented in Tables 6.2–6.5.

TABLE 6.2 Concrete Mixture Proportions Used in First Stage of This Study

Sample code	Fiber (F) or Nano-Sio$_2$ (NS) content
P1	1.5% F*, 0.75% NS**
P2	3% F, 3% NS
P3	3% F, 1.5% NS
P4	2.25% F, 2.25% NS
P5	0% F, 0% NS
P6	0% F, 3% NS
P7	1.5% F, 0% NS
P8	0.75% F, 2.25% NS
P9	0% F, 1.5% NS
P10	1.5% F, 3% NS
P11	3% F, 0% NS

*F refers to the weight percent of cement for PP fiber.
** NS refers to the weight percent of cement for Nano-SiO$_2$.

TABLE 6.3 Result of Concrete Mixtures Used in First Stage of This Study

Specimens code	P1	P2	P3	P4	P5	P6	P7	P8	P9	P10	P11
Compression Strength (MPa) (7 days)	25	27	33	36	36	27	34	31	31	34	23
Compression Strength (MPa) (28 days)	38	38	41	44	33	41	38	40	42	36	30

TABLE 6.4 Concrete Mixture Proportions Used in Second Stage of This Study

Sample code	Weight percentage of cement in mixture for Fiber (F), Nano-Sio$_2$ (NS), and Nano-TiO$_2$ (NT) contents in
a	2.25% F, 0.75% NT, 0% NS
b	2.25% F, 0.75% NT, 0.75% NS
c	2.25% F, 2.25% NT, 2.25% NS

TABLE 6.4   *(Continued)*

Sample code	Weight percentage of cement in mixture for Fiber (F), Nano-Sio$_2$ (NS), and Nano-TiO$_2$ (NT) contents in
d-1	2.25% F, 2.25% NS
d-2	2.25% F, 2.25% NS
e-1	0% F, 0% NS
e-2	0% F, 0% NS
f	2.25% F, 0% NS

TABLE 6.5   Results of Compression and Flexural Strength and Water Absorption for Second Stage Specimens

Specimens code	a	b	c	d-1	d-2	e-1	e-2	f
Compression Strength (MPa) (28 days)	28.8	40.9	38	45.4	45.7	31.6	31.6	32.4
Flexural Strength (MPa) (28 days)	—	—	—	10.5	12	8.5	11.1	10.7
Water absorption	Plain (2.4%)				Optimum (1.4%)			

6.3.1.1   COMPRESSIVE AND FLEXURAL STRENGTHS

Result showed that fibers and Nano-particles have effectively improved concrete's compressive and flexural strengths, (see Table 6.3 and 6.5). For example, in P4 mixture that contains 2.25% of PP fiber and 2.25% of nano-SiO$_2$, flexural strength is 13% higher, and compressive strength is 25% higher than those of plain sample (P5 sample).

It also shows that both compressive and flexural strengths increase to its maximum amount for P4 sample; a good relation about these two factors can be confirmed. This sample also selected as the optimum sample at both age of curing. It also showed that both compressive and flexural strengths decrease with increase in fibers or nano particles content. For example, P11 mixture with more fiber content than others even has lower strengths than p5 mixture. It is due to the random contribution of fibers and lack of nano particles' good dispersion.

By adding the nano-SiO_2 to (b) and (c) mixtures, compressive strength is increased, significantly. In other words, nano-SiO_2 is more effective than nano-TiO_2 to improve the compressive strength. The reason may be attributed to the fact that the specific surface of nano-SiO_2 is much larger than nano-TiO_2, so nano-TiO_2 is more difficult to uniformly disperse than nano-SiO_2 in concrete microstructure. Results of 28-days compressive strength is more than 7-days for all specimens as it was expected. Also by adding PP fibers in the mixture, flexural strength is improved significantly for all specimens. It seems that increase in flexural strength of concrete due to adding PP fiber is more than increase in compressive strength of samples. For p6 and p9 samples that include only nano-SiO_2, both 7-days and 28-days compressive strengths are more than those for p7 and p11 samples that were contained only PP fibers. This shows that nano-SiO_2 is more effective in compressive strength improvement than PP fiber.

RSM plots (Figs. 6.1 and 6.2) confirm the results of this test in first stage, since the values were statistically analyzed by the ANOVA technique (variances analyzes tables due to design of experiments' method) at the 95% confidence level. The coefficient of determination (R2) calculated in 7-days and 28-days compressive strength, was 0.9620 and 0.9288, respectively, showed that this model is acceptable.

6.3.1.2   WATER PENETRATION

Water penetration result (Table 6.5) shows that the water penetration depth of concrete has significantly reduced after adding nano-SiO_2 (2.25% of PP fiber and 2.25% of nano-SiO_2). Water penetration for P4 sample (1.4%) is much less than that of plain sample (2.4%). Therefore, by adding fibers and nano-particles to concrete samples, water penetration was decreased. The reason for this could be due to decreasing in porosity and inhibiting the formation of micro cracks in concrete microstructure, which can consequently enhance durability of concrete.

6.3.2   SEM RESULTS AND ANALYSIS ABOUT MICROSTRUCTURE OF CONCRETE

The primary phase compositions of hydrated concrete include the amorphous and poorly crystalline calcium silicate hydrate (C–S–H), the layered calcium hydroxide (CH, or Ca(OH)$_2$), and the calcium aluminate/aluminoferrite hydrate phases (AFm and AFt-type phases) [13]. C–S–H is the primary bind phase in the hardened concrete. CH is also an important component occupying about 20–25% of hydrate products by volume [14]. Figure 6.3 shows the microstructures of the plain and optimum modified concrete samples. It seems that C1 mixture (plain concrete) contains many crystalline with the shape like fiber, snow flower, or needle, which interweave together among C–S–H. However, the CH crystalline presents in a layered structure with high orientation. Meanwhile, many voids are noticeable among these crystalline regions. In contrast, optimum sample contains many gel components and have much less voids than C1 sample. Its crystalline size is obviously smaller than that of C1. The size of the needle-shaped crystallines decreases from 8–10 µm (C1 mixture) to 2–4 µm (F2 mixture). It even seems hard to identify the interfaces between different crystalline layers. Especially F2 mixture contains very few voids among crystalline, and its gels look very integral and continuous. Figure 6.3 also shows the microstructures of the aggregate-cement interfacial transition zone for the plain and NF reinforced concretes with a magnifying ratio of 3000. It shows that C1 contains many C–S–H gel and hexagon sill CH crystalline with relatively large particle sizes, and it has noticeable voids and microcracking at the interfacial transition zone. However, the microstructures at the interfacial transition zone of the optimum mixtures especially F2 look much denser with more uniform particle sizes than that of C1, the voids at its interfacial transition zone are very minimal (F2 mixture in Fig. 6.3).

6.3.3   MECHANISM OF FIBER AND NANO PARTICLES MODIFICATION

It has expected that fiber and nano particles performs a kind of network, that make a bridge between cracks that cause dense microstructure and toughens the reinforce concrete, thus fiber and nano particles significantly improves the toughness of cementation matrices [2, 3, 11, 12].

It seems that fibers and nano particles network's formation is restricted the growth of CH crystalline and reduced the micro voids, and therefore decreases the size and orientation of CH. As a result, the microstructures and the aggregate-cement interfacial transition zone of NF are much denser with less microcracking than that of plain concrete, and the size and orientation of CH have significantly reduced. Therefore, NF's compressive and flexural strengths values, have significantly increased, and the water penetration have obviously decreased. Furthermore, fibers and nano particles create the networking effect to hold concrete aggregates together and also prevent crack propagation. Fibers would also perform the bridging crack function to take part of the internal tensile stress and thus resist the crack propagation after appearing microcracks. In addition, fiber reduces the shrinkage contractions and thus reduces the interfacial relative slides between concrete and other substrate like hardened concrete, which contribute to part of the improved interfacial bonding strength discussed previously. However, when fibers and nano particles contents are relatively up to (3%), fibers and nano particles distribution in concrete mixture becomes uniform weekly due to the reduced workability.

As a result, concrete density decreases with increasing fibers and nano particles contents and the aggregated fibers and nano particles may form weak points and induce relatively large voids in the concrete structure. Also, the material strengths and water penetration resistance decrease with increasing fiber and nano particles content. However, other research showed that additives like silica fume may improve the dispersion of fiber and nano particles in concrete, and thus help improve the performance of concrete samples.

6.4   CONCLUSION

This research endeavors to study how fiber and nano particles modify the physical and mechanical properties of concrete. The primary findings are summarized as follows:

1. simultaneous application of the fiber and nano particles can significantly alter the microstructure of concrete as it reduces the amount, size, and orientation of CH crystals and micro voids, and it reduces the voids and microcracking at the interfacial transition zone between aggregate and cement;

2. simultaneous application of the fiber and nano particles effectively improves the engineering properties of concrete including material strengths, and water penetration resistance, and fiber specifically reduces the shrinkage contraction of concrete and increases the flexural strength;

3. simultaneous application of the fiber and nano particles forms a network to restrict the growth of CH crystallines and thus condenses the microstructure of concrete and improves physical and mechanical properties of reinforcement concrete;

4. increase the nano particles and pp fiber content over to 3% weight in mixture, showed inverse influence to mechanical properties of concrete. Also it is concluded that nano-SiO_2 is more effective than nano-TiO_2 and PP fiber in improving physical and mechanical properties of concrete. The results indicated that simultaneously adding of nano particles and PP fibers can significantly improve the physical and mechanical properties of concrete.

KEYWORDS

- **concrete**
- **fiber reinforced concrete**
- **ITZ**
- **nanoparticle**
- **physical and mechanical properties**

REFERENCES

1. Mazaheripour, H., & Ghanbarpour, S. (2011). "The Effect of PP Fibers on the Properties of Fresh and Hardened Light Weight Self Compacting Concrete" Construction and Building Material 25(1), 351–358.

2. Raki, L., Beaudion, J., Alizadeh, R., Maker, J., & Sate, T. (2010). "Cement and Concrete Nano Science and Nano Technology", Materials, 3, 918–942.
3. Bagherzadeh, R., Sadeghi, A. H., & Latifi, M. (2012). "Utilizing Polypropylene Fibers to Improve Physical and Mechanical Properties of Concrete", Textile Research Journal, 82(1), 88–96.
4. Bagherzadeh, R., Pakravan, H. R., Sadeghi, A. H., Latifi, M., & Merati, A. A. (2012). "An Investigation on Adding Polypropylene Fibers to Reinforce Lightweight Cement Composites (LWC)", Journal of Engineered Fibers & Fabrics, 7(4), 13–21.
5. Sun, Z., & Xu, Q. (2009). Materials Science and Engineering A, 527 (1–2), 198–204.
6. Behnood, A., & Ghandehari, M. (2009). "Comparison of Compressive and Splitting Tensile Strength of High Strength Concrete with and Without PP Fibers Heated To High Temperature" Fire Safety Journal 44 (8), 1015–1022.
7. Diamanti, M. V., Ormellese, M., & Pedeforri, M. P. (2008). "Characterization of Photo Catalytic and Super Hydrophilic Properties of Mortars Containing TiO_2", Cement and Concrete Research 38, 1349–1353.
8. Hui, Li., Hua, Z. M., & Ou Jing-Ping (2006). "Abrasion Resistances of Concrete containing Nanoparticles", Wear, 260 (11–12), 1262–1266.
9. Kodur, V. K. R., Cheng, F. P., Wang, T. C., & Sultan, M. A. (2003). "Effects of Strength and Fiber Reinforcement on the Fire Resistance of High Strength Concrete Columns", Journal of Structural Engineering, 129(2), 253–259.
10. Hui, Li., Yuon-Jie, & Jing-Ping, Ou. (2004). "Microstructure of Cement Mortar with Nanoparticles", Composites Part B Engineering, 35, 185–189.
11. Jye-Chyi Lu, Kaibo Wang, & Shuen-Lin Jeng (2009). "A Review of Statistical Methods for Quality Improvement and Control in Nanotechnology", Journal of Quality Technology, 41(2).
12. Hui, Li., Hua, Z. M., & Ou Jing Ping, (2007). "Flexural Fatigue Performance of Concrete Containing Nano Particles for Pavement", International Journal of Fatigue, 29, 1292–1301.
13. Girao, A. V., Richardson, I. G., Porteneuve, C. B., & Brydson, R. M. D. (2007). "Composition, Morphology and Nanostructure of C–S–H in White Portland Cement Pastes hydrated at 55°C", Cement and Concrete Research, 37, 1571–1582.
14. Richardson, I. G. (2004). "Tobermorite Jennite and Tobermorite Calcium Hydroxide-Based Models for the Structure of C-S-H: Applicability to Hardened Pastes of Tricalcium Silicate, β-Dicalcium Silicate, Portland Cement, and blends of Portland Cement with Blast-Furnace Slag, Metakaolin, or Silica Fume", Cement and Concrete Research, 34, 1733–1777.

CHAPTER 7

UV-PROTECTIVE NANOCOMPOSITES FILMS BASED ON POLYETHYLENE

A. A. OLKHOV[1], M. A. GOLDSHTRAKH[1], G. NYSZKO[2],
K. MAJEWSKI[2], J. PIELICHOWSKI[3], and G. E. ZAIKOV[4]

[1]N.N. Semenov Institute of Chemical physics Russian Academy of Sciences
119991 Moscow, street Kosygina, 4, E-mail: *aolkhov72@yandex.ru*

[2]Military Institute of Chemistry and Radiometry, 105 Allea of General A. Chrusciela,
00-910 Warsaw, Poland, E-mail: Grzegorz.Nyszko@wichir.waw.pl; K.Majewski@
wichir.waw.pl

[3]Cracow University of Technology, Department of Polymer Science and Technology,
Warszawska str., 31-155 Krakow, Poland; E-mail: Pielich@pk.edu.pl

[4]N. M. Emanuel Institute of Biochemical physics Russian Academy of Sciences
119334 Moscow, street Kosygina, 4; E-mail: chembio@sky.chph.ras.ru

CONTENTS

Abstract 74
7.1   Introduction 74
7.2   Objects of Research 76
7.3   Experimental Methods 77
7.4   Results and Discussion 78
Keywords 84
References...85

ABSTRACT

High-strength polyethylene films containing 0.5–1.0 wt.% of nanocrystalline silicon (nc-Si) were synthesized. Samples of nc-Si with an average core diameter of 7–10 nm were produced by plasmochemical method and by laser-induced decomposition of monosilane. Spectral studies revealed almost complete (up to ~95%) absorption of UV radiation in 200–400 nm spectral region by 85 micron thick film if the nc-Si content approaches to 1.0 wt.%. The density function of particle size in the starting powders and polymer films containing immobilized silicon nanocrystallites were obtained using the modeling a complete profile of X-ray diffraction patterns, assuming spherical grains and the lognormal distribution. The results of X-ray analysis shown that the crystallite size distribution function remains almost unchanged and the crystallinity of the original polymer increases to about 10% with the implantation of the initial nc-Si samples in the polymer matrix.

7.1   INTRODUCTION

In recent years, considerable efforts have been devoted for search new functional nanocomposite materials with unique properties that are lacking in their traditional analogs. Control of these properties is an important fundamental problem. The use of nanocrystals as one of the elements of a polymer composite opens up new possibilities for targeted modification of its optical properties because of a strong dependence of the electronic structure of nanocrystals on their sizes and geometric shapes. An increase in the number of nanocrystals in the bulk of composites is expected to enhance long-range correlation effects on their properties. Among the known nanocrystals, nanocrystalline silicon (nc-Si) attracts high attention due to its extraordinary optoelectronic properties and manifestation of quantum size effects. Therefore, it is widely used for designing new generation functional materials for nanoelectronics and information technologies. The use of nc-Si in polymer composites calls for a knowledge of the processes of its interaction with polymeric media. Solid nanoparticles can be combined into aggregates (clusters), and, when the percolation threshold is achieved, a continuous cluster is formed.

An orderly arrangement of interacting nanocrystals in a long-range potential minimum leads to formation of periodic structures. Because of the well-developed interface, an important role in such systems belongs to adsorption processes, which are determined by the structure of the nanocrystal surface. In a polymer medium, nanocrystals are surrounded by an adsorption layer consisting of polymer, which may change the electronic properties of the nanocrystals. The structure of the adsorption layer has an effect on the processes of self-organization of solid-phase particles, as well as on the size, shape, and optical properties of resulting aggregates. According to data obtained for metallic [1] and semiconducting [2] clusters, aggregation and adsorption in three-phase systems with nanocrystals have an effect on the optical properties of the whole system. In this context, it is important to reveal the structural features of systems containing nanocrystals, characterizing aggregation and adsorption processes in these systems, which will make it possible to establish a correlation between the structural and the optical properties of functional nano-composite systems.

Silicon nanoclusters embedded in various transparent media are a new, interesting object for physicochemical investigation. For example, for particles smaller than 4 nm in size, quantum size effects become significant. It makes possible to control the luminescence and absorption characteristics of materials based on such particles using of these effects [3, 4]. For nanoparticles about 10 nm in size or larger (containing $\sim10^4$ Si atoms), the absorption characteristics in the UV and visible ranges are determined in many respects by properties typical of massive crystalline or amorphous silicon samples. These characteristics depend on a number of factors: the presence of structural defects and impurities, the phase state, etc. [5, 6]. For effective practical application and creation on a basis nc-Si the new polymeric materials possessing useful properties: sun-protection films [7] and the coverings [8] photoluminescent and electroluminescent composites [9, 10], stable to light dyes [11], embedding of these nanosized particles in polymeric matrixes becomes an important synthetic problem.

The method of manufacture of silicon nanoparticles in the form of a powder by plasma chemical deposition, which was used in this study, makes possible to vary the chemical composition of their surface layers. As a result, another possibility of controlling their spectral characteristics arises, which is absent in conventional methods of manufacture of nanocrystalline silicon in solid matrices (e.g., in α-SiO_2) by implantation of charged silicon particles [5] or radio frequency deposition of silicon [2]. Polymer composites based on silicon nanopowder are a new object for comprehensive spectral investigation. At the same time, detailed spectral analysis has been performed for silicon nanopowder prepared by laser-induced decomposition of gaseous SiH_4 (see, e.g., [6, 12]). It is of interest to consider the possibility of designing new effective UV protectors based on polymer containing silicon nanoparticles [13]. An advantage of this nanocomposite in comparison with other known UV protectors is its environmental safety, that is, ability to hinder the formation of biologically harmful compounds during UV-induced degradation of components of commercial materials. In addition, changing the size distribution of nanoparticles and their concentration in a polymer and correspondingly modifying the state of their surface, one can deliberately change the spectral characteristics of nanocomposite as a whole. In this case, it is necessary to minimize the transmission in the wavelength range below 400 nm (which determines the properties of UV-protectors [13]) by changing the characteristics of the silicon powder.

7.2   OBJECTS OF RESEARCH

In this study, the possibilities of using polymers containing silicon nanoparticles as effective UV protectors are considered. First, the structure of nc-Si obtained under different conditions and its aggregates, their adsorption and optical properties was studied in order to find ways of control the UV spectral characteristics of multi-phase polymer composites containing nanocrystalline silicon. Also, the purpose of this work was to investigate the effect of the concentration of silicon nanoparticles embedded in polymer matrix and the methods of preparation of these nanoparticles on the spectral characteristics of such nanocomposites. On the basis of the data obtained, recommendations for designing UV protectors based on these nanocom-posites were formulated.

nc-Si consists of core–shell nanoparticles in which the core is crystalline silicon coated with a shell formed in the course of passivation of nc-Si with oxygen and/or nitrogen. nc-Si samples were synthesized by an original procedure in an argon plas-ma in a closed gas loop. To do this, we used a plasma vaporizer/condenser operating in a low-frequency arc discharge. A special consideration was given to the formation of a nanocrystalline core of specified size. The initial reagent was a silicon powder, which was fed into a reactor with a gas flow from a dosing pump. In the reactor, the powder vaporized at 7000–10,000°C. At the outlet of the high-temperature plasma zone, the resulting gas–vapor mixture was sharply cooled by gas jets, which resulted in condensation of silicon vapor to form an aerosol. The synthesis of nc-Si in a low-frequency arc discharge was described in detail in Ref. [3].

The microstructure of nc-Si was studied by transmission electron microscopy (TEM) on a Philips NED microscope. X-ray powder diffraction analysis was car-ried out on a Shimadzu Lab XRD-6000 diffractometer. The degree of crystallinity of nc-Si was calculated from the integrated intensity of the most characteristic peak at $2\theta = 28°$. Low-temperature adsorption isotherms at 77.3 K were measured with a Gravimat-4303 automated vacuum adsorption apparatus. FTIR spectra were re-corded on in the region of 400–5000 cm^{-1} with resolution of about 1 cm^{-1}.

Three samples of nc-Si powders with specific surfaces of 55, 60, and 110 m^2/g were studied. The D values for these samples calculated by Eq. (2) are 1.71, 1.85, and 1.95, respectively; that is, they are lower than the limiting values for rough objects. The corresponding D values calculated by Eq. (3) are 2.57, 2.62, and 2.65, respectively. Hence, the adsorption of nitrogen on nc-Si at 77.3 K is determined by capillary forces acting at the liquid–gas interface. Thus, in argon plasma with addi-tion of oxygen or nitrogen, ultra disperse silicon particles are formed, which consist of a crystalline core coated with a silicon oxide or oxynitride shell. This shell pre-vents the degradation or uncontrollable transformation of the electronic properties of nc-Si upon its integration into polymer media. Solid structural elements (threads or nanowires) are structurally similar, which stimulates self-organization leading to fractal clusters. The surface fractal dimension of the clusters determined from the nitrogen adsorption isotherm at 77.3 K is a structurally sensitive parameter, which characterizes both the structure of clusters and the morphology of particles and aggregates of nanocrystalline silicon.

As the origin materials for preparation film nanocomposites served polyethylene of low density (LDPE) marks 10803–020 and ultra disperse crystal silicon. Silicon powders have been received by a method plazmochemical recondensation of coarse-crystalline silicon in nanocrystalline powder. Synthesis nc-Si was carried out in argon plasma in the closed gas cycle in the plasma evaporator the condenser working in the arc low-frequency category. After particle synthesis nc-Si were exposed microcapsulating at which on their surfaces the protective cover from SiO_2, protecting a powder from atmospheric influence and doing it steady was created at storage. In the given work powders of silicon from two parties were used: nc-Si-36 with a specific surface of particles ~36 m^2/g and nc-Si-97 with a specific surface ~97 m^2/g.

Preliminary mixture of polyethylene with a powder nc-Si firms "Brabender" (Germany) carried out by means of closed hummer chambers at temperature $135\pm5°C$, within 10 min and speed of rotation of a rotor of 100 min^{-1}. Two compositions LDPE + nc-Si have been prepared: (i) composition PE + 0.5% ncSi-97 on a basis ncSi-97, containing 0.5 weights silicon %; (ii) composition PE + 1% ncSi-36 on a basis ncSi-36, containing 1.0 weights silicon %.

Formation of films by thickness 85 ± 5 micron was spent on semiindustrial extrusion unit ARP-20–150 (Russia) for producing the sleeve film. The temperature was 120–190°C on zones extruder and extrusion die. The speed of auger was 120 min^{-1}. Technological parameters of the nanocomposites choose, proceeding from conditions of thermostability and the characteristic viscosity recommended for processing polymer melting.

7.3 EXPERIMENTAL METHODS

Mechanical properties and an optical transparency of polymer films, their phase structure and crystallinity, and also communication of mechanical and optical properties with a microstructure of polyethylene and granulometric structure of modifying powders nc-Si were observed.

Physicomechanical properties of films at a stretching (extrusion) measured in a direction by means of universal tensile machine EZ-40 (Germany) in accordance with Russian State Standard GOST-14236–71. Tests are spent on rectangular samples in width of 10 mm, and a working site of 50 mm. The speed of movement of a clip was 240 mm/min. The 5 parallel samples were tested.

Optical transparency of films was estimated on absorption spectra. Spectra of absorption of the obtained films were measured on spectrophotometer SF-104 (Russia) in a range of wavelengths 200–800 nanometers. Samples of films of polyethylene and composite films PE + 0.5% ncSi-36 and PE + 1% ncSi-36 in the size 3×3 cm were investigated. The special holder was used for maintenance uniform a film tension.

X-ray diffraction analysis by wide-angle scattering of monochromatic X-rays data was applied for research phase structure of materials, degree of crystallinity of a polymeric matrix, the size of single-crystal blocks in powders nc-Si and in a polymeric matrix, and also functions of density of distribution of the size crystalline particles in initial powders nc-Si.

X-ray diffraction measurements were observed on Guinier diffractometer: chamber G670 Huber [14] with bent Ge (111) monochromator of a primary beam which are cutting out line $K\alpha_1$ (length of wave $l = 1.5405981$ Å) characteristic radiation of *x*-ray tube with the copper anode. The diffraction picture in a range of corners 2q from 3° to 100° was registered by the plate with optical memory (IP-detector) of the camera bent on a circle. Measurements were spent on original powders nc-Si-36 and nc-Si-97, on the pure film LDPE further marked as PE, and on composite films PE + 0.5% ncSi-97 and PE + 1.0% nc-Si-36. For elimination of tool distortions effect diffractogram standard SRM660a NIST from the crystal powder LaB_6 certificated for these purposes by Institute of standards of the USA was measured. Further it was used as diffractometer tool function.

Samples of initial powders nc-Si-36 and nc-Si-97 for X-ray diffraction measurements were prepared by drawing of a thin layer of a powder on a substrate from a special film in the thickness 6 microns (MYLAR, Chemplex Industries Inc., Cat. No: 250, Lot No: 011671). Film samples LDPE and its composites were established in the diffractometer holder without any substrate, but for minimization of structure effect two layers of a film focused by directions extrusion perpendicular each other were used.

Phase analysis and granulometric analysis was spent by interpretation of the X-ray diffraction data. For these purposes the two different full-crest analysis methods [15, 16] were applied: (i) method of approximation of a profile diffractogram using analytical functions, polynoms and splines with diffractogram decomposition on making parts; (ii) method of diffractogram modeling on the basis of physical principles of scattering of X-rays. The package of computer programs WinXPOW was applied to approximation and profile decomposition diffractogram ver. 2.02 (Stoe, Germany) [17], and diffractogram modeling at the analysis of distribution of particles in the sizes was spent by means of program PM2K (version 2009) [18].

7.4   RESULTS AND DISCUSSION

Results of mechanical tests of the prepared materials are presented to Table 7.1 from which it is visible that additives of particles nc-Si have improved mechanical characteristics of polyethylene.

TABLE 7.1 Mechanical Characteristics of Nanocomposite Films Based of LDPE and nc-Si

Sample	Tensile strength, kg/cm²	Relative elongation-at-break, %
PE	100 ± 12	200–450
PE + 1% ncSi-36	122 ± 12	250–390
PE + 0.5% ncSi-97	118 ± 12	380–500

The results presented in the table show that additives of powders of silicon raise mechanical characteristics of films, and the effect of improvement of mechanical properties is more expressed in case of composite PE + 0.5% ncSi-97 at which in comparison with pure polyethylene relative elongation-at-break has essentially grown.

Transmittance spectra of the investigated films are shown on Fig. 7.1.

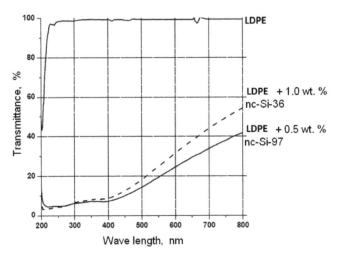

FIGURE 7.1 Transmittance spectra of the investigated films LDPE and nanocomposite films PE + 0.5% ncSi-97 and PE + 1.0% ncSi-36.

It is visible that additives of powders nc-Si reduce a transparency of films in all investigated range of wavelengths, but especially strong decrease transmittance (almost in 20 times) is observed in a range of lengths of waves of 220–400 nanometers, that is, in UV areas.

The wide-angle scattering of X-rays data were used for the observing phase structure of materials and their component. Measured X-ray diffractograms of initial powders ncSi-36 and ncSi-97 on intensity and Bragg peaks position completely corresponded to a phase of pure crystal silicon (a cubic elementary cell of type of diamond – spatial group $Fd\overline{3}m$, cell parameter $a_{Si} = 0.5435$ nanometers).

For the present research granulometric structure of initial powders nc-Si is of interest. Density function of particle size in a powder was restored on X-ray diffractogram a powder by means of computer program PM2K [18] in which the method [19] modelings of a full profile diffractogram based on the theory of physical processes of diffraction of X-rays is realized. Modeling was spent in the assumption of the spherical form of crystalline particles and logarithmically normal distributions of their sizes. Deformation effects from flat and linear defects of a crystal lattice were considered. Received function of density of distribution of the size crystalline particles for initial powders nc-Si are represented graphically on Fig. 7.2, in the signature to which statistical parameters of the found distributions are resulted. These distributions are characterized by such important parameters as, $Mo(d)$ – position of maximum (a distribution mode); $<d>_V$ – average size of crystalline particles based on volume of the sample (the average arithmetic size) and $Me(d)$ – the median of distribution defining the size d, specifying that particles with diameters less than this size make half of volume of a powder.

The results represented on Fig. 7.2 show that initial powders nc-Si in the structure have particles with the sizes less than 10 nanometers which especially effectively absorb UV radiation. The both powders modes of density function of particle size are very close, but median of density function of particle size of a powder ncSi-36 it is essential more than at a powder ncSi-97. It suggests that the number of crystalline particles with diameters is less 10 nanometers in unit of volume of a powder ncSi-36 much less, than in unit of volume of a powder ncSi-97. As a part of a powder ncSi-36 it is a lot of particles with a diameter more than 100 nanometers and even there are particles more largely 300 nanometers whereas the sizes of particles in a powder ncSi-97 don't exceed 150 nanometers and the basic part of crystalline particles has diameter less than 100 nanometers.

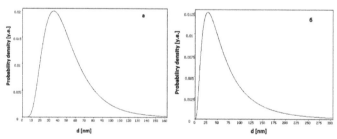

FIGURE 7.2 Density function of particle size in powders ncSi, received from x-ray diffractogram by means of program PM2K:

(a) – ncSi-97	Mo(d) = 35 nm	Me(d) = 45 nm	$<d>_V$ = 51 nm;
(b) – ncSi-36	Mo(d) = 30 nm	Me(d) = 54 nm	$<d>_V$ = 76 nm.

The phase structure of the obtained films was estimated on wide-angle scattering diffractogram only qualitatively. Complexity of diffraction pictures of scattering and structure don't poses the quantitative phase analysis of polymeric films [20]. At the phase analysis of polymers often it is necessary to be content with the comparative qualitative analysis, which allows watching evolution of structure depending on certain parameters of technology of production. Measured wide-angle X-rays scattering diffractograms of investigated films are shown on Fig. 7.3. Diffractograms have a typical form for polymers. As a rule, polymers are the two-phase systems consisting of an amorphous phase and areas with distant order, conditionally named crystals. Their diffractograms represent [20] superposition of intensity of scattering by the amorphous phase, which is looking like wide halo on the small-angle area (in this case in area 2q between 10° and 30°), and intensity Bragg peaks scattering by a crystal phase.

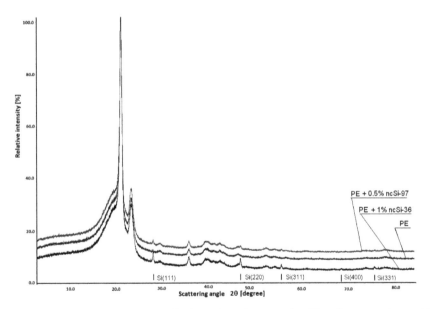

FIGURE 7.3 Diffractograms of the investigated composite films in comparison with diffractogram of pure polyethylene. Below vertical strokes specify reference positions of diffraction lines of silicon with their interference indexes (hkl).

Data on Fig. 7.3 is presented in a scale of relative intensities (intensity of the highest peak is accepted equal 100%). For convenience of consideration curves are represented with displacement on an axis of ordinates. The scattering plots without displacement represented completely overlapping of diffractogram profiles of composite films with diffractogram of a pure LDPE film, except peaks of crystal silicon, which weren't present on PE diffractogram. It testifies that additives of powders nc-Si practically haven't changed crystal structure of polymer.

The peaks of crystal silicon are well distinguishable on diffractograms of films with silicon (the reference positions with Miller's corresponding indexes are pointed below). Heights of the peaks of silicon with the same name (i.e., peaks with identical indexes) on diffractograms of the composite films PE + 0.5% ncSi-97 and PE + 1.0% ncSi-36 differ approximately twice that corresponds to a parity of mass concentration Si set at their manufacturing.

Degree of crystallinity of polymer films (a volume fraction of the crystal ordered areas in a material) in this research was defined by diffractograms Fig. 7.3 for a series of samples only semiquantitative (more/less). The essence of the method of crystallinity definition consists in analytical division of a diffractogram profile on the Bragg peaks from crystal areas and diffusion peak of an amorphous phase [20], as is shown in Fig. 7.4.

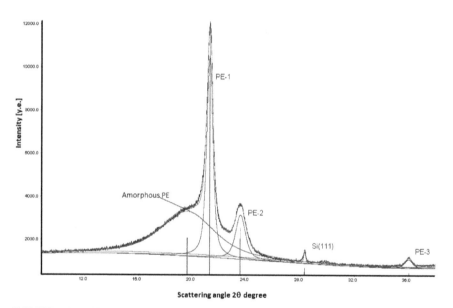

Scattering angle 2θ degree

FIGURE 7.4 Diffractogram decomposition on separate peaks and a background by means of approximation of a full profile by analytical functions on an example of the data from sample PE+1%ncSi-36 (Fig. 7.3). PE-n designate Bragg peaks of crystal polyethylene with serial numbers n from left to right. Si (111) – Bragg silicon peak ncSi-36. Vertical strokes specify positions of maxima of peaks.

Peaks profiles of including peak of an amorphous phase, were approximated by function pseudo-Foigt, a background 4 order polynoms of Chebysheva. The nonlinear method of the least squares minimized a difference between intensity of points experimental and approximating curves. The width and height of approximating functions, positions of their maxima and the integrated areas, and also background parameters were thus specified. The relation of integrated intensity of a scattering profile by an amorphous phase to full integrated intensity of scattering by all phases except for particles of crystal silicon gives a share of amorphy of the sample, and crystallinity degree turns out as a difference between unit and an amorphy fraction.

It was supposed that one technology of film obtaining allowed an identical structure. It proved to be true by coincidence relative intensities of all peaks on diffractograms Fig. 7.3, and samples consist only crystal and amorphous phases of the same chemical compound. Therefore, received values of degree of crystallinity should reflect correctly a tendency of its change at modification polyethylene by powders nc-Si though because of a structure of films they can quantitatively differ considerably from the valid concentration of crystal areas in the given material. The found values of degree of crystallinity are represented in Table 7.2.

TABLE 7.2 Characteristics of the Ordered (Crystal) Areas in Polyethylene and Its Composites with nc-Si

PE			PE + 1% ncSi-36			PE + 0.5% ncSi-97		
Crystallinity	**46%**		**47.5%**			**48%**		
2q [°]	d [E]	e	2q [°]	d [E]	e	2q [°]	d [E]	e
21.274	276	8.9	21.285	229	7.7	21.282	220	7.9
23.566	151	12.8	23.582	128	11.2	23.567	123	11.6
36.038	191	6.8	36.035	165	5.8	36.038	162	5.8
Average values	206	9.5 $\times 10^{-3}$		174	8.2 $\times 10^{-3}$		168	8.4 $\times 10^{-3}$

One more important characteristic of crystallinity of polymer is the size d of the ordered areas in it. For definition of the size of crystalline particles and their maximum deformation e *in* X-ray diffraction analysis [21] Bragg peaks width on half of maximum intensity (Bragg lines half-width) is often used. In the given research the sizes of crystalline particles in a polyethylene matrix calculated on three well-expressed diffractogram peaks Fig. 7.3. The peaks of polyethylene located at corners 2q approximately equal 21.28°, 23.57° and 36.03° (peaks PE-1, PE-2 and PE-3 on Fig. 7.4) were used. The ordered areas size d and the maximum relative deformation e of their lattice were calculated by the joint decision of the equations of Sherrera and Wilson [21] with use of half-width of the peaks defined as a result of approximation by analytical functions, and taking into account experimentally measured diffractometer tool function. Calculations were spent by means of program $WinX^{POW}$

size/strain. Received *d* and *e*, and also their average values for investigated films are presented in Table 7.2. The updated positions of maxima of diffraction peaks used at calculations are specified in the table.

The offered technology allowed the obtaining of films LDPE and composite films LDPE + 1% ncSi-36 and LDPE + 0.5% ncSi-97 an identical thickness (85 microns). Thus concentration of modifying additives ncSi in composite films corresponded to the set structure that is confirmed by the X-ray phase analysis.

By direct measurements it is established that additives of powders ncSi have reduced a polyethylene transparency in all investigated range of lengths of waves, but especially strong transmittance decrease (almost in 20 times) is observed in a range of lengths of waves of 220–400 nanometers, that is, in UV areas. Especially strongly effect of suppression UV of radiation is expressed in LDPE film + 0.5% ncSi-97 though concentration of an additive of silicon in this material is less. It is possible to explain this fact to that according to experimentally received function of density of distribution of the size the quantity of particles with the sizes is less 10 nanometers on volume/weight unit in a powder ncSi-97 more than in a powder ncSi-36.

Direct measurements define mechanical characteristics of the received films – durability at a stretching and relative lengthening at disrupture (Table 7.1). The received results show that additives of powders of silicon raise durability of films approximately on 20% in comparison with pure polyethylene. Composite films in comparison with pure polyethylene also have higher lengthening at disrupture, especially this improvement is expressed in case of composite PE + 0.5% ncSi-97. Observable improvement of mechanical properties correlates with degree of crystallinity of films and the average sizes of crystal blocks in them (Table 7.2). By results of the X-ray analysis the highest crystallinity at LDPE film + 0.5% ncSi-97, and at it the smallest size the crystal ordered areas that should promote durability and plasticity increase.

This work is supported by grants RFBR № 10–02–92000 and RFBR № 11–02–00868 also by grants FCP "Scientific and scientific and pedagogical shots of innovative Russia," contract № 2353 from 17.11.09 and contract № 2352 from 13.11.09.

KEYWORDS

- **nanocrystalline silicon**
- **polyethylene**
- **polymer nanocomposites**
- **spectroscopy**
- **UV-protective film**
- **X-ray diffraction analysis**

REFERENCES

1. Karpov, S. V., & Slabko, V. V. (2003). Optical and Photo Physical Properties of Fractally Structured Metal Sols, Novosibirsk: Sib. Otd. Ross. Akad Nauk.
2. Varfolomeev, A. E., Volkov, A. V., Yu, D., Godovskii, et al. (1995). Pis'ma Zh Eksp, Teor Fiz, *62*, 344.
3. Delerue, C., Allan, G., & Lannoo, M. (1999). J. Lumin, *80*, 65.
4. Soni, R. K., Fonseca, L. F., Resto, O. et al. (1999). J. Lumin, *83–84*, 187.
5. Altman, I. S., Lee, D., Chung, J. D. et al. (2001). Phys. Rev. B: Condens. Matter Material Physics *63*, 161402.
6. Knief, S., & Von Niessen, W. (1999). Phys. Rev. B: Condensation. Matter Mater Phys, *59*, 12940.
7. Olkhov, A. A., Goldschtrakh, M. A., & Ischenko, A. A. (2009). RU Patent №: 2009145013.
8. Bagratashvili, V. N., Tutorskii, I. A., Belogorokhov, A. I. et al. (2005). Reports of Academy of Science, Physical Chemistry, *405*, 360.
9. Kumar, V. (Ed.) (2008). Nano Silicon Elsevier Ltd. XIII 368p.
10. Carl, C., & Koch. N. Y. Eds. (2009). Nanostructured Materials. Processing, Properties, and Applications. William Andrew Publishing 752.
11. Ischenko, A. A., Dorofeev, S. G., Kononov, N. N. et al. (2009). RU Patent №: 2009146715.
12. Kuzmin, G. P., Karasev, M. E., Khokhlov, E. M. et al. (2000). Laser Physics, *10*, 939.
13. Beckman, J., & Ischenko, A. A. (2003). RU Patent No: 2 227 015.
14. Stehl, K. (2000). The Huber G670 Imaging-Plate Guinier Camera Tested on Beamline I711 at the MAX II Synchrotron. J. Appl. Cryst., *33*, 394–396.
15. Fetisov, G. V. (2010). The X-ray phase analysis. Chapter 11, 153–184. Analytical Chemistry and Physical and Chemical Methods of the Analysis T 2 Red Ischenko, A. A., M: ITc Academy, 416p.
16. Scardi, P., & Leoni, M. (2006). Line Profile Analysis: Pattern Modeling Versus Profile Fitting. J. Appl. Crystal. *39*, 24–31.
17. WINX^POW Version 1.06, STOE & CIE GmbH Darmstadt Germany (1999).
18. Leoni, M., Confente, T., & Scardi, P. (2006). PM2K: A Flexible Program Implementing Whole Powder Pattern Modelling Z. Kristallogr. Suppl. *23*, 249–254.
19. Scardi, P. (2008). Recent Advancements in Whole Powder Pattern Modeling Z Kristallogr Suppl. *27,* 101–111.
20. Strbeck, N. (2007). X-ray Scattering of Soft Matter, Springer-Verlag Berlin Heidelberg. 238p.
21. Iveronova, V. I., & Revkevich, U. P. (1978). The Theory of Scattering of X-Rays. M: MGU 278 p.

CHAPTER 8

NANOCOMPOSITE FOILS BASED ON SILICATE PRECURSOR WITH SURFACE-ACTIVE COMPOUNDS: A RESEARCH NOTE

L. O. ZASKOKINA, V. V. OSIPOVA, Y. G. GALYAMETDINOV, and A. A. MOMZYAKOV

CONTENTS

Abstract 88
8.1   Introduction 88
8.2   Experimental Part 88
8.3   Implications and Discussion 89
8.4   Conclusion 91
Keywords 92
References 92

ABSTRACT

Obtained film composites containing complex of Eu (DBM) 3 bpy. It is shown that the using of liquid-crystal matrix results to a more uniform and orderly arrangement of the complex and silicate substrate.

8.1   INTRODUCTION

Liquid crystals are extremely flexible that is driven under the influence of a relatively weak external factors leading to changes in macroscopic physical properties of the sample. Therefore, materials based on them have found practical application in the most advanced fields of science and technology [1].

Great interest is the creation based on lyotropic liquid crystals of nanoscale materials with improved physical properties by doping silicate matrices and lanthanide ions and other metals into them [2–3]. Adding silicate matrix to LCD systems allows to stabilize nanoorganized structure [4].

One of the most important areas of modern material engineering aligned with the obtaining of nanostructures with desired properties. For this widely used approach of obtaining composite nanomaterials, that is, particles of prisoners in a chemically inert matrix. In many cases, by way of such matrices we use different porous materials In these pores, various compounds may be administered and then, after chemical modification provide particles of the desired material, size and shape are the same shape as the cavities of the matrix, and its walls prevent their aggregation and to protect against environmental influences [5]. The aim of this work was to obtain nanocomposites lanthanide complexes with silicate matrices.

8.2   EXPERIMENTAL PART

Silicate matrices were prepared by the sol-gel process from reaction mixtures consisting of vinyl trimethoxysilane (VMOS)/phenyl trimethoxysilane (PhTEOS), ethanol and water [6]. Firstly, we measured the weight of precursor and then we added ethanol plus water. The samples obtained were placed on a magnetic stirrer at a rate of 450 rev/min at temperature of 50°C for time duration of 2–4 days. Periodically we checked pH value (2–3 times), time by time the water phase was transformed into a gel. A few hours later, we could observe a condensation into a solid continuous network. Cooked system is characterized by composition of the initial components, that is the molar and weight ratio.

Lyomesophases synthesis was carried out as described in article [8], based on nonionic surfactants (C12EO4, C12EO9, C12EO10). Identification of liquid-crystalline properties conducted according to the polarization-optical microscopy (POM) (Olympus BX 51) – the observed textures set the type of mesophase and the phase transition temperature.

Silicate was built in liquid crystal matrix between micellar aggregates. Incorporation of small amounts of hydrochloric acid necessary sought pH (2 ÷ 3) system, wherein the water phase is gelled. Few hours later we could observe the condensation into solid continuous network. [9]

Using spin coating (SPIN COATER LAU-TELL WS – 400–6NPP – LITE) we obtained films of nanocomposites including Si/C12EOn/H_2O. Following this method, we should introduce a system for glass and rotate it into device at a rate of 1000 rev/min.

Luminescent characteristics of multicomponent films by focusing mesophases become better, which allows passing from the supramolecular organization of the sample to streamline the entire volume.

8.3   IMPLICATIONS AND DISCUSSION

Control of the LCD systems synthesis process completeness performed by fixing constant temperature of mesophase transition, as a result, we got isotropic liquid in the whole volume of the sample.

Exploring LCD system textures, we found out that one system has a lamellar mesophase (C12EO14/H_2O), and two others have hexagonal one (C12EO10/H_2O, C12EO9/H_2O).

System were obtained with silicate matrix based on two different precursors – vinyl trimethoxysilane and phenyl triethoxysilane by sol-gel technology. During the synthesis of the silicate matrix, molar ratios, aging time and drying time were the most important characteristics, so they were experimentally optimized (Table 8.1).

TABLE 8.1   Controlled Parameters During the Sol-Gel Synthesis

Factor	Range*	Optimum
pH	1–7	2–3
		If the pH is too high (>pH 3.5), adding some amount of water, initial solution is turbid enough and the gel time is rapid
Sol temperature	18–100°C	50°C
		Samples are thickened quickly (a few days)
Aging time	1–20 days	2 days
		Drying time and temperature are the most important parameters

TABLE 8.1 *(Continued)*

Drying temperature	18–100°C	18°C
		All samples are dried at elevated temperatures, cracked
Drying time	1 h–6 months	> 45 days for 4 mL of sol
		It takes a long drying time.

* Published data.

After the admission of silicate into LCD, synthesized systems were investigated by polarized light. As in the original liquid crystal systems, appropriate texture observed that to characterize the supramolecular organization of the molecules in the mesophase.

Nanocomposite films were prepared by the method of spin coating. This method allows control of the film thickness, but this is a uniform distribution of the sample on the substrate.

Annealed porous films placed in a solution of the complex tris [1,3-diphenyl-1,3-propanedione] – [2,2'-bipyridyl] Europium Eu (DBM) 3 bpy in toluene for adsorption equation for 24 h. Then the samples were washed in toluene to remove the surfactant and water out of the surface and dried to remove the solvent [10]. After the process, the systems under study were examined by fluorescent spectroscopy (Fig. 8.1).

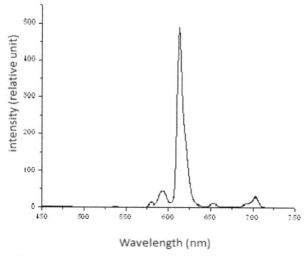

FIGURE 8.1 The Luminescence Spectrum and Kinetics of Luminescence System SiO_2/ $C12EO_4$/Eu (DBM) 3 bpy.

Lack of main energy transition signal splitting looks as shown: 5D0 → 7F1, 5D0 → 7F2 and 5D0 → 7F3 and indicates a low symmetry of the ligand environment of ion Eu^{3+}, which in principle expected in lyotropic systems containing a large amount of solvent. Small ratio of the peak areas of transitions 5D0 → 7F2 and 5D0 → 7F1 equal to 5.9 indicates a weak energy transfer from the ligand to the ion.

In films obtained on the glass, there is a uniform and orderly arrangement of the complex Eu (DBM) 3 bpy silicate and SiO_2, which increases the luminescence lifetime (Fig. 8.2). Found out that oriented systems compared to disordered systems have more intense photoluminescence.

It should be emphasized the role of the orientation of mesophases as a necessary step in the organization of a liquid crystal template. Orientation mesophases switches from supramolecular organization in the domains of the sample to streamline the entire volume, which is especially important for the practical design of new functional materials when used in molecular electronics and laser optics.

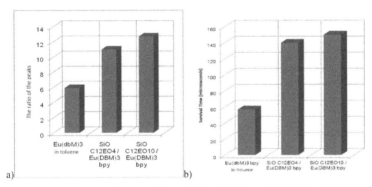

FIGURE 8.2 (a) The ratio of the peak intensities of 5D0 → 7F2/5D0 → 7F1; (b) Survival Time the glow luminescent complexes.

8.4 CONCLUSION

We obtained multicomponent $Si/C12EOn/H_2O$ systems and films containing complex of Eu (DBM) 3 bpy. In films, there is a uniform and orderly arrangement of Eu (DBM) 3 bpy silicate and SiO2, which increases the intensity and luminescence lifetime.

The study was supported by RFBR under research project number 12–08–31405 mol_a.

KEYWORDS

- **aging time of luminescence**
- **composite films**
- **liquid crystals**
- **luminescence**
- **multicomponent systems**
- **nanocomposites**
- **silicate matrix**

REFERENCES

1. Blackstock, J. J., Donley, C. L., Stickle, W. F., Ohlberg, D. A. A., Yang, J. J., Stewart, D. R., & Williams, R. S. (2008). J. Am. Chem. Soc., 130(12), 4041–4047.
2. Koen Binnemans, Yury, G. Galyametdinov, Rik Van Deun, & Duncan, W. Bruce, (2000). J. Am Chem. Soc. 122, 4335–4344.
3. Duncan, W. Bruce (2000). Acc. Chem. Res., 33, 831–840.
4. Ohtake, T., Ogasawara, M., Ito-Akita, K., Nishina, N., Ujiie, S., Ohno, H., & Kato, T. (2000). Chem. Mater, 12, 782.
5. Eliseev, A. A., Kolesnik, I. V., Lukashin, A. V., & Tretyakov, Y. D. (2005). Adv. Eng, Mater, 7(4), 213–217.
6. Margolin, V. I., Zharbeev, V. A., & Tupik, V. A. (2008). Physical Basis of Microelectronics of Academy, 400p.
7. Osipova, V. V., Zaskokina, L. O., Gumerov, F. M., & Galyametdinov, J. G. (2012). Vestnik of KNRTU, 17, 29–30.
8. Osipova, V. V., Selivanov, N. M., Danovski, D. E., & Galyametdinov, J. G. (2007). Vestnik of KNRTU, 6, 30–35.
9. Osipova, V. V., Yarullin, L. Y., Gumerov, F. M., & Galyametdinov, J. G. (2010). Vestnik of KNRTU, 9, 879–881.
10. Ostapenko, N. I., Kozlova, N. V., Frolov, E., Ostapenko, V., Pekus, D., Gulbinas V., Eremenko, A. M., Smirnov, N., Surovtceva, N. I., Suto, S., & Watanabe, A. (2011). Journal of Applied Spectroscopy, 78(1), 82–88.

CHAPTER 9

EXPERIMENTAL INVESTIGATION ON THE EFFECTS OF NANO CLAY ON MECHANICAL PROPERTIES OF AGED ASPHALT MIXTURE

M. ARABANI[1], A. K. HAGHI[2], and R. TANZADEH[3]

[1]Professor, Faculty of Engineering, University of Guilan, Rasht, Postal code: 3756, I. R. Iran.Tel: +98(131)6690270, Fax: +98 (131) 6690270. Email:arabani@guilan.ac.ir

[2]Professor, Faculty of Engineering, University of Guilan, Rasht, Postal code: 3756, I. R. Iran.Tel: +98(131)6690270, Fax: +98 (131) 6690270. Email:Haghi@guilan.ac.ir

[3]Graduated student. Department of Civil Engineering.University of Guilan, Rasht, Iran.Tel: +98(131)3229883, Fax: +98 (131) 3231116. Email: rashidtanzadeh@yahoo.com

CONTENTS

Abstract 94
9.1   Introduction 94
9.2   Materials and Test Program 95
9.3   Results and Discussions 97
9.4   Conclusions 104
Keywords 104
References 104

ABSTRACT

Bitumen is a very important adhesive part of the mixture that affects properties of asphalt mixtures. SBS is one of the most successful polymers, which could improve resistance to high temperature rutting. For further reinforced polymer and application of new material in bitumen, nano clay particles were used. Nano is a new additive used for modifying bitumen to obtain high resistance asphalt mixture. In this study, different percentages of Iranian nano clay with 5% SBS polymer (nano clay composite) were added to aged binder using thin film oven test. Creep tests were performed on the modified aged asphalt mixtures at 40°C according to the research requirements in Iran. The results showed that nano clay composite could modify aged asphalt mixture's creep parameters. Strain percent of the asphalt samples with modified aged bitumen with a specified nano clay composite value, decreased under static loads.

9.1   INTRODUCTION

Nanotechnology has opened a new world in nano-scale while civil engineering infrastructure is focused on macro scale. Despite the fact that good pavement is constructed using the existing materials, there are significant applications of nanotechnology to improve the pavement performance. Scientists have anticipated that nanotechnology may provide a great potential in the fields of material design, manufacturing, properties, monitoring and modeling in order to advance in asphalt pavement technology [1]. One of the most successful polymers is the styrene–butadiene–styrene block copolymer (SBS), which can retard low-temperature thermal stress cracking and improve resistance to high temperature rutting [2]. Nano material as a new molecular level size significantly improves properties at mechanical and related levels. Nano clay composite in the field of nanotechnology field provides a new insight into fundamental properties of clays [3]. The first polymer–clay nanocomposites (PCN) of Nylon 6-clay hybrid was invented in 1985 [4]. A number of bitumen physical properties enhanced successfully when as a polymer was modified with small amount of nanoclay [5]. Styrene–Butadiene–Styrene tri block copolymer used for modifying physical, mechanical and rheological properties of bitumen [6]. Asphalt concrete is a mixture of bitumen and aggregates and engineers attempt to improve performance of asphalt pavements. One of the most conventional ways to improve pavement performance is bitumen reinforcement with different additives but the most popular bitumen modification technique is polymer modification [7–9]. In this study, bitumen was aged using TFOT and modified by different percent's of nano clay composite. Marshal samples were made by nano composite modified aged bitumen and put under static loading. The results showed improvement in creep parameters and according to images of scanning electron microscope, nano clay composite was uniformly distributed in the bitumen.

9.2 MATERIALS AND TEST PROGRAM

9.2.1 MATERIALS

According to 101 Iranian Issue [10], middle range of dense graded HMA in Topeka and Binder was used, the specifications of which are given in Table 9.1.

TABLE 9.1   The Gradation of the Aggregates Used in This Research

Sieve size	Passing Percentage Topeka & Binder
1"	-
3/4"	100
1/2"	80
3/8"	-
#4	59
# 8	43
#50	13
#200	6

The bitumen used for this study was kindly supplied by courtesy of refinery in Tehran. The polymer SBS (Styrene – Butadiene – Styrene) with specified 5% with different nano clay percent was used. Engineering properties of the bitumen, the nano clay and SBS are presented in Tables 9.2, 9.3 and 9.4, respectively.

TABLE 9.2   Properties of the Bitumen Used in This Research

Specific gravity (at 25°C)	Penetration grade (mm/10)	Soften-ing point (°C)	Viscos-ity Pa.s	Ductility (cm)	Flash point (°C)	Purity grade (%)
1.013	67	50	289	112	308	99.6

TABLE 9.3   Properties of the Nano Clay Used in This Research

PM	Appearance	Purity grade (%)	SSA (m²/cm³)	APS (nm)
Nearly Spherical	White powder	99	750	10–25

TABLE 9.4   Properties of the SBS Used in This Research

Polymer	Type	Styrene Percent	Molecular structure
SBS	Carl Prene 501	31	Linear

9.2.2 SAMPLE PREPARATION

To simulate short-term aging of bitumen with respect to the existing facilities, initial bitumen in the laboratory of Guilan University was aged using thin film oven test (TFOT). In this case, bitumen was heated for 5 h at 163°C. Afterwards, the obtained aged bitumen used for Marshal samples. In the Marshal samples construction of reinforce aged bitumen with nano clay composite, first, nano clay composite percentage with 2, 4, 6 and 8% bitumen weight was added to the asphalt mixture. The modified bitumen was mixed with different percentages of nano clay composite for 20 min at temperatures varying from 120 to 150 degrees at 28,000 rpm; this mixer was built by the authors of this article. Then, Marshal mixed design was carried out for both modified and unmodified mixtures.

9.2.3 STATIC CREEP TEST

This test determined the resistance to permanent deformation of asphalt mixtures at temperatures and loads similar to those experienced by these mixtures in the actual field. Creep properties including stiffness, permanent strain and slop could be determined. Nottingham device test (NDT) was used in order not to destroying experimental tests. In this study, stress was considered as 150 KPa and the test was performed at constant temperature of 40°C.

9.2.4 SCANNING ELECTRON MICROSCOPE (SEM)

Observation of the bitumen structure revealed that bitumen could be also described as a complex binder with two important parts, in which that asphaltenes as a nanometers solid part was surrounded with by an oily liquid matrix (maltenes) [11]. To investigate nano clay composite effect on bitumen structure and evaluate bitumen distribution, scanning electron microscope (SEM) was used. SEM produces images by scanning a sample using a focused beam of electrons.

9.2.5   FINE PARTICLE HOMOGENEOUS SCATTER

Uniform nano composite distributing in the bitumen is an important factor for obtaining stable bitumen [12]. When nano composite particles are accumulated, bitumen nano structure would change the bitumen behavior. With respect to the existing facilities, a new device, as shown in Fig. 9.1, was made. A microingredient blender was also made to uniformly distribute nano composite-ingredients in the bitumen considering the activities being executed.

FIGURE 9.1   Fine particle homogeneous scatter

9.3   RESULTS AND DISCUSSIONS

9.3.1   MARSHAL TEST RESULTS

In order to determine optimum bitumen percent, Marshal samples were made according to ASTM D6926. Based on these tests, it was found that different nano clay composite percentages did not significantly affect optimum bitumen value. Thus, to observe the overall evaluation of changes, Marshal test diagrams of aged sample and modified aged sample with 2% nano clay composite are given in Fig. 9.2.

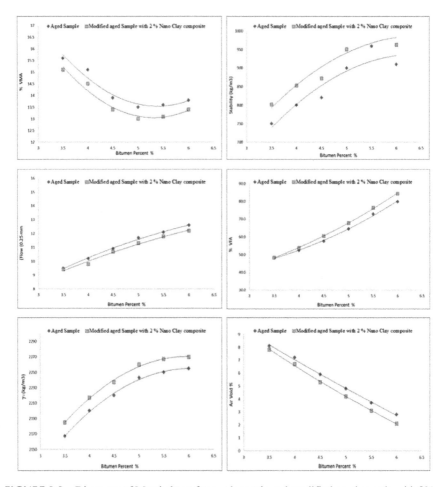

FIGURE 9.2 Diagrams of Marshal test for aged sample and modified aged sample with 2% nano clay composite.

Results of the Marshall experiments were used on aged and modified nano clay composite ingredients to determine percentage of optimized bitumen. As an example, graphs of the modified bitumen using 2% nano clay composite were compared with the aged bitumen. Special weight and stability of bitumen samples modified by nano clay composites increased due to the existence of nano ingredients and increase in bitumen viscosity. Considering the calculation of actual special weight of the asphaltic mixture, by increasing nano-ingredient, not only sample weight of air would increase but also saturated samples' weight with dry level in air would decrease due to filling of empty spaces of asphaltic mixtures. Using nano clay composite ingredients and lack of water penetration to the asphaltic mixtures and according to the results of Marshal experiments, special weight of asphaltic mixture was increased to a certain level. This value remained almost unchanged by increasing percentage of nano ingredients. By increasing nano composite ingredients' weight of dry asphalt in air, decreasing sample weight of asphalt in water after being put under vacuum pressure and increasing actual special weight of asphalt, empty space percentage of asphaltic samples was compressed and diminished; by increasing percentage of nano composite-ingredients, this value did not dramatically change. Marshal stability is maximum resistance against deformation, which can be calculated for a constant loading rate. Value of Marshal stability changes with type and gradation of aggregates and type, amount and calibration of bitumen. Different institutions have various criteria for Marshal stability. In fact, stability of asphaltic mixture is the maximum force that can be tolerated by the sample right before it breaks [13]. The samples are loaded by loading jacks or loading machines with constant rate of 50 ± 5 per minute until the load indicator gauge shows decreased in load rate. The maximum applied load by the machine or the maximum force calculated from the conversion of maximum gauge record is considered to be Marshal stability. According to Marshal experiment, by adding nano clay composite to old bitumen, actual special weight of the compressed sample increased and special weight of the aggregates mixture decreased; therefore, empty space of the aggregates decreased. By adding more percentage of nano ingredients, these changes continued. Marshal flow is determination (elastic to extra plastic) of asphaltic mixture attained during the experiment. If rate of flow in the chosen optimized bitumen was higher than the upper limit, the mixture would become too plastic and would be considered instable; if it was less than the lower limit, it would be considered too fragile (upper and lower limits were attained according to Iranian Pavement Regulations of 234 Publication). In other words, Marshal flow which was obtained using the flow measurement was deformation of the whole sample between the point at which the load was not applied and the point at which the maximum load started to diminish. According to Marshal experiment, by adding nano clay composite to old bitumen, due to increase in bitumen volume, the empty space filled with bitumen would increases. By adding more percentage of nano ingredients, these changes would continue. Optimum bitumen percent changed from 5.51 to 5.98 by adding 8% nano clay composite to aged bitumen.

9.3.2   STATIC CREEP ON MODIFIED AGED ASPHALT SAMPLES

Results of static creep tests results are determined in Fig. 9.3. These tests measure a specimen's permanent deformation and are conducted by applying a static load to modified aged bitumen and then measuring the specimen's permanent deformation after unloading. High rutting potential is associated with a large amount of permanent deformation. In this study, stress of 150 KPa and temperature of 400°C determined. Creep loading was placed for 1000 sec and then unloading was carried out to 200 sec according to Figs. 9.3 to 9.6.

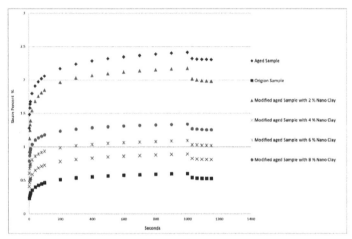

FIGURE 9.3   Static creep test on modified aged samples with different nano clay composite percent.

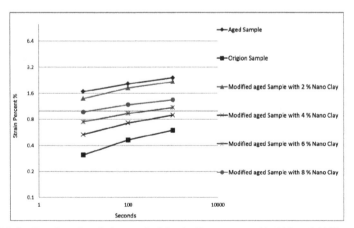

FIGURE 9.4   Results of asphalt samples' Static Creep test at 10, 100 and 1000 sec on the modified aged samples with different nano clay composite percentages.

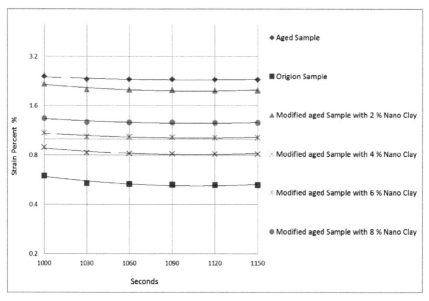

FIGURE 9.5   Asphalt samples' elastic static creep test after unloading on the modified aged samples with different nano clay composite percentages.

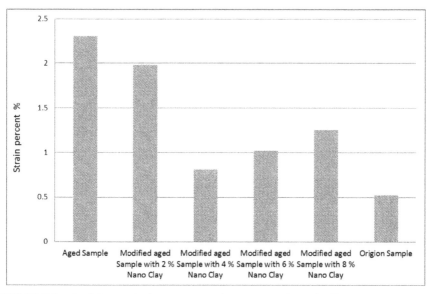

FIGURE 9.6   Asphalt samples' permanent static creep test after unloading.

As can be observed, static creep of the asphaltic mixture decreased as percentage of clay nano-ingredients added to the bitumen increased; this process continued to the 4% increase. After this value, it diminished due to the inappropriate connection between bitumen and aggregate caused by accumulation of nano composite-ingredients. Considering the above results, it is clarified that total and permanent strain decreased after adding nano clay composite in value of 4% because nano clay composites had the maximum value. After adding 4% nano clay composite ingredients, asphalt mixture's strain increased due to shriveling of the bitumen fall in its flexibility. For better view about the strain values, Fig. 9.4 is demonstrated in the logarithmic plot.

Also, in the logarithmic plot above the static creep, changes in the time of 10,100, 1000 sec from the starting point of the static loadings were depicted. In Fig. 9.5, elastic creep of the samples made of Nano clay composite is clear. As can be seen in the figure, the sample made with 4% nano clay composite had more ramp (returned with more ramp). In fact, for better evaluation of elastic creep, Fig. 9.5 shows these values after unloading. It is clear that the modified aged sample with 4% nano clay composite had the best performance. Results of asphalt samples' permanent static creep test after unloading are given in Fig. 9.6.

According to Fig. 9.6, modified aged sample with 4% nano clay composite, had the lowest strain percentage compared with aged sample and asphalt strain increased after adding nano clay composite.

9.3.3 MODULUS TEST RESULTS

Modulus experiment on the resilience of the asphaltic mixture samples including initial bitumen and modified bitumen was performed 50°C. Based on the ASTM-D4123 standard methods in these experiments, loading pattern of semisinusoidal with loading frequency of one hertz, loading cycle time period of one sec, loading time period of 0.1 sec and relaxation time period of 0.9 sec in each loading cycle were used. Five pulses for determining asphalt sample's resilient modulus are shown in Table 9.5.

TABLE 9.5   Resilient Modulus Results of Asphalt Samples

Aged samples	Pals1	Pals2	Pals3	Pals4	Pals5	Average Modulus (MPa)
	230	242	222	233	265	238/4
Modified aged sample with 4% Nano clay composite	Pals1	Pals2	Pals3	Pals4	Pals5	Average Modulus (MPa)
	248	261	243	241	284	255/4

9.3.4. FATIGUE TEST RESULTS

According to the results taken from the Nottingham experiment and by determining the optimized percentage of nano materials, it was clarified that 4% nano clay composite had a better impact on improving. Functional properties of asphaltic mixtures, these samples were made under 150 Kpa stress at 25°C to be put under exhaustion. The purpose was to analyze exhaustion operation of the samples which had the best creep operation. In Fig. 9.7, the modified aged samples with 4% nano clay composite compared with aged samples had higher strain against the aged samples in the nearly same loading cycles. These results confirmed that nano clay composite did not have a very good effect on fatigue performance [14].

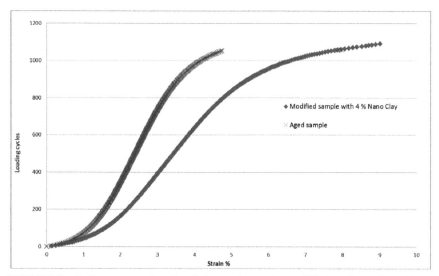

FIGURE 9.7   Fatigue test results of asphalt samples on the modified aged samples with 4% nano clay composite Percent and aged sample.

9.3.5   SEM RESULTS

About 20 years ago, this microscope was invented and scientists observed atoms. This microscope works through assessing very small flows made between the microscope tip and the sample. The wire charged with positive loads by the thin tip in the microscope which is in the distance of about one nanometer from surface of the object causes a weak flow and throw of the surface electron to the detector (microscope wire). These images were taken from the bitumen in the thin layer laboratory in Iran University of Science and Technology. Due to high thermal sensitivity of bitumen and its ruining probability in the case of increased system voltage, quality of the image was not so good.

The nano composite-ingredients used in this research were almost spherical in shape, which is shown in the figures related to the ingredient. SEM images for identifying the size of nano composite can be observed in Figs. 9.8 and 9.9. Nano composite's uniform distribution can be also notified in bitumen. As seen in these images, mechanical mixing method had good agreement with the innovative device developed in this article.

9.4 CONCLUSIONS

In this study, bitumen was aged using TFOT and then modified by different nano composite percentages. Marshal samples were made using modified aged bitumen with different nano composite percentages and aggregates with Topeka and Binder gradation. These samples were put under static creep loading and those with high creep resistance were placed under fatigue loading and defined resilient modulus. According to the conventional tests, adding 4% nano composite in bitumen returned the missed properties of bitumen by strengthening the bond between bitumen components. After a few years, bonding between aggregates and bitumen decreased and, with creating a gap, aged asphalt mixture's creep increased. In this study, static creep test results on the asphalt mixtures made by modified aged bitumen were observed. Nano clay composite decreased total strain of aged asphalt mixtures made by modified bitumen and, according to creep test results, 4% nano clay composite was a decent amount. This research was one of the pioneers on the asphaltic samples made of aged bitumen and aged bitumen modified by nano composite ingredients of clay. Generally, nano ingredients decrease thermal sensitivity of bitumen and the whole mixture. Asphalt samples' resistance against static loading also increased. Resilient modulus increased and it seems that nano clay composite ingredients did not have a positive impact on the mixture operation under the fatigue load.

KEYWORDS

- aging
- creep test
- homogeneous scatter
- nanoclay composite

REFERENCES

1. Steyn Wynand, J. V. D. M. (2011). "Applications of Nanotechnology in Road Pavement Engineering", Springer-Verlag Berlin Heidelberg, Nanotechnology in Civil Infrastructure, 49–83.

2. Sabbagh, A. B., & Lesser, A. J. (1998). Effect of Particle Morphology on the Emulsion Stability and Mechanical Performance of Polyolefin Modified Asphalts Polymer Eng. Science, *38(5)*, 707–715.

3. Chong, K. P. (2003). "Nanotechnology in Civil Engineering" In: Proc. 1st Int. Symp on Nanotechnology in Construction, Paisley, Scotland 13–21.

4. Gang Liu, Martin van de Ven, & Andre Molenaar, (2011). "Influence of Organo-Montmorillonites on Fatigue Properties of Bitumen and Mortar", Journal Homepage: www.elsevier.com/locates/ijfatigue, International Journal of Fatigue *33*, 1574–1582.

5. Saeed Ghaffarpour Jahromi, & Ali Khodaii (2009). "Effects of Nano clay on Rheological Properties of Bitumen Binder", Journal Homepage: www.elsevier.com/locate/conbuildmat, Construction and Building Materials *23*, 2894–2904.

6. Saeed Sadeghpour Galooyak, Bahram Dabir, Ali Ehsan Nazarbeygi, & Alireza Moeini (2010). "Rheological Properties and Storage Stability of Bitumen SBS Montmorillonite Composites", Journal Homepage: www.elsevier.com/locate/conbuildmat, Construction and Building Materials *24*, 300–307.

7. Sayyed Mahdi Abtahi, Mohammad Sheikhzadeh, & Sayyed Mahdi Hejazi (2010). "Fiber-Reinforced Asphalt-Concrete a Review", Construction and Building Materials *24*, 871–877.

8. Shen, J., Amirkhanian, S., & Miller, J. A. (2007). Effects of Rejuvenator on Performance-Based Properties of Rejuvenated Asphalt Binder and Mixtures Construction Build Mater *21(5)*, 958–964.

9. Haghi, A. K., Arabani, M., Shakeri, M., Haj jafari, M., & Mobasher, B. (2005). Strength Modifications of Asphalt Pavement using Waste Tires, 7th International Fracture Conference, University of Kocaeli, Kocaeli, Turkey.

10. Iranian Issues, (2010). General Technical specification of Roads, *101*, http://tec.mporg.ir.

11. Zhanping You, Julian Mills-Beale, Justin, M. Foley, Samit Roy, Gregory, M. Odegard, Qingli Dai, & Shu Wei Goh (2011). Nanoclay Modified Asphalt Materials: Preparation and Characterization, Construction and Building Materials *25*, 1072–1078.

12. Jian-Ying Yu, Peng-Cheng Feng, Heng-Long Zhang, & Shao-Peng Wu (2009). Effect of Organo Montmorillonite on Aging Properties of Asphalt, Construction and Building Materials, *23*, 2636–2640.

13. ASTM D6927, Standard Test Method for Marshall Stability and Flow of Bituminous Mixtures, (2007).

14. VandeVen, M. F. C., & Molenaar, A. A. A. (2009). "NanoClay for binder Modification of Asphalt Mixtures", Taylor & Francis Group, London, ISBN 978-0-415-55854-9.

CHAPTER 10

NANOSTRUCTURAL ELEMENTS AND THE MOLECULAR MECHANICS AND DYNAMICS INTERACTIONS: A SYSTEMATIC STUDY

A. V. VAKHRUSHEV, and A. M. LIPANOV

Institute of Mechanics, Ural Branch of the Russian Academy of Sciences, T. Baramsinoy 34, Izhevsk, Russia; E-mail: postmaster@ntm.udm.ru

CONTENTS

Abstract 108
10.1   Introduction 108
10.2   Problem Statement 110
10.3   Problem Formulation for Interaction of Several Nanoelements 115
10.4   Numerical Procedures and Simulation Techniques 115
10.5   Results and Discussions 117
10.6   Conclusions 134
Acknowledgments 135
Keywords 136
References 136

ABSTRACT

In this chapter, a systematic investigation on application of theoretical methods for calculation of internal structure and the equilibrium configuration (shape) of separate noninteracting nanoparticles by the molecular mechanics and dynamics Interactions of nanostructural elements presented.

10.1   INTRODUCTION

The properties of a nanocomposite are determined by the structure and properties of the nanoelements, which form it. One of the main tasks in making nanocomposites is building the dependence of the structure and shape of the nanoelements forming the basis of the composite on their sizes. This is because with an increase or a decrease in the specific size of nanoelements (nanofibers, nanotubes, nanoparticles, etc.), their physical-mechanical properties such as coefficient of elasticity, strength, deformation parameter, etc. are varying over one order [1–5].

The calculations and experiments show that this is primarily due to a significant rearrangement (which is not necessarily monotonous) of the atomic structure and the shape of the nanoelement. The experimental investigation of the above parameters of the nanoelements is technically complicated and laborious because of their small sizes. In addition, the experimental results are often inconsistent. In particular, some authors have pointed to an increase in the distance between the atoms adjacent to the surface in contrast to the atoms inside the nanoelement, while others observe a decrease in the aforementioned distance [6].

Thus, further detailed systematic investigations of the problem with the use of theoretical methods, that is, mathematical modeling, are required.

The atomic structure and the shape of nanoelements depend both on their sizes and on the methods of obtaining, which can be divided into two main groups:

1. Obtaining nanoelements in the atomic coalescence process by "assembling" the atoms and by stopping the process when the nanoparticles grow to a desired size (the so-called "bottom-up" processes). The process of the particle growth is stopped by the change of physical or chemical conditions of the particle formation, by cutting off supplies of the substances that are necessary to form particles, or because of the limitations of the space where nanoelements form.

2. Obtaining nanoelements by breaking or destruction of more massive (coarse) formations to the fragments of the desired size (the so-called "up down" processes).

In fact, there are many publications describing the modeling of the "bottom-up" processes [7, 8], while the "up down" processes have been studied very little. Therefore, the objective of this work is the investigation of the regularities of the changes in the structure and shape of nanoparticles formed in the destruction ("up down") processes depending on the nanoparticle sizes, and building up theoretical dependences describing the above parameters of nanoparticles.

When the characteristics of powder nanocomposites are calculated it is also very important to take into account the interaction of the nanoelements since the changes in their original shapes and sizes in the interaction process and during the formation (or usage) of the nanocomposite can lead to a significant change in its properties and a cardinal structural rearrangement. In addition, the experimental investigations show the appearance of the processes of ordering and self-assembling leading to a more organized form of a nanosystems [9–15]. In general, three main processes can be distinguished: the first process is due to the regular structure formation at the interaction of the nanostructural elements with the surface where they are situated; the second one arises from the interaction of the nanostructural elements with one another; the third process takes place because of the influence of the ambient medium surrounding the nanostructural elements. The ambient medium influence can have "isotropic distribution" in the space or it can be presented by the action of separate active molecules connecting nanoelements to one another in a certain order. The external action significantly changes the original shape of the structures formed by the nanoelements. For example, the application of the external tensile stress leads to the "stretch" of the nanoelement system in the direction of the maximal tensile stress action; the rise in temperature, vice versa, promotes a decrease in the spatial anisotropy of the nanostructures [10]. Note that in the self-organizing process, parallel with the linear moving, the nanoelements are in rotary movement. The latter can be explained by the action of moment of forces caused by the asymmetry of the interaction force fields of the nanoelements, by the presence of the "attraction" and "repulsion" local regions on the nanoelement surface, and by the "nonisotropic" action of the ambient as well.

The above phenomena play an important role in nanotechnological processes. They allow developing nanotechnologies for the formation of nanostructures by the self-assembling method (which is based on self-organizing processes) and building up complex spatial nanostructures consisting of different nanoelements (nanoparticles, nanotubes, fullerenes, super-molecules, etc.) [15]. However, in a number of cases, the tendency towards self-organization interferes with the formation of a desired nanostructure. Thus, the nanostructure arising from the self-organizing process is, as a rule, "rigid" and stable against external actions. For example, the "adhesion" of nanoparticles interferes with the use of separate nanoparticles in various nanotechnological processes, the uniform mixing of the nanoparticles from different materials and the formation of nanocomposite with desired properties. In connection with this, it is important to model the processes of static and dynamic interaction of the nanostructure elements. In this case, it is essential to take into consideration the interaction force moments of the nanostructure elements, which causes the mutual rotation of the nanoelements.

The investigation of the above dependences based on the mathematical modeling methods requires the solution of the aforementioned problem on the atomic level. This requires large computational aids and computational time, which makes the development of economical calculation methods urgent. The objective of this work was the development of such a technique.

This chapter gives results of the studies of problems of numeric modeling within the framework of molecular mechanics and dynamics for investigating the regularities of the amorphous phase formation and the nucleation and spread of the crystalline or hypocrystalline phases over the entire nanoparticle volume depending on the process parameters, nanoparticles sizes and thermodynamic conditions of the ambient. Also the method for calculating the interactions of nanostructural elements is offered, which is based on the potential built up with the help of the approximation of the numerical calculation results using the method of molecular dynamics of the pairwise static interaction of nanoparticles. Based on the potential of the pairwise interaction of the nanostructure elements, which takes into account forces and moments of forces, the method for calculating the ordering and self-organizing processes has been developed. The investigation results on the self-organization of the system consisting of two or more particles are presented and the analysis of the equilibrium stability of various types of nanostructures has been carried out. These results are a generalization of the authors' research in Refs. [16–24]. A more detailed description of the problem you can obtain in these works.

10.2   PROBLEM STATEMENT

The problem on calculating the internal structure and the equilibrium configuration (shape) of separate noninteracting nanoparticles by the molecular mechanics and dynamics methods has two main stages:

1. The "initiation" of the task, that is, the determination of the conditions under which the process of the nanoparticle shape and structure formation begins.
2. The process of the nanoparticle formation.

Note that the original coordinates and initial velocities of the nanoparticle atoms should be determined from the calculation of the macroscopic parameters of the destructive processes at static and dynamic loadings taking place both on the nanoscale and on the macroscale. Therefore, in the general case, the coordinates and velocities are the result of solving the problem of modeling physical-mechanical destruction processes at different structural levels. This problem due to its enormity and complexity is not considered in this paper. The detailed description of its statement and the numerical results of its solution are given in the works of the authors [16–19].

The problem of calculating the interaction of ordering and self-organization of the nanostructure elements includes three main stages: the first stage is building the internal structure and the equilibrium configuration (shape) of each separate noninteracting nanostructure element; the second stage is calculating the pairwise interaction of two nanostructure elements; and the third stage is establishing the regularities of the spatial structure and evolution with time of the nanostructure as a whole.

Let us consider the above problems in sequence.

10.2.1   THE CALCULATION OF THE INTERNAL STRUCTURE AND THE SHAPE OF THE NONINTERACTING NANOELEMENT

The initialization of the problem is in giving the initial coordinates and velocities of the nanoparticle atoms

$$\vec{\mathbf{x}}_i = \vec{\mathbf{x}}_{i0}, \vec{\mathbf{V}}_i = \vec{\mathbf{V}}_{i0}, \ t = 0, \vec{\mathbf{x}}_i \subset \Omega_k \, , \tag{1}$$

where $\vec{\mathbf{x}}_{i0}, \vec{\mathbf{x}}_i$ are original and current coordinates of the i-th atom; $\vec{\mathbf{V}}_{i0}, \vec{\mathbf{V}}_i$ are initial and current velocities of the i-th atom, respectively; Ω_k is an area occupied by the nanoelement.

The problem of calculating the structure and the equilibrium configuration of the nanoelement will be carried out with the use of the molecular dynamics method taking into consideration the interaction of all the atoms forming the nanoelement. Since, at the first stage of the solution, the nanoelement is not exposed to the action of external forces, it is taking the equilibrium configuration with time, which is further used for the next stage of calculations.

At the first stage, the movement of the atoms forming the nanoparticle is determined by the set of Langevin differential equations at the boundary conditions (1) [25]

$$m_i \cdot \frac{d\vec{\mathbf{V}}_i}{dt} = \sum_{j=1}^{N_k} \vec{\mathbf{F}}_{ij} + \vec{\mathbf{F}}_i\left(t\right) - \alpha_i m_i \vec{\mathbf{V}}_i, \qquad i = 1, 2, .., N_k,$$

$$\frac{d\vec{\mathbf{x}}_i}{dt} = \vec{\mathbf{V}}_i, \tag{2}$$

where N_k is the number of atoms forming each nanoparticle; m_i is the mass of the i-th atom; α_i is the "friction" coefficient in the atomic structure; $\vec{\mathbf{F}}_i(t)$ is a random set of forces at a given temperature which is given by Gaussian distribution.

The interatomic interaction forces usually are potential and determined by the relation

$$\vec{F}_{ij} = -\sum_{1}^{n} \frac{\partial \Phi(\vec{\rho}_{ij})}{\partial \vec{\rho}_{ij}} \, , \ i = 1, 2, ..., N_k, \ j = 1, 2, ..., N_k, \tag{3}$$

where $\bar{\rho}_{ij}$ is a radius-vector determining the position of the l-th atom relative to the j-th atom; $\Phi(\bar{\rho}_{ij})$ is a potential depending on the mutual positions of all the atoms; n is the number of interatomic interaction types.

In the general case, the potential $\Phi(\bar{\rho}_{ij})$ is given in the form of the sum of several components corresponding to different interaction types:

$$\Phi(\vec{\rho}_{ij}) = \Phi_{cb} + \Phi_{va} + \Phi_{ta} + \Phi_{pg} + \Phi_{vv} + \Phi_{es} + \Phi_{hb}. \tag{4}$$

Here the following potentials are implied: Φ_{cb} – of chemical bonds; Φ_{va} – of valence angles; Φ_{ta} – of torsion angles; Φ_{pg} – of flat groups; Φ_{vv} – of Van der Waals contacts; Φ_{es} – of electrostatics; Φ_{hb}- of hydrogen bonds.

The above addends have different functional forms. The parameter values for the interaction potentials are determined based on the experiments (crystallography, spectral, calorimetric, etc.) and quantum calculations [25].

Giving original coordinates (and forces of atomic interactions) and velocities of all the atoms of each nanoparticle in accordance with Eq. (2), at the start time, we find the change of the coordinates and the velocities of each nanoparticle atoms with time from the equation of motion (1). Since the nanoparticles are not exposed to the action of external forces, they take some atomic equilibrium configuration with time that we will use for the next calculation stage.

1.2.2 THE CALCULATION OF THE PAIRWISE INTERACTION OF THE TWO NANOSTRUCTURE ELEMENTS

At this stage of solving the problem, we consider two interacting nanoelements. First, let us consider the problem statement for symmetric nanoelements, and then for arbitrary shaped nanoelements.

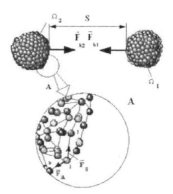

FIGURE 10.1 The scheme of the nanoparticle interaction; A – an enlarged view of the nanoparticle fragment.

First of all, let us consider two symmetric nanoelements situated at the distance S from one another (Fig. 10.1) at the initial conditions

$$\vec{x}_i = \vec{x}_{i0}, \vec{V}_i = 0, \ t = 0, \ \vec{x}_i \subset \Omega_1 \bigcup \Omega_2 \ , \tag{5}$$

where Ω_1, Ω_2 are the areas occupied by the first and the second nanoparticle, respectively.

We obtain the coordinates $\vec{x}_{i0}$ from Eq. (2) solution at initial conditions of Eq. (1). It allows calculating the combined interaction forces of the nanoelements,

$$\vec{F}_{b1} = -\vec{F}_{b2} = \sum_{i=1}^{N_1} \sum_{j=1}^{N_2} \vec{F}_{ij} \ , \tag{6}$$

where i, j are the atoms and N_1, N_2 are the numbers of atoms in the first and in the second nanoparticle, respectively.

Forces $\vec{F}_{ij}$ are defined from Eq. (3).

In the general case, the force magnitude of the nanoparticle interaction $\left|\vec{F}_{bi}\right|$ can be written as product of functions depending on the sizes of the nanoelements and the distance between them:

$$\left|\vec{F}_{bi}\right| = \Phi_{11}(S_c) \cdot \Phi_{12}(D) \tag{7}$$

The $\vec{F}_{bi}$ vector direction is determined by the direction cosines of a vector connecting the centers of the nanoelements.

Now, let us consider two interacting asymmetric nanoelements situated at the distance S_c between their centers of mass (Fig. 10.2) and oriented at certain specified angles relative to each other.

In contrast to the previous problem, the interatomic interaction of the nanoelements leads not only to the relative displacement of the nanoelements but to their rotation as well. Consequently, in the general case, the sum of all the forces of the interatomic interactions of the nanoelements is brought to the principal vector of forces $\vec{F}_c$ and the principal moment $\vec{M}_c$

$$\vec{F}_c = \vec{F}_{b1} = -\vec{F}_{b2} = \sum_{i=1}^{N_1} \sum_{j=1}^{N_2} \vec{F}_{ij} \ , \tag{8}$$

$$\vec{M}_c = \vec{M}_{c1} = -\vec{M}_{c2} = \sum_{i=1}^{N_1} \sum_{j=1}^{N_2} \vec{\rho}_{cj} \times \vec{F}_{ij} \ , \tag{9}$$

where $\vec{\rho}_{cj}$ is a vector connecting points c and j.

The main objective of this calculation stage is building the dependences of the forces and moments of the nanostructure nanoelement interactions on the distance S_c between the centers of mass of the nanostructure nanoelements, on the angles of mutual orientation of the nanoelements $\Theta_1, \Theta_2, \Theta_3$ (shapes of the nanoelements) and on the characteristic size D of the nanoelement. In the general case, these dependences can be given in the form

$$\vec{F}_{bi} = \vec{\Phi}_F(S_c, \Theta_1, \Theta_2, \Theta_3, D), \qquad (10)$$

$$\vec{M}_{bi} = \vec{\Phi}_M(S_c, \Theta_1, \Theta_2, \Theta_3, D), \qquad (11)$$

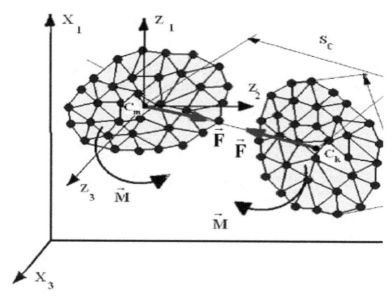

FIGURE 10.2 Two interacting nanoelements; $\vec{M}, \vec{F}$, are the principal moment and the principal vector of the forces, respectively.

For spherical nanoelements, the angles of the mutual orientation do not influence the force of their interaction; therefore, in Eq. (12), the moment is zero.

In the general case, Eqs. (11) and (12) can be approximated by analogy with Eq. (8), as the product of functions $S_0, \Theta_1, \Theta_2, \Theta_3, D$, respectively. For the further numerical solution of the problem of the self-organization of nanoelements, it is sufficient to give the above functions in their tabular form and to use the linear (or nonlinear) interpolation of them in space.

10.3   PROBLEM FORMULATION FOR INTERACTION OF SEVERAL NANOELEMENTS

When the evolution of the nanosystem as whole (including the processes of ordering and self-organization of the nanostructure nanoelements) is investigated, the movement of each system nanoelement is considered as the movement of a single whole. In this case, the translational motion of the center of mass of each nanoelement is given in the coordinate system X_1, X_2, X_3, and the nanoelement rotation is described in the coordinate system Z_1, Z_2, Z_3, which is related to the center of mass of the nanoelement (Fig. 10.2). The system of equations describing the above processes has the form

$$
\begin{cases}
M_k \dfrac{d^2 X_1^k}{dt^2} = \displaystyle\sum_{j=1}^{N_e} F_{X_1}^{kj} + F_{X_1}^{ke}, \\[2mm]
M_k \dfrac{d^2 X_2^k}{dt^2} = \displaystyle\sum_{j=1}^{N_e} F_{X_2}^{kj} + F_{X_2}^{ke}, \\[2mm]
M_k \dfrac{d^2 X_3^k}{dt^2} = \displaystyle\sum_{j=1}^{N_e} F_{X_3}^{kj} + F_{X_3}^{ke}, \\[2mm]
J_{Z_1}^k \dfrac{d^2 \Theta_1^k}{dt^2} + \dfrac{d\Theta_2^k}{dt}\cdot\dfrac{d\Theta_3^k}{dt}(J_{Z_3}^k - J_{Z_2}^k) = \displaystyle\sum_{j=1}^{N_e} M_{Z_1}^{kj} + M_{Z_1}^{ke}, \\[2mm]
J_{Z_2}^k \dfrac{d^2 \Theta_2^k}{dt^2} + \dfrac{d\Theta_1^k}{dt}\cdot\dfrac{d\Theta_3^k}{dt}(J_{Z_1}^k - J_{Z_3}^k) = \displaystyle\sum_{j=1}^{N_e} M_{Z_2}^{kj} + M_{Z_2}^{ke}, \\[2mm]
J_{Z_3}^k \dfrac{d^2 \Theta_3^k}{dt^2} + \dfrac{d\Theta_2^k}{dt}\cdot\dfrac{d\Theta_1^k}{dt}(J_{Z_2}^k - J_{Z_1}^k) = \displaystyle\sum_{j=1}^{N_e} M_{Z_3}^{kj} + M_{Z_3}^{ke},
\end{cases}
\tag{12}
$$

where x_i^k, Θ_i^k are coordinates of the centers of mass and angles of the spatial orientation of the principal axes Z_1, Z_2, Z_3 of nanoelements; $F_{X_1}^{kj}, F_{X_2}^{kj}, F_{X_3}^{kj}$ are the interaction forces of nanoelements; $F_{X_1}^{ke}, F_{X_2}^{ke}, F_{X_3}^{ke}$ are external forces acting on nanoelements; N_e is the number of nanoelements; M_k is a mass of a nanoelement; $M_{Z_1}^{kj}, M_{Z_2}^{kj}, M_{Z_3}^{kj}$ is the moment of forces of the nanoelement interaction; $M_{Z_1}^{ke}, M_{Z_2}^{ke}, M_{Z_3}^{ke}$ are external moments acting on nanoelements; $J_{Z_1}^k, J_{Z_2}^k, J_{Z_3}^k$ are moments of inertia of a nanoelement.

The initial conditions for the system of Eqs. (13) and (14) have the form

$$
\vec{X}^k = \vec{X}_0^k; \quad \Theta^k = \Theta_0^k; \quad \vec{V}^k = \vec{V}_0^k; \quad \frac{d\Theta^k}{dt} = \frac{d\Theta_0^k}{dt}; \quad t = 0,
\tag{13}
$$

10.4   NUMERICAL PROCEDURES AND SIMULATION TECHNIQUES

In the general case, the problem formulated in the previous sections has no analytical solution at each stage; therefore, numerical methods for solving are used, as a rule. In this work, for the first stages, the numerical integration of the equation of motion of the nanoparticle atoms in the relaxation process are used in accordance with Verlet scheme [26]:

$$\vec{x}_i^{n+1} = \vec{x}_i^n + \Delta t \, \vec{V}_i^n + \left((\Delta t)^2 / 2m_i\right) \left(\sum_{j=1}^{N_k} \vec{F}_{ij} + \vec{F}_i - \alpha_i m_i \vec{V}_i^n \right)^n \tag{14}$$

$$\vec{V}_i^{n+1} = (1 - \Delta t \alpha_i)\vec{V}_i^n + (\Delta t / 2m_i)((\sum_{j=1}^{N_k} \vec{F}_{ij} + \vec{F}_i)^n + (\sum_{j=1}^{N_k} \vec{F}_{ij} + \vec{F}_i)^{n+1}), \tag{15}$$

where $\vec{x}_i^n, \vec{v}_i^n$, are a coordinate and a velocity of the i-th atom at the n-th step with respect to the time; Δt is a step with respect to the time.

The solution of the Eq. (13) also requires the application of numerical methods of integration. In the present work, Runge–Kutta method [27] is used for solving Eq. (13).

$$(X_i^k)_{n+1} = (X_i^k)_n + (V_i^k)_n \Delta t + \frac{1}{6}(\mu_{1i}^k + \mu_{2i}^k + \mu_{3i}^k)\Delta t \tag{16}$$

$$(V_i^k)_{n+1} = (V_i^k)_n + \frac{1}{6}(\mu_{1i}^k + 2\mu_{2i}^k + 2\mu_{3i}^k + \mu_{4i}^k). \tag{17}$$

$$\mu_{1i}^k = \Phi_i^k(t_n;(X_i^k)_n,\dots;(V_i^k)_n\dots)\Delta t,$$

$$\mu_{2i}^k = \Phi_i^k(t_n + \frac{\Delta t}{2};(X_i^k + V_i^k\frac{\Delta t}{2})_n,\dots;(V_i^k)_n + \frac{\mu_{1i}^k}{2},\dots)\Delta t,$$

$$\mu_{3i}^k = \Phi_i^k(t_n + \frac{\Delta t}{2};(X_i^k + V_i^k\frac{\Delta t}{2} + \mu_{1i}^k\frac{\Delta t}{4})_n,\dots;(V_i^k)_n + \frac{\mu_{2i}^k}{2},\dots)\Delta t, \tag{18}$$

$$\mu_{4i}^k = \Phi_i^k(t_n + \Delta t;(X_i^k + V_i^k\Delta t + \mu_{2i}^k\frac{\Delta t}{2})_n,\dots;(V_i^k)_n + \mu_{2i}^k,\dots)\Delta t$$

$$\Phi_i^k = \frac{1}{M_k}(\sum_{j=1}^{N_e} F_{X_3}^{kj} + F_{X_3}^{ke}) \tag{19}$$

$$(\Theta_i^k)_{n+1} = (\Theta_i^k)_n + (\frac{d\Theta_i^k}{dt})_n \Delta t + \frac{1}{6}(\lambda_{1i}^k + \lambda_{2i}^k + \lambda_{3i}^k)\Delta t \tag{20}$$

$$(\frac{d\Theta_i^k}{dt})_{n+1} = (\frac{d\Theta_i^k}{dt})_n + \frac{1}{6}(\lambda_{1i}^k + 2\lambda_{2i}^k + 2\lambda_{3i}^k + \lambda_{4i}^k) \qquad (21)$$

$$\lambda_{1i}^k = \Psi_i^k(t_n; (\Theta_i^k)_n, ...; (\frac{d\Theta_i^k}{dt})_n ...)\Delta t,$$

$$\lambda_{2i}^k = \Psi_i^k(t_n + \frac{\Delta t}{2}; (\Theta_i^k + \frac{d\Theta_i^k}{dt}\frac{\Delta t}{2})_n, ...; (\frac{d\Theta_i^k}{dt})_n + \frac{\lambda_{1i}^k}{2}, ...)\Delta t$$

$$\lambda_{3i}^k = \Psi_i^k(t_n + \frac{\Delta t}{2}; (\Theta_i^k + \frac{d\Theta_i^k}{dt}\frac{\Delta t}{2} + \lambda_{1i}^k\frac{\Delta t}{4})_n, ...; (\frac{d\Theta_i^k}{dt})_n + \frac{\lambda_{2i}^k}{2}, ...)\Delta t \qquad (22)$$

$$\lambda_{4i}^k = \Psi_i^k(t_n + \Delta t; (\Theta_i^k + \frac{d\Theta_i^k}{dt}\Delta t + \lambda_{2i}^k\frac{\Delta t}{2})_n, ...; (\frac{d\Theta_i^k}{dt})_n + \lambda_{2i}^k, ...)\Delta t$$

$$\Psi_1^k = \frac{1}{J_{Z_1}^k}(-\frac{d\Theta_2^k}{dt} \times \frac{d\Theta_3^k}{dt}(J_{Z_3}^k - J_{Z_2}^k) + \sum_{j=1}^{N_e} M_{Z_1}^{kj} + M_{Z_1}^{ke}),$$

$$\Psi_2^k = \frac{1}{J_{Z_2}^k}(-\frac{d\Theta_1^k}{dt} \cdot \frac{d\Theta_3^k}{dt}(J_{Z_1}^k - J_{Z_3}^k) + \sum_{j=1}^{N_e} M_{Z_2}^{kj} + M_{Z_2}^{ke}), \qquad (23\text{-}a)$$

$$\Psi_3^k = \frac{1}{J_{Z_3}^k}(-\frac{d\Theta_1^k}{dt} \cdot \frac{d\Theta_2^k}{dt}(J_{Z_1}^k - J_{Z_2}^k) + \sum_{j=1}^{N_e} M_{Z_3}^{kj} + M_{Z_3}^{ke}),$$

were $i = 1,2,3$; $k=1,2, ...N_e$

10.5 RESULTS AND DISCUSSIONS

Let us consider the realization of the above procedure taking as an example the calculation of the metal nanoparticle.

The potentials of the atomic interaction of Morse (23-b) and Lennard-Johns (24) were used in the following calculations

$$\Phi(\vec{\rho}_{ij})_m = D_m (\exp(-2\lambda_m(|\vec{\rho}_{ij}| - \rho_0)) - 2\exp(-\lambda_m(|\vec{\rho}_{ij}| - \rho_0))), \qquad (23\text{-}b)$$

$$\Phi(\bar{\boldsymbol{\rho}}_{ij})_{LD} = 4\varepsilon \left[\left(\frac{\sigma}{\left|\vec{\boldsymbol{\rho}}_{ij}\right|} \right)^{12} - \left(\frac{\sigma}{\left|\vec{\boldsymbol{\rho}}_{ij}\right|} \right)^{6} \right], \tag{24}$$

where $D_m.\lambda_m.\rho_0.\varepsilon.\sigma$ are the constants of the materials studied.

For sequential and parallel solving the molecular dynamics equations, the program package developed at Applied Mechanics Institute, the Ural Branch of the Russian Academy of Sciences, and the advanced program package NAMD developed at the University of Illinois and Beckman Institute (USA) by the Theoretical Biophysics Group were used. The graphic imaging of the nanoparticle calculation results was carried out with the use of the program package VMD.

10.5.1   STRUCTURE AND FORMS OF NANOPARTICLES

At the first stage of the problem, the coordinates of the atoms positioned at the ordinary material lattice points (Fig. 10.3, (1)) were taken as the original coordinates. During the relaxation process, the initial atomic system is rearranged into a new "equilibrium" configuration (Fig. 10.3, (2)) in accordance with the calculations based on Eqs. (6)–(9), which satisfies the condition when the system potential energy is approaching the minimum (Fig. 10.3, the plot).

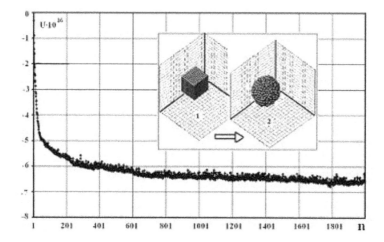

FIGURE 10.3   The initial crystalline (1) and cluster (2) structures of the nanoparticle consisting of 1331 atoms after relaxation; the plot of the potential energy $U[J]$ variations for this atomic system in the relaxation process (n – number of iterations with respect to the time).

After the relaxation, the nanoparticles can have quite diverse shapes: globe-like, spherical centered, spherical eccentric, spherical icosahedral nanoparticles and asymmetric nanoparticles (Fig. 10.4).

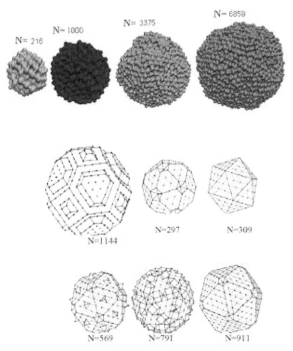

FIGURE 10.4   Nanoparticles of diverse shapes, depending on the number of atoms they consists.

In this case, the number of atoms N significantly determines the shape of a nanoparticle. Note, that symmetric nanoparticles are formed only at a certain number of atoms. As a rule, in the general case, the nanoparticle deviates from the symmetric shape in the form of irregular raised portions on the surface. Besides, there are several different equilibrium shapes for the same number of atoms. The plot of the nanoparticle potential energy change in the relaxation process (Fig. 10.5) illustrates it.

As it follows from Fig. 10.5, the curve has two areas: the area of the decrease of the potential energy and the area of its stabilization promoting the formation of the first nanoparticle equilibrium shape (1). Then, a repeated decrease in the nanoparticle potential energy and the stabilization area corresponding to the formation of the second nanoparticle equilibrium shape are observed (2). Between them, there is a region of the transition from the first shape to the second one (P). The second equilibrium shape is more stable due to the lesser nanoparticle potential energy. However, the first equilibrium shape also "exists" rather long in the calculation process. The change of the equilibrium shapes is especially characteristic of the nanoparticles with an "irregular" shape. The internal structure of the nanoparticles is of importance since their atomic structure significantly differs from the crystalline structure of the bulk materials: the distance between the atoms and the angles change, and the surface formations of different types appear. In Fig. 10.6, the change of the structure of a two-dimensional nanoparticle in the relaxation process is shown.

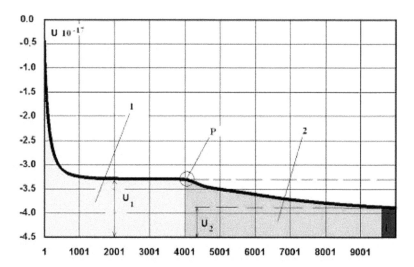

FIGURE 10.5   The plot of the potential energy change of the nanoparticle in the relaxation process: 1 – a region of the stabilization of the first nanoparticle equilibrium shape; 2 – a region of the stabilization of the second nanoparticle equilibrium shape; P – a region of the transition of the first nanoparticle equilibrium shape into the second one.

The Fig. 10.6 shows how the initial nanoparticle crystalline structure (1) is successively rearranging with time in the relaxation process (positions 2, 3, 4). Note that the resultant shape of the nanoparticle is not round, that is, it has "remembered" the initial atomic structure. It is also of interest that in the relaxation process, in the nanoparticle, the defects in the form of pores (designation "p" in the figure) and the density fluctuation regions (designation "c" in the figure) have been formed, which are absent in the final structure.

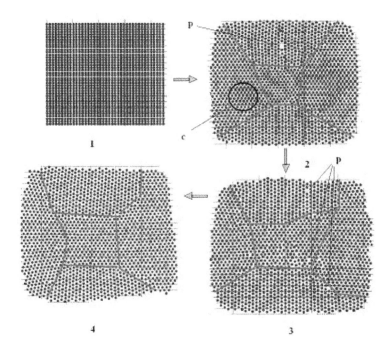

FIGURE 10.6 The change of the structure of a two-dimensional nanoparticle in the relaxation process: 1 – the initial crystalline structure; 2, 3, 4 – the nanoparticles structures which change in the relaxation process; p – pores; c – the region of compression.

10.5.2 NANOPARTICLES INTERACTION

Let us consider some examples of nanoparticles interaction. Figure 10.7 shows the calculation results demonstrating the influence of the sizes of the nanoparticles on their interaction force. One can see from the plot that the larger nanoparticles are attracted stronger, that is, the maximal interaction force increases with the size growth of the particle. Let us divide the interaction force of the nanoparticles by its maximal value for each nanoparticle size, respectively. The obtained plot of the "relative" (dimensionless) force (Fig. 10.8) shows that the value does not practically depend on the nanoparticle size since all the curves come close and can be approximated to one line.

Figure 10.9 displays the dependence of the maximal attraction force between the nanoparticles on their diameter that is characterized by nonlinearity and a general tendency towards the growth of the maximal force with the nanoparticle size growth.

The total force of the interaction between the nanoparticles is determined by multiplying of the two plots (Figs. 10.8 and 10.9).

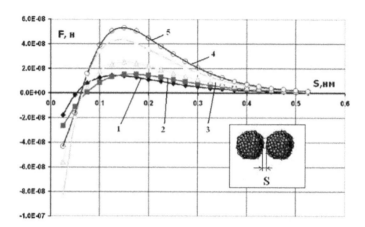

FIGURE 7   The dependence of the interaction force F [N] of the nanoparticles on the distance S [nm] between them and on the nanoparticle size: $1 - d = 2.04$; $2 - d = 2.40$; $3 - d = 3.05$; $4 - d = 3.69$; $5 - d = 4.09$ [nm].

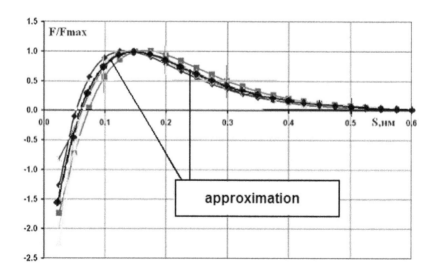

FIGURE 8   The dependence of the "relative" force $\bar{F}$ of the interaction of the nanoparticles on the distance S [nm] between them.

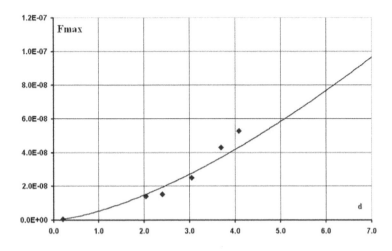

FIGURE 9   The dependence of the maximal attraction force F_{max} [N] the nanoparticles on the nanoparticle diameter d [nm].

Using the polynomial approximation of the curve in Fig. 10.5 and the power mode approximation of the curve in Fig. 10.6, we obtain

$$\overline{F} = (-1.13S^6 + 3.08S^5 - 3.41S^4 - 0.58S^3 + 0.82S - 0.00335)10^3 , \tag{25}$$

$$F_{max} \cdot = 0.5 \cdot 10^{-9} \cdot d^{1.499} , \tag{26}$$

$$F = F_{max} \cdot \overline{F} , \tag{27}$$

where d and S are the diameter of the nanoparticles and the distance between them [nm], respectively; F_{max} is the maximal force of the interaction of the nanoparticles [N].

Dependences (25)–(27) were used for the calculation of the nanocomposite ultimate strength for different patterns of nanoparticles' "packing" in the composite (Fig. 10.10).

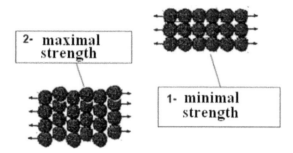

FIGURE 10   Different types of the nanoparticles' "packing" in the composite.

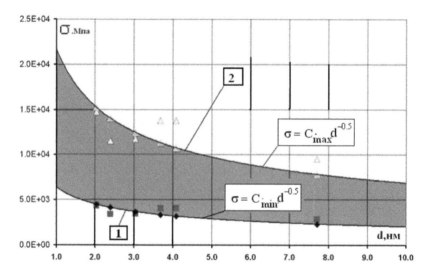

FIGURE 11   The dependence of the ultimate strength σ [MPa] of the nanocomposite formed by monodisperse nanoparticles on the nanoparticle sizes d [nm].

Figure 11 shows the dependence of the ultimate strength of the nanocomposite formed by monodisperse nanoparticles on the nanoparticle sizes. One can see that with the decrease of the nanoparticle sizes, the ultimate strength of the nanomaterial increases, and vice versa. The calculations have shown that the nanocomposite strength properties are significantly influenced by the nanoparticles' "packing" type in the material. The material strength grows when the packing density of nanoparticles increases. It should be specially noted that the material strength changes in inverse proportion to the nanoparticle diameter in the degree of 0.5, which agrees with the experimentally established law of strength change of nanomaterials (the law by Hall-Petch) [18]:

$$\sigma = C \cdot d^{-0.5},\tag{28}$$

where $C = C_{max} = 2.17 \times 10^4$ is for the maximal packing density; $C = C_{min} = 6.4 \times 10^3$ is for the minimal packing density.

The electrostatic forces can strongly change force of interaction of nanoparticles. For example, numerical simulation of charged sodium (NaCl) nanoparticles system (Fig. 10.12) has been carried out. Considered ensemble consists of eight separate nanoparticles. The nanoparticles interact due to Van-der-Waals and electrostatic forces.

Results of particles center of masses motion are introduced at Fig. 10.13 representing trajectories of all nanoparticles included into system. It shows the dependence of the modulus of displacement vector $|R|$ on time. One can see that nanoparticle moves intensively at first stage of calculation process. At the end of numerical calculation, all particles have got new stable locations, and the graphs of the radius vector $|R|$ become stationary. However, the nanoparticles continue to "vibrate" even at the final stage of numerical calculations. Nevertheless, despite of "vibration," the system of nanoparticles occupies steady position.

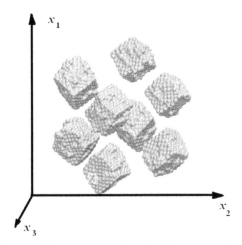

FIGURE 10.12 Nanoparticles system consists of eight nanoparticles NaCl.

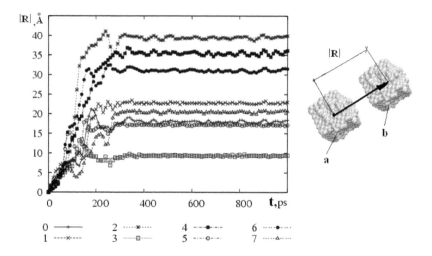

FIGURE 10.13   The dependence of nanoparticle centers of masses motion $|R|$ on the time t; a, b–the nanoparticle positions at time 0 and t, accordingly; 1–8 are the numbers of the nanoparticles.

However, one can observe a number of other situations. Let us consider, for example, the self-organization calculation for the system consisting of 125 cubic nanoparticles, the atomic interaction of which is determined by Morse potential (Fig. 10.14).

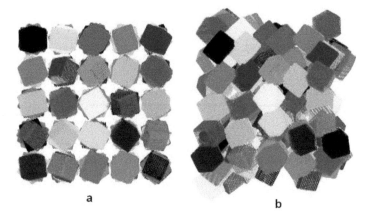

FIGURE 10.14   The positions of the 125 cubic nanoparticles: a – initial configuration; b – final configuration of nanoparticles.

As you see, the nanoparticles are moving and rotating in the self-organization process forming the structure with minimal potential energy. Let us consider, for example, the calculation of the self-organization of the system consisting of two cubic nanoparticles, the atomic interaction of which is determined by Morse potential [12]. Figure 10.15 displays possible mutual positions of these nanoparticles. The positions, where the principal moment of forces is zero, corresponds to pairs of the nanoparticles 2–3; 3–4; 2–5 (Fig. 10.15) and defines the possible positions of their equilibrium.

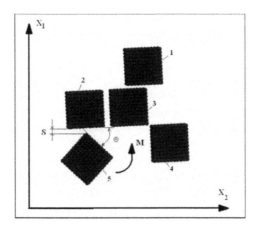

FIGURE 10.15   Characteristic positions of the cubic nanoparticles.

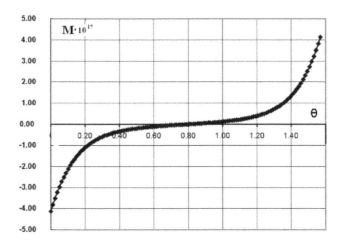

FIGURE 10.16   The dependence of the moment M [Nm] of the interaction force between cubic nanoparticles 1–3 (see in Fig. 10.9) on the angle of their relative rotation θ [rad].

Figure 10.16 presents the dependence of the moment of the interaction force between the cubic nanoparticles 1–3 (Fig. 10.15) on the angle of their relative rotation. From the plot follows that when the rotation angle of particle 1 relative to particle 3 is $\pi/4$, the force moment of their interaction is zero. At an increase or a decrease in the angle the force moment appears. In the range of $\pi/8 < \theta < 3\pi/4$ the moment is small. The force moment rapidly grows outside of this range. The distance S between the nanoparticles plays a significant role in establishing their equilibrium. If $S > S_0$ (where S_0 is the distance, where the interaction forces of the nanoparticles are zero), then the particles are attracted to one another. In this case, the sign of the moment corresponds to the sign of the angle θ deviation from $\pi/4$. At $S < S_0$ (the repulsion of the nanoparticles), the sign of the moment is opposite to the sign of the angle deviation. In other words, in the first case, the increase of the angle deviation causes the increase of the moment promoting the movement of the nanoelement in the given direction, and in the second case, the angle deviation causes the increase of the moment hindering the movement of the nanoelement in the given direction. Thus, the first case corresponds to the unstable equilibrium of nanoparticles, and the second case – to their stable equilibrium. The potential energy change plots for the system of the interaction of two cubic nanoparticles (Fig. 10.17) illustrate the influence of the parameter S. Here, curve 1 corresponds to the condition $S < S_0$ and it has a well-expressed minimum in the $0.3 < \theta < 1.3$ region. At $\theta < 0.3$ and $\theta > 1.3$, the interaction potential energy sharply increases, which leads to the return of the system into the initial equilibrium position. At $S > S_0$ (curves 2–5), the potential energy plot has a maximum at the $\theta = 0$ point, which corresponds to the unstable position.

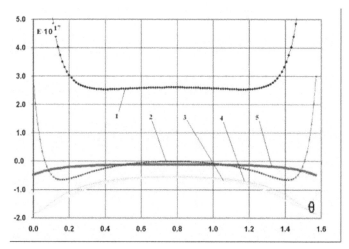

FIGURE 10.17   The plots of the change of the potential energy E [Nm] for the interaction of two cubic nanoparticles depending on the angle of their relative rotation θ [rad] and the distance between them (positions of the nanoparticles 1–3, Fig. 10.9).

The carried out theoretical analysis is confirmed by the works of the scientists from New Jersey University and California University in Berkeley who experimentally found the self-organization of the cubic microparticles of plumbum zirconate-titanate (PZT) [28]: the ordered groups of cubic microcrystals from PZT obtained by hydrothermal synthesis formed a flat layer of particles on the air-water interface, where the particle occupied the more stable position corresponding to position 2–3 in Fig. 10.15.

Thus, the analysis of the interaction of two cubic nanoparticles has shown that different variants of their final stationary state of equilibrium are possible, in which the principal vectors of forces and moments are zero. However, there are both stable and unstable stationary states of this system: nanoparticle positions 2–3 are stable, and positions 3–4 and 2–5 have limited stability or they are unstable depending on the distance between the nanoparticles.

Note that for the structures consisting of a large number of nanoparticles, there can be a quantity of stable stationary and unstable forms of equilibrium. Accordingly, the stable and unstable nanostructures of composite materials can appear. The search and analysis of the parameters determining the formation of stable nanosystems is an urgent task.

It is necessary to note, that the method offered has restrictions. This is explained by change of the nanoparticles form and accordingly variation of interaction pair potential during nanoparticles coming together at certain conditions.

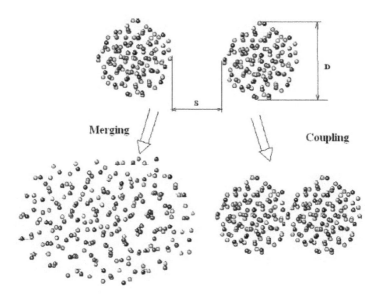

FIGURE 10.18   Different type of nanoparticles connection (merging and coupling).

The merge (accretion [4]) of two or several nanoparticles into a single whole is possible (Fig. 10.18). Change of a kind of connection cooperating nanoparticles (merging or coupling in larger particles) depending on its sizes, it is possible to explain on the basis of the analysis of the energy change graph of connection nanoparticles (Fig. 10.19). From Fig.10.19 follows that though with the size increasing of a particle energy of nanoparticles connection E_{np} grows also, its size in comparison with superficial energy E_S of a particle sharply increases at reduction of the sizes nanoparticles. Hence, for finer particles energy of connection can appear sufficient for destruction of their configuration under action of a mutual attraction and merging in larger particle.

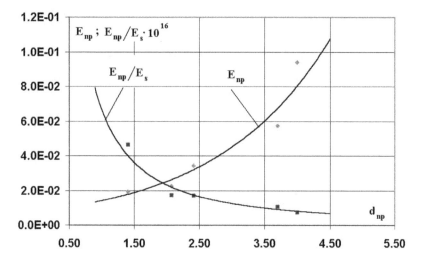

FIGURE 10.19 Change of energy of nanoparticles connection E_{np} [Nm] and E_{np} ration to superficial energy E_s depending on nanoparticles diameter d [nm]. Points designate the calculated values, and the continuous lines are approximations.

Spatial distribution of particles influences on rate of the forces holding nanostructures, formed from several nanoparticles, also. On Fig. 10.20 the chain nanoparticles, formed is resulted at coupling of three nanoparticles (from 512 atoms everyone), located in the initial moment on one line. Calculations have shown, that in this case nanoparticles form a stable chain. Thus, particles practically do not change the form and cooperate on "small platforms."

In the same figure the result of connection of three nanoparticles, located in the initial moment on a circle and consisting of 256 atoms everyone is submitted. In this case particles incorporate among themselves "densely," contacting on a significant part of the external surface.

Distance between particles at which they are in balance it is much less for the particles collected in group ($L_{3np}^0 < L_{2np}^0$) It confirms also the graph of forces from which it is visible, that the maximal force of an attraction between particles in this case (is designated by a continuous line) in some times more, than at an arrangement of particles in a chain (dashed line) $F_{3np} > F_{2np}$.

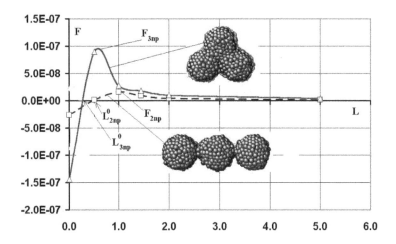

FIGURE 10.20   Change of force F [N] of 3 nanoparticles interaction, consisting of 512 atoms everyone, and connected among themselves on a line and on the beams missing under a corner of 120 degrees, accordingly, depending on distance between them L [nm].

Experimental investigation of the spatial structures formed by nanoparticles [4], confirm that nanoparticles gather to compact objects. Thus the internal nuclear structure of the connections area of nanoparticles considerably differs from structure of a free nanoparticle.

Nanoelements kind of interaction depends strongly on the temperature. In Fig. 10.21 shows the picture of the interaction of nanoparticles at different temperatures (Fig. 10.22). It is seen that with increasing temperature the interaction of changes in sequence: coupling (1,2), merging (3,4). With further increase in temperature the nanoparticles dispersed.

FIGURE 10.21   Change of nanoparticles connection at increase in temperature

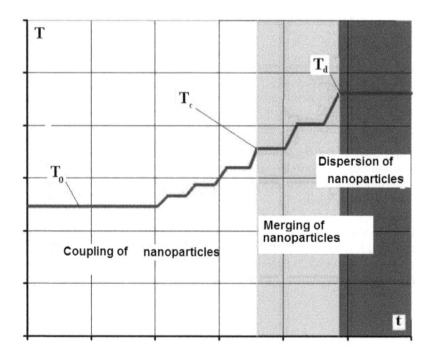

FIGURE 10.22   Curve of temperature change.

In the conclusion to the section we will consider problems of dynamics of nanoparticles. The analysis of interaction of nanoparticles among themselves also allows to draw a conclusion on an essential role in this process of energy of initial movement of particles. Various processes at interaction of the nanoparticles, moving with different speed, are observed: the processes of agglomerate formation, formation of larger particles at merge of the smaller size particles, absorption by large particles of the smaller ones, dispersion of particles on separate smaller ones or atoms.

For example, in Fig. 10.23 the interactions of two particles are moving towards each other with different speed are shown. At small speed of moving is formed steady agglomerate (Fig. 10.23).

In Fig. 10.23 (left) is submitted interaction of two particles moving towards each other with the large speed. It is visible, that steady formation in this case is not appearing and the particles collapse.

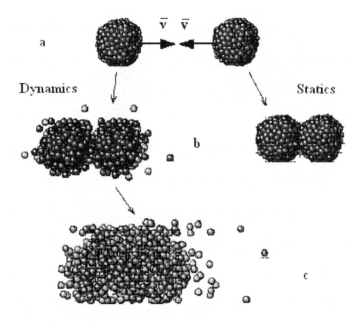

FIGURE 10.23 Pictures of dynamic interaction of two nanoparticles: (a) an initial configuration nanoparticles; (b) nanoparticles at dynamic interaction, (c) the "cloud" of atoms formed because of dynamic destruction two nanoparticles.

Nature of interaction of nanoparticles, along with speed of their movement, essentially depends on a ratio of their sizes. In Fig. 10.24 pictures of interaction of two nanoparticles of zinc of the different size are presented. As well as in the previous case of a nanoparticle move from initial situation (1) towards each other. At small initial speed, nanoparticles incorporate at contact and form a steady conglomerate (2).

FIGURE 10.24 Pictures of interaction of two nanoparticles of zinc: 1 – initial configuration of nanoparticles; 2 – connection of nanoparticles, 3, 4 – absorption by a large nanoparticle of a particle of the smaller size, 5 – destruction of nanoparticles at blow.

At increase in speed of movement larger nanoparticle absorbs smaller, and the uniform nanoparticle (3,4) is formed. At further increase in speed of movement of nanoparticles, owing to blow, the smaller particle intensively takes root in big and destroys its.

The given examples show that use of dynamic processes of pressing for formation of nanocomposites demands a right choice of a mode of the loading providing integrity of nanoparticles. At big energy of dynamic loading, instead of a nanocomposite with dispersion corresponding to the initial size of nanoparticles, the nanocomposite with much larger grain that will essentially change properties of a composite can be obtained.

10.6 CONCLUSIONS

In conclusion, the following basic regularities of the nanoparticle formation and self-organization should be noted.

1. The existence of several types of the forms and structures of nanoparticles is possible depending on the thermodynamic conditions.
2. The absence of the crystal nucleus in small nanoparticles.
3. The formation of a single (ideal) crystal nucleus defects on the nucleus surface connected to the amorphous shell.
4. The formation of the polycrystal nucleus with defects distributed among the crystal grains with low atomic density and change interatomic distances. In addition, the grain boundaries are nonequilibrium and contain a great number of grain-boundary defects.
5. When there is an increase in sizes, the structure of nanoparticles is changing from amorphous to roentgen-amorphous and then into the crystalline structure.
6. The formation of the defect structures of different types on the boundaries and the surface of a nanoparticle.
7. The nanoparticle transition from the globe-shaped to the crystal-like shape.
8. The formation of the "regular" and "irregular" shapes of nanoparticles depending on the number of atoms forming the nanoparticle and the relaxation conditions (the rate of cooling, first of all).
9. The structure of a nanoparticle is strained because of the different distances between the atoms inside the nanoparticle and in its surface layers.
10. The systems of nanoparticles can to form stable and unstable nanostructures.

ACKNOWLEDGMENTS

This work was carried out with financial support from the Research Program of the Ural Branch of the Russian Academy of Sciences: the projects 12-P-12-2010, 12-C-1-1004, 12-T-1-1009 and was supported by the grants of Russian Foundation for Basic Research (RFFI) 04-01-96017-r2004ural_a; 05-08-50090-a; 07-01-96015-r_ural_a; 08-08-12082-ofi; 11-03-00571-a.

The author is grateful to his young colleagues doctor A.A. Vakhrushev and doctor A.Yu. Fedotov for active participation in the development of the software complex, calculations and analyzing of numerous calculation results.

The calculations were performed at the Joint Supercomputer Center of the Russian Academy of Sciences.

KEYWORDS

- **dynamics interactions**
- **molecular mechanics**
- **nanocompoites**
- **nanoparticles**
- **nanostructural elements**
- **systematic study**

REFERENCES

1. Qing-Qing Ni, Yaqin Fu, & Masaharu Iwamoto, (2004). "Evaluation of Elastic Modulus of Nano Particles in PMMA/Silica Nanocomposites", Journal of the Society of Materials Science, Japan, *53(9)*, 956–961.

2. Ruoff, R. S., & Nicola, M. Pugno (2004). "Strength of Nanostructures", Mechanics of the 21st Century. Proceedings of the 21th International Congress of Theoretical and Applied Mechanics Warsaw: Springer, 303–311.

3. Diao, J., Gall, K., & Dunn, M. L. (2004). "Atomistic Simulation of the Structure and Elastic Properties of Gold Nanowires" Journal of the Mechanics and Physics of Solids, *52(9)*, 1935–1962.

4. Dingreville, R., Qu, J., & Cherkaoui, M. (2004). "Surface Free Energy and Its Effect on the Elastic Behavior of Nano-Sized Particles, Wires and Films", Journal of the Mechanics and Physics of Solids, *53(8)*, 1827–1854.

5. Duan, H. L., Wang, J., Huang, Z. P., & Karihaloo, B. L. (2005). "Size-Dependent Effective Elastic Constants of Solids Containing Nano-Inhomogeneities with Interface Stress", Journal of the Mechanics and Physics of Solids, *53(7)*, 1574–1596.

6. Gusev, A. I., & Rempel, A. A. (2001). "Nanocrystalline Materials" Moscow: Physical Mathematical literature, (in Russian).

7. Hoare, M. R. (1987). "Structure and Dynamics of Simple Micro Clusters" Ach Chem. Phys, *40*, 49–135.

8. *Brooks, B. R., Bruccoleri, R. E., Olafson, B. D., States, D. J., Swaminathan, S., & Karplus, M. (1983). "Charmm: "A Program for Macromolecular Energy Minimization and Dynamics Calculations" J Comput Chemistry, 4(2), 187–217.*

9. Friedlander, S. K. (1999). "Polymer-Like Behavior of Inorganic Nanoparticle Chain Aggregates", Journal of Nanoparticle Research, *1*, 9–15.

10. Grzegorczyk, M., Rybaczuk, M., & Maruszewski, K. (2004). "Ballistic Aggregation: An Alternative Approach to Modeling of Silica Sol–Gel Structures" Chaos, Solitons and Fractals, *19*, 1003–1100.

11. Shevchenko, E. V., Talapin, D. V., Kotov, N. A., O'Brien, S., & Murray, C. B. (2006). "Structural Diversity in Binary Nanoparticle Superlattices," Nature Letter, *439*, 55–59.

12. Kang, Z. C., & Wang, Z. L. (1996). "On Accretion of Nanosize Carbon Spheres", Journal Physical Chemistry, *100*, 5163–5165.

13. Melikhov, I. V., & Bozhevol'nov, V. E. (2003). "Variability and Self-Organization in Nanosystems", Journal of Nanoparticle Research, *5*, 465–472.

14. Kim, D., & Lu, W. (2004). "Self-Organized Nanostructures in Multi-Phase Epilayers" Nano-technology, *15,* 667–674.

15. Kurt, E. Geckeler (2005). "Novel Super molecular Nanomaterials: From Design To Reality", Proceeding of the 12 Annual International Conference on Composites Nano Engineering, Tenerife, Spain, August 1–6, CD Rom Edition.

16. Vakhrouchev, A. V. & Lipanov, A. M. (1992). "A Numerical Analysis of the Rupture of Powder Materials under the Power Impact Influence", Computer & Structures, *1/2(44)*, 481–486.

17. Vakhrouchev, A. V. (2004). "Modeling of Static and Dynamic processes of Nanoparticles Interaction" CD–ROM Proceeding of the 21th International Congress of Theoretical and Applied Mechanics ID12054 Warsaw Poland.

18. Vakhrouchev, A. V. (2005). "Simulation of Nanoparticles Interaction", Proceeding of the 12 Annual International Conferences on Composites/Nano Engineering, Tenerife, Spain, August 1–6, CD Rom Edition.

19. Vakhrouchev, A. V. (2006). "Simulation of Nano-Elements Interactions and Self-Assembling", Modeling and Simulation in Materials Science and Engineering, *14*, 975–991.

20. Vakhrouchev, A. V., & Lipanov, A. M. (2007). Numerical Analysis of the Atomic Structure and Shape of Metal Nanoparticles Computational Mathematics and Mathematical Physics, *47(10)*, 1702–1711.

21. Vakhrouchev, A. V. (2008). Modeling of the Process of Formation and Use of Powder Nano-composites Composites with Micro and Nano-Structures Computational Modeling and Experiments Computational Methods in Applied Sciences Series Barcelona, Spain: Springer Science, *9,* 107–136.

22. Vakhrouchev, A. V. (2009). Modeling of the Nano Systems Formation by the Molecular Dynamics, Mesodynamics and Continuum Mechanics Methods Multidiscipline Modeling in Material and Structures, *5(N2),* 99–118.

23. Vakhrouchev, A. V. (2008). Theoretical Bases of Nanotechnology Application to Thermal Engines and Equipment Izhevsk, Institute of Applied Mechanics Ural Branch of the Russian Academy of Sciences, 212p (in Russian).

24. Alikin, V. N., Vakhrouchev A. V., Golubchikov, V. B., Lipanov, A. M., & Serebrennikov, S. Y. (2010). Development and Investigation of the Aerosol Nanotechnology Moscow: Mashinostroenie, 196p, (in Russian).

25. Heerman, W. D. (1986). "Computer simulation methods in theoretical physics" Berlin: Springer-Verlag.

26. Verlet, L. (1967). "Computer "Experiments" On Classical Fluids I. Thermo Dynamical Properties of Lennard-Jones Molecules", Phys. Rev. *159*, 98–103.

27. Korn, G. A., & Korn, M. T. (1968). "Mathematical Handbook", New York: McGraw-Hill Book Company.

28. Self-Organizing of Micro particles Piezoelectric Materials, News of Chemistry, date news php.htm.

CHAPTER 11

A STUDY ON BIOLOGICAL APPLICATION OF AG AND CO/AG NANOPARTICLES CYTOTOXICITY AND GENOTOXICITY

IMAN E. GOMAA[1*], SAMARTH BHATT[2], THOMAS LIEHR[2], MONA BAKR[3], and TAREK A. EL-TAYEB[3]

[1*]German University in Cairo, Egypt; *E-mail: iman.gomaa@guc.edu.eg

[2]Jena University Hospital, Friedrich Schiller University, Institute of Human Genetics, Kollegiengasse 10, D-07743 Jena, Germany.

[3]The National Institute for Laser Enhanced Sciences, Cairo University, Egypt

CONTENTS

Abstract 140
11.1   Introduction 140
11.2   Materials and Methods 141
11.3   Results 144
11.4   Discussion 149
11.5   Conclusions and Recommendations 151
Acknowledgement 151
Keywords 152
References 152

ABSTRACT

The use of nanoparticles (NPs) in cancer treatment has drawn more attention in different scientific fields. This is due to their altered physical, chemical and biological properties from their macrostructures. The downsizing of the materials to the nano-scale size increases their ability to interact with biological size-equivalent macromolecules.

The present work is directed to investigate and compare between the photothermal cytotoxic effect of silver "Ag" nanoparticles and cobalt core – silver shell (Co/Ag) nanoparticles at in vitro level for determination of the efficacy as well as the safety borders of cancer cells killing using photo thermal therapy (PTT).

In-vitro study using HEp-2 human laryngeal carcinoma cells was undergone in order to determine the most reliable concentrations of either Ag or Co/Ag particles required for tumor cytotoxicity. Monochromatic blue light exposure using a LED source of 460 nm and 200 mW was applied at 5 and 72 h post-NPs incubation. Light exposure of 2 to 15 min, has been tested in order to exert the most effective therapy. Ag and Co/Ag nanoparticles provide considerable efficacy when up taken by laryngeal cancer cells. However, both types of particles showed considerable genetic side effects both on DNA and chromosomal levels.

11.1   INTRODUCTION

Hyperthermia is a state where cells absorb more heat than they can dissipate, which can be lethal to the cells. Such phenomenon has been found as a promising approach for cancer therapy because it directly kills cancer cells and indirectly activates anticancer immunity [1]. Since focusing the heat on an intended region without damaging the healthy tissue was an important problem that arose by the treatment with hyperthermia, targeting of a specific region was important. If nanoparticles can target a specific malignant tissue, hyperthermia can be directed to this specific tissue. Unlike the conventional hyper thermic techniques, some studies showed that using nanoparticles for hyperthermia helped cancer cells to reach the lethal temperature without damaging the surrounding tissue [2].

The choice of a specific light delivery mode in clinical settings is usually based on the nature and location of the disease. The optimal light dose can be obtained by adjusting the fluency rate and fluency element. The characterization of light penetration and distribution in solid tumors is important since it will influence choosing a light source with an appropriate wavelength. From the light sources used in photo thermal therapy, the Light Emitting Diode "LED" generating a desired high energy of specific wavelengths and can be assembled in a range of geometries and sizes [3].

Metal nanoparticles with their wide range of applications, such as catalytic systems with optimized selectivity and efficiency, optical components, targeted thermal agents for exploitation in drug delivery and medical therapies, as well as surface-enhanced Raman spectral probing, have attracted research attention during the last decade. The size-dependent and shape-dependent properties of nanoparticles render them different from their corresponding bulk materials with macroscopic dimensions [4].

Due to their special physicochemical properties, metal nanoparticles showed a great progress in the bioanalytical and medical applications, such as multiplexed bioassays, biomedicine, ultrasensitive bio detection [5], and bioimaging [6]. They have several biological applications including; drug delivery, magnetic resonance imaging enhancement "MRI" [6], as well as in cancer treatment [7].

Silver nanoparticles are among the noble metallic nanomaterials that have received considerable attention due to their attractive physicochemical properties. The surface Plasmon resonance and the large effective scattering cross section of individual nanoparticles make them ideal candidates for biomedical applications [8]. On the other hand, other studies were directed towards synthesizing silver nano shells of 40–50 nm outer surface diameter and 20–30 nm inner diameter using cobalt (Co) nanoparticles as sacrificial templates. In this case, the thermal reaction deriving force comes from the large reduction potential gap between the Ag^+/Ag and the Co^{+2}/Co redox couples which results in the consumption of Co cores and the formation of a hollow cavity of Ag nano shells. The UV spectrum of this nanostructure exhibits a distinct difference from that of solid nanoparticles, which makes it a good candidate for application in photo thermal materials [9].

As little is known about their biological applications in cancer treatment, and relying on the fact of being novel candidates providing high thermal effect, this study was directed towards investigation of the photo thermal cancer therapy using silver nanoparticles and cobalt core silver shell nanoparticles in HEp-2 laryngeal cell carcinoma in vitro, as well as the side genetic effects both on DNA and chromosomal levels.

11.2 MATERIALS AND METHODS

11.2.1 SYNTHESIS OF AGNPS AND CO/AGNPS

The Ag and Co/Ag nanoparticles were obtained from the Nanotechnology Lab, the National Institute of Laser Enhanced Sciences (NILES), Cairo University. Spectrophotometric analysis was done to determine the Absorption wavelengths range of Ag and Co/Ag NPs. Transmission Electron Microscope was used to investigate the shape and size range of these nanoparticles before using for application in cancer hyperthermia.

11.2.2 CELL CULTURE CONDITIONS

An established Human Epidermoid cancer cells (HEp-2) isolated from the larynx was obtained from American Type Culture Collection (CCL-23, ATCC, Rockville,

MD) have been used for application of the in vitro tumor viability assay. Cells were routinely maintained in Eagle's minimal essential medium (EMEM) supplemented with 10% FBS and 1% antibiotic solution (including 10,000 U penicillin and 10 mg streptomycin) at 37°C in a humidified atmosphere of 5% CO_2 in air. Cells (100 cells/mm²) were seeded in 24 well plates and grown to 75% confluence level overnight.

11.2.3 IN VITRO VIABILITY ASSAY

In order to ensure highest cytotoxic levels of tumor cells, the optimum light doses and nanoparticles concentrations were first determined by application of series of dark and light control experiments. Three replicates of HEp-2 laryngeal carcinoma cells were seeded overnight in 24 wells culture plates. Cells were incubated with different concentrations of the drugs of interest (Ag or Co/Ag) NPs in 5% FBS containing medium and incubated at the standard culture conditions. On the next day, they were subjected to different monochromatic light doses from light emitting diode (LED) of 460 nm and 200 mW, while being incubated in 2% FBS medium. The distance between the light source and cell line surface was adjusted to 1 cm² spot surface area. Duration of light exposure was calculated with the following general formula; Time (s) = [No. of Joules × Well surface area (m²)]/irradiance of LED (W/m²).

Cells were washed with Phosphate Buffer Saline (PBS) to get rid of the excess unabsorbed drug, and incubated with 0.0075% neutral red viability assay solution (Sigma Aldrich, N7005–1G) diluted in 2% FBS medium for 3 h. Excess dye was washed off with 0.9% NaCl, and the stained viable cells were lysed in lysis buffer (1% absolute ethanol: 1% distilled water: 0.02% glacial acetic acid) for 20 min with continuous shaking. Spectrophotometric measurement of cellular viability was performed in a 96 wells plate. Absorbance of the red color released after lyses of viable cells was measured using a plate reader (Victor 3 V-1420, Germany) at 572 nm where the absorbance values indicate cell viability.

11.2.4 GENOTOXICITY TEST (COMET ASSAY)

The LC50of both AgNPs as well as Co/AgNPs mediated PTT were used to evaluate the genotoxicity in freshly isolated peripheral blood lymphocytes. Blood lymphocytes were separated according to Ref. [10]. Lymphocytes layer seen as a buffy coat was aspirated and washed with PBS then incubated overnight in RPMI growth medium containing 0.001% phytohemagglutinin. Equal numbers of cells were seeded in 24 wells plate for testing each experimental condition (treated cells with either AgNPs or Co/AgNPs at 50 J/cm², dark control and light control). These were compared to cells treated with the strong genotoxic chemotherapeutic drug Bleomycin, as a positive control. Viability of cells was monitored at 70% with reference to the control cells using Trypan blue stain in order to exclude the cytotoxic effects of either drug upon interpretation of the genotoxicity results [11]. Comet assay was applied according to Ref. [12]. During electrophoresis the broken DNA moves towards the anode forming a Comet tail. The greater the extent of DNA damage, the longer the tail length. Finally, 150 cells per slide were randomly selected for quantification of genotoxic effect of the tested drugs as a percentage of DNA damage using (Comet Imager, metasystem, version 2.2, GmbH) software.

11.2.5 MOLECULAR CYTOGENETIC TEST: MULTIPLEX FLUORESCENCE IN SITU HYBRIDIZATION (MFISH)

Blood collected from a healthy volunteer was cultured for 72 h in RPMI medium containing 1% penicillin/streptomycin, 15% FBS and 2% phytohaemagglutinin. Duplicate whole blood cultures were treated with the same LC50 conditions used for the comet assay, before being harvested. Both dark and light control experiments were also included, in order to detect the sole effect of either the drug or the monochromatic light on exerting chromosomal aberrations. In-vitro human lymphocyte assay for evaluating the chromosomal anomalies caused by chemical agents was applied on metaphases of duplicate slides from each sample referring to Ref. [13].

Slides prepared from lymphocyte cultures treated with either of the two drugs of interest and containing metaphase chromosomes were used for investigation of chromosomal aberrations by M-FISH. The M-FISH probe was prepared at the Institute of Human Genetics in Jena, Germany. After hybridization of two slides from each duplicate culture, slide washes and signal detection steps were applied as described in Refs. [14, 15]. Counterstain and anti-fade solution was applied using 4',6-diamino-2-phenylindole (DAPI) solution (Invitrogen, H-1200). The slides were stored in the dark at 4°C for 15 min before fluorescent microscopic examination. A total of 50 metaphases were analyzed using a fluorescent microscope (Axio Imager. Z1 mot; Zeiss) equipped with the appropriate filter sets to discriminate between the five fluorochromes and the DAPI counterstain using a XC77 CCD camera with on-chip integration (Sony, Vienna, Austria). Image capturing and processing were carried out using an ISIS imaging system (MetaSystems, Altlussheim, Germany). Chromosomal gaps or breaks were estimated in comparison to a normal karyotype.

A number of 100 metaphases of each sample were analyzed by standard karyotyping using ISIS software (Metasystems GmbH, Altlussheim, Germany, 0017) for detection of the presence of any chromosomal aberrations. Karyotypes nomenclature was determined according to the International System for human Cytogenetic Nomenclature (ISCN). Both numerical and structural chromosomal aberrations were recorded for either AgNPs or Co/AgNPs treated cells as compared to the control non-treated ones. Finally, the mitotic index per 1000 blast cells for each treated sample with either drug of interest in comparison to the (untreated) cells was also assessed.

11.2.6 STATISTICAL ANALYSIS

Statistical analysis was applied using the GraphPad Prism 5.0 software. Results were obtained from three independent experiments and data were expressed as the mean ± standard error of the mean (SEM), using One-way ANOVA Turkey's Multiple Comparison Test, and 95% confidence interval levels. Data were considered significant at p-value < 0.05.

11.3 RESULTS

11.3.1 PHYSICOCHEMICAL CHARACTERIZATION OF THE PRODUCED NPS

The prepared particles show absorption in the visible range due to surface plasmon resonance at 405 nm and the TEM images show that, the particles are spherical with homogenous size distribution (Fig. 11.1a and 11.1b). As confirmed by the TEM images, the absorption spectra show one narrow band indicating the spherical shape of the obtained particles (Fig. 11.2a and 11.2b).

(a)   **(b)**

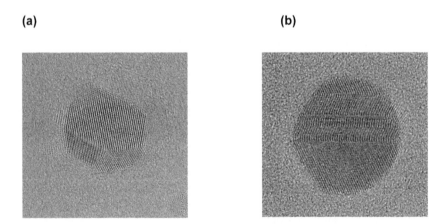

FIGURE 11.1   TEM micrographs of AgNPs (a) & Co/AgNPs (b).

(a) (b)

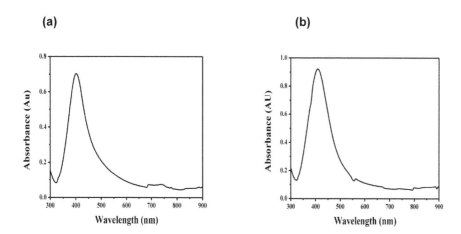

FIGURE 11.2 Absorption spectrum of AgNPs (a) & Co/AgNPs (b).

11.3.2 IN VITRO VIABILITY ASSAY

Our in vitro study was based on evaluation of the photo thermal cytotoxic effect of Ag and Co/Ag core shell nanoparticles upon treatment of HEp-2 laryngeal cancer cells.

A dark toxicity test was performed in order to investigate the effect of both Ag-NPs and Co/AgNPs on the viability of HEp-2 Laryngeal cell carcinoma incubated with either type of nanoparticles in the absence of monochromatic light activation. Two different concentrations have been examined for their cytotoxic effect on HEp-2 tumor cells (10^{-6} M/L and 5×10^{-5} M/L) both at 5 h and 24 h incubation times. Evaluation of cellular viability using NR assay resulted into absolutely no cell death at 10^{-6} M/L with either concentrations of both AgNPs or Co/AgNPs at 5 h incubation time. Longer incubation periods at 24 h, resulted into a slight decrease in cell viability as indicated by 3.13% and 5.29%, at 10^{-6} M/L of AgNPs and Co/AgNPs respectively, and insignificant decrease in cell survival as shown by 13.56% and 15.84% at 5×10^{-5} M/L of AgNPs and Co/AgNPs respectively. These results of dark toxicity test prove that both AgNPs and Co/AgNPs are not significantly toxic at either 10^{-6} M/L or 5×10^{-5} M/L (Fig. 11.3a). Meanwhile, exposing HEp-2 laryngeal carcinoma cells to monochromatic blue light of 460 nm at different exposure times of 5, 10, and 15 min showed slight and insignificant decrease in cell viability as indicated by 0.88%, 4.83% and 7.8% respectively (Fig. 11.3b).

Treatment of HEp-2 cells with 5×10^{-5} M/L of either particles and 200 mW exposure to monochromatic blue light resulted into dramatic decrease in HEp-2 cells viability. The LC50 conditions of HEp-2 cells were exceeded and could not be detected upon incubation with either AgNPs, or Co/AgNPs, as complete cell death was obtained at 10 min light exposure of AgNPs, and Co/AgNPs respectively (data not shown). On the other hand, HEp-2 laryngeal carcinoma cells that have been incubated for 24 h with 10^{-6} M/L concentration of either AgNPs or Co/AgNPs while exposed to monochromatic blue light of 200 mW and 460 nm for 5, 10, and 15 min resulted into gradual decrease in cell viability from 84.1%, 75.3%, 51.3% in case of cells incubated with AgNPs, and 77.3%, 49.5, 37.8% in those incubated with Co/AgNPs. This means, conditions required for reaching the LC50 (50% cell death) at 10^{-6} M/L of either type of particles were 15 and 10 min light exposure for AgNPs and Co/AgNPs, respectively (Fig. 11.4a and 11.4b).

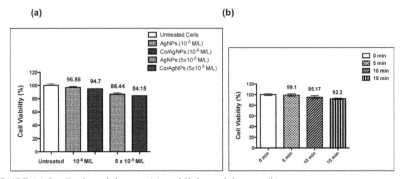

FIGURE 11.3   Dark toxicity test (a) and light toxicity test (b).

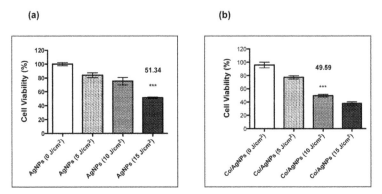

FIGURE 11.4   Cell viability test of HEp-2 laryngeal carcinoma cells incubated with 10^{-6} M/L AgNPs (a) and Co/AgNPs (b) and exposed to 200 mW monochromatic blue light of 460 nm.

11.3.3   EVALUATION OF GENOTOXICITY

In the presented study, an alkaline single-cell gel electrophoresis assay (Comet assay) was applied on peripheral blood lymphocytes in order to measure the single strand breaks (SSB), double strand breaks (DSB), as well as alkali labile sites as indicated by the number of DNA stand breaks for each therapeutic modality of either chemotherapy or PTT. The threshold level for cytotoxicity after application of each therapeutic modality was set at 70% cell viability, in order to minimize cytotoxicity-induced genotoxic effects [16]. The LC50 conditions of both AgNPs and Co/AgNPs concluded from the in vitro tumor viability assay in HEp-2 laryngeal carcinoma cells exerted less than 70% lymphocyte cell death as proved by Trypan Blue exclusion test. Accordingly, the genotoxicity test by alkaline comet assay was applied on those conditions (10^{-6} M/L of either type of NPs for 24 h incubation, and 15 min monochromatic blue light exposure) and compared to the untreated control cells (Fig. 11.5, Panel A).

On one hand, blood lymphocytes treated with LC50 conditions of AgNPs gave significant percentage (p < 0.05) of DNA damage (49.85%) as compared to the untreated control cells [Fig. 11.5, Panel B (a)]. On the other hand, lymphocytes treated with LC50 conditions of Co/AgNPs mediated PTT, resulted into a highly significant increase (p = 0.01) of DNA damage as indicated by 51.2% compared to the untreated control cells [Fig. 11.5, Panel B (b)]. Also, the sole effect of particles concentration gave 28.2% and 29.4% DNA damage for AgNPs and Co/AgNPs, respectively, while the 460 nm monochromatic blue light exposure showed 26.5% and 27.3% at 10 and 15 min, respectively. Such levels of both dark control, as well as light control tests, proved to have insignificant levels (p > 0.05) of DNA damage compared to that exhibited by the untreated cells (23.4%). Bleomycin is a standard chemotherapy that has exerted 99.98% level of DNA damage proving a very highly significant level of DNA damage (p = 0.001) in comparison to the untreated cells [Fig. 11.5, Panel B (a and b)].

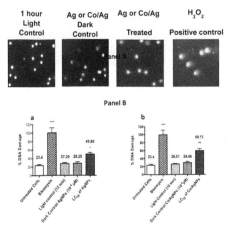

FIGURE 11.5 Genotoxicity "Comet Assay." Percentage of DNA damage exerted by LC50 conditions of AgNPs, panel [B (a)], and Co/AgNPs, panel [B (b)] on freshly isolated peripheral blood lymphocytes.

11.3.4 EVALUATION OF MUTAGENICITY

Chromosomal mutations resulted from treatment of peripheral blood lymphocytes with LC50 of either AgNPs or Co/AgNPs was investigated using standard karyotyping by GTG banding technique. Data of the light control and dark control, lymphocytes with LC50 conditions of either type of particles showed normal karyotype (46, XX) with no chromosomal aberrations (data not shown). On the other hand, metaphase analysis of lymphocytes treated with LC50 conditions of AgNPs and Co/AgNPs exhibited different types of mutations including chromatid type breaks as well as hypoploidy such as 45, XX, −18 and 45, XX, −18 in case of AgNPs treated lymphocytes, while 40, XX, −6, −10, −11, −12–17, −20 chtb(1), and 45, XX, −22, chtb(9)in case of the Co/AgNPs treated lymphocytes (Fig. 11.6a and 11.6b, respectively).

Additionally, the mitotic index was investigated in order to evaluate the lethality magnitude of the LC50 of each drug on peripheral blood lymphocytes. Results proved significant decrease (53.3%) in the mitotic index of cells incubated with LC50 of Ag-NPs, and a highly significant decrease (40.37%) in mitotic index of cells treated with Co/AgNPs mediated PTT, in comparison to the that of control cells, which have not been exposed to PTT. On the other hand, there has been a major drop in the mitotic index of blood lymphocytes treated for 24 h with the strong chemotherapeutic drug Bleomycine (4.33%) showing a very highly statistical significance (p = 0.001) compared to the control untreated cells (Fig. 11.7).

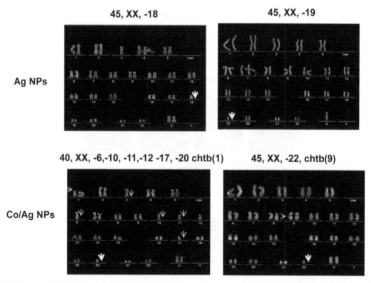

FIGURE 11.6 M-FISH analysis of human karyotype treated with LC50 conditions of either AgNPs or Co/AgNPs mediated PTT: Cells incubated with 10^{-6} M/L of either AgNPs, or Co/AgNPs mediated PTT, and exposed to monochromatic blue light of 460 nm and 200 mW.

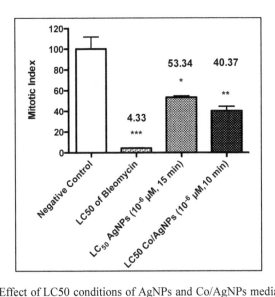

FIGURE 11.7   Effect of LC50 conditions of AgNPs and Co/AgNPs mediated PTT on the mitotic index of peripheral blood lymphocytes.

11.4   DISCUSSION

Silver nanoparticles were applied in wound dressings, catheters, and various household products due to their antimicrobial activity [17]. The achievement of different sizes and shapes of silver nanoparticles can be obtained by minor modifications during their synthesis parameters [18, 19]. Therefore, in this study, the toxicity of AgNPs and Co/AgNPs on HEp-2 laryngeal cell line was evaluated using cell viability, DNA damage and chromosomal aberrations.

Spherical AgNPs and Co/AgNPs of average size 20 nm were used in this study. The optical absorption of both nanoparticles covers the spectral absorption range of 350 nm – 600 nm. The maximum emission wavelength of light emitting diode (LED) used in this work (460 nm) is found in the AgNPs and Co/AgNPs absorption range. Both HEp-2 cells dark and light controls showed insignificant change when they were exposed to AgNPs and Co/AgNPs concentrations of 10^{-6} and 5×10^{-5} M/l or 5 J/Cm^2, 10 J/Cm^2 and 15 J/Cm^2 of light irradiance, respectively. This means that the reduction in the percentage of cell viability at Fig. 11.4, compared to the untreated control cells, is due to the activation of AgNPs and Co/AgNPs with the blue light photons (460 nm). Absorbed light increases the plasmon resonance of these nanoparticles, and by particles relaxation process the excess energy is dissipated in form of heat. The heat energy surrounding each nanoparticle causes cellular localized effect called hyperthermia, which is a state where cells absorb more heat than they can dissipate. Such a heat can be lethal to cancer cells [1].

Different light fluencies rates (5 J/Cm², 10 J/Cm² and 15 J/Cm²) were applied in this work to have suitable LC50 of AgNPs and Co/AgNPs using the same above concentrations to be used in the cytotoxicity and genotoxicity studies in blood lymphocytes. Light fluence rate of 15 J/Cm²) in case of AgNPs treatment have the same reduction effect of the 10 J/Cm² in case of Co/AgNPs. This may reflect that Co/Ag-NPs is more sensitive than AgNPs to this kind of treatment. This sensitivity may be due to the fact that the thermal reaction deriving force in case of Co/AgNPs comes from the large reduction potential gap between the Ag^+/Ag and the Co^{+2}/Co redox couples which results in the consumption of Co cores and the formation of a hollow cavity of Ag nano shells. The UV spectrum of this nanostructure exhibits a distinct difference from that of solid nanoparticles, which makes it a good candidate for application in photo thermal materials [9].

Cellular uptake of nanoparticles via diffusion or endocytosis, is followed by their random distributed everywhere in the cells. It has been reported that in case of normal human lung fibroblast cells (IMR-90) and human glioblastoma cells (U251), the transmission electron microscopic (TEM) analysis indicated the presence of Ag-NPs inside the mitochondria and nucleus, implicating their direct involvement in the mitochondrial toxicity and DNA damage [17]. This fact has encouraged us to study the side effect of the photo thermal process of AgNPs and Co/AgNPs treated blood lymphocytes.

In the current study, we are interested in the genotoxic effects caused by the LC50 conditions (10^{-6} M/L NPs and 200 mW blue light) of either AgNPs or Co/AgNPs resulting from the viability assay. Genotoxicity was evaluated and showed percentage of DNA damage of a significant and highly significant level for the AgNPs and Co/AgNPs, respectively. This result is in agreement with the study made on Balb/3T3 cell model at which, statistically significant DNA damage of $AgNO_3$ (7–10 μM), and (> 1 μM) Co nano particles was obtained indicating their strong genotoxic potential by Ref. [20]. In another study [21], the comet assay showed a statistically significant dose-related increase in % Tail DNA for (10^{-5} to 10^{-4} M) of CoNP ($p < 0.001$) upon 24 h incubation with peripheral blood lymphocytes. Meanwhile, a dose dependent increase in genotoxicity has been observed when AgNPs were incubated with CHO-K1 cells causing 450% increase in DNA breakage at a concentration of 9×10^{-6} M, as compared to control cells (p<0.01) [22]. In all cases it is anticipated that DNA damage is augmented by deposition, followed by interactions of either silver or cobalt silver nanoparticles to the DNA leading to cell cycle arrest in the G_2/M phase [23].

Investigation of the effect of both AgNPs and Co/AgNPs on chromosomal aberrations of human peripheral blood lymphocytes indicated the exhibition of numerical (hypoploidy) chromosomal aberrations in case of AgNPs, while both numerical (hypoploidy) and structural (Chromatid type breaks) aberrations in case of Co/AgNPs. Although no available studies have been found for the mutagenic effect of either AgNPs or Co/AgNPs on the level of human chromosomes, but several studies have proved the mutagenic effect of both types of particles using the Ames muta-genicity test, as well as micronucleus test [22, 24].

Meanwhile, the genetic investigations showed significant drop (53.3%) in the mitotic index of cells incubated with LC50 of AgNPs, and a highly significant de-crease (40.37%) in mitotic index of cells treated with Co/AgNPs mediated PTT, in comparison to the that of control cells, which have not been exposed to PTT. This suggests that PTT tumor cells killing could happen through a molecular mechanism involving DNA double strands breaks or chromosomal mutations. A possible mech-anism could involve inhibition of extracellular signal-regulated kinases (ERKs) that are responsible for cell proliferation and differentiation, or down regulation of cy-clin D1 and cyclin E which lead to antiproliferative effect of chlorophyll derivative [25, 26].

11.5 CONCLUSIONS AND RECOMMENDATIONS

In conclusion, AgNPs and Co/AgNPs mediated PTT can be used in the medical field offering a promising tool for cancer cells death. It provides considerable cytotoxic effect when administered by laryngeal cancer cells. However, both types of particles showed considerable genetic side effects both on DNA and chromosomal levels. Therefore, AgNPs and Co/AgNPs mediated PTT needs further optimization as well as in vivo investigations for possible future clinical applications.

The efficacy of AgNPs and Co/AgNPs should also be tested in other cancer cell lines, and their cellular uptake should be examined by confocal microscopy in order to specify the time required for their exocytosis that might influence their efficacy at tumor cell killing. Additionally, higher selectivity to tumor cells could be examined via in vivo applications, taking into consideration, that the surface modifications of both types of nanoparticles might increase their specificity at tumor cell targeting.

ACKNOWLEDGEMENT

The authors thank Marwa Ramadan at the "National Institute for Laser Enhanced Sciences – NILES" for the synthesis of both AgNPs and Co/AgNPs. This work has been partially supported by The German Academic Exchange Service, Deutsche Akademische Austausch Dienst (DAAD).

KEYWORDS

- **Ag and Co/Ag nanoparticles**
- **genotoxicity**
- **in vitro study**
- **mutagenicity**
- **photo thermal therapy**

REFERENCES

1. Van der Zee, J. (2002). Heating the Patient: A Promising Approach? Annuals of Oncology, 13(8), 1173–1184.
2. Kikumori, T., Kobayashi, T., & Sawaki, M. (2009). Anti-Cancer Effect of Hyperthermia on Breast Cancer by Magnetite Nanoparticle Loaded Anti-HER2 Immunoliposomes Breast Cancer Res. Treat, 113(3), 435–41.
3. Zheng, Y., Hunting, D. J., Ayotte, P., & Sanche, L. (2008). Radio Sensitization of DNA by Gold Nanoparticles Irradiated with High Energy Electrons Radiation. Res, 169(1), 19–27.
4. Zamiri, R., Zakaria, A., Husin, M. S., Wahab, Z. A., & Nazarpour, F. K. (2011). Formation of Silver Micro belt Structures by Laser Irradiation of Silver Nanoparticles in Ethanol. Int J Nanomedicine. 6, 2221–2224.
5. Nam, Y. S., Kang, H. S., Park, J. Y., Park, T. G., Han, S. H., & Chang, I. S. (2003). New Micelle-Like Polymer Aggregates Made From PEI-PLGA Diblock Copolymers: Micellar Characteristics and Cellular Uptake. Biomaterials, 24(12), 2053–2059.
6. Iida, H. (2008). Chemical synthesis of Nanoparticles and their Applications to Bioanalysis and Medical Care, PhD Dissertation, Waseda University, Shinjuku Tokyo, Japan.
7. Ruddon, R. W. (2007). Cancer Biology, 4th ed, Oxford University Press, Inc, Oxford.
8. Liau, S. Y., Read, D. C., Pugh, W. J., Furr, J. R., & Russell, A. D. (1997). Interaction of Silver Nitrate with Readily Identifiable Groups: Relationship to the Antibacterial Action of Silver Ions, Lett. Appl. Microbiol., 25(4), 279–283.
9. Guildford, A. L., Poletti, T., Osbourne, L. H., Di Cerbo, A., Gatti, A. M., & Santin, M. (2009). Nanoparticles of a Different Source Induce Different Patterns of Activation in Key Biochemical and Cellular Components of the Host Response. J. R. Soc. Interface, 6, 6(41), 1213–1221.
10. Singh, N. P., McCoy, M. T., Tice, R. R., & Schneider, E. L. (1988). A Simple Technique 458 for Quantitation of Low Levels of DNA damage in Individual Cells Experimental Cell Research, 184–91.
11. Aardema, M. J., Albertini, S., Arni, P., Henderson, L. M., Kirsch-Volders, M., Mackay, J. M., Sarrif, A. M., Stringer, D. A., & Taalman, R. D. (1998). Aneuploidy: A Report of an ECETOC Task Force. Mutat Res. 410(1), 73–79.
12. Anderson, D., & Plewa, M. J. (1998). The International Comet Assay Workshop Mutagenesis, 13(1), 67–73.
13. Preston, R. J., San Sebastian, J. R., & McFee, A. F. (1987). The In Vitro Human Lymphocyte Assay for Assessing the Clastogenicity of Chemical Agents Mutat Res, 189(2), 175–183.
14. Liehr, T., Thoma, K., Kammler, K., Gehring, C., Grahi, H., Ekici, A., Bathice, K. D., & Rautenstrauss, B. (1995). Direct Preparation of Uncultured EDTA-treated or Heparinized Blood for Interphase Fish analysis Applied Cytogenetic, 21(6), 185–188.

15. Chudoba, I., Plesch, A., Lörch, T., Lemke, J., Claussen, U., & Senger, G. (1999). High Resolution Multicolor-Banding: A New Technique for Refined Fish Analysis of Human Chromosomes, Cytogenet Cell Genet, 84(3–4), 156–160.
16. Hartmann, A., Plappert, U., Poetter, F., & Suter, W. (2003). Comparative Study with the Alkaline Comet Assay and the Chromosome Aberration Test Mut Res, 536(1–2), 27–38.
17. Pal, S., Tak, Y. K., & Song, J. M. (2007). Does the Antibacterial Activity of Silver Nanoparticles Depend on the Shape of the Nanoparticle? A Study of the Gram Negative Bacterium Escherichia Coli, Applied and Environmental Microbiology, 73(6), 1712–1720.
18. Bogle, K. A., Dhole, S. D., & Bhoraskar, V. N. (2006). Silver Nanoparticles: Synthesis and Size control by Electron Irradiation, Nanotechnology, 17(13), 3204.
19. Panáček, A., Kvitek, L., Prucek, R., Kolar, M., Vecerova, R., Pizurova, N., & Zboril, R. (2006). Silver Colloid Nanoparticles: Synthesis, Characterization, and their Antibacterial Activity. The Journal of Physical Chemistry B, 110(33), 16248–16253.
20. Munaro, B. (2009). Mechanistic in Vitro Tests for Genotoxicity and Carcinogenicity of Heavy Metals and their Nanoparticles PhD dissertation Konstanz University, Germany.
21. Colognato, Bonelli A., Ponti, J., Farina, M., Bergamaschi, E., & Sabbioni, E. (2008). Comparative Genotoxicity of Cobalt Nanoparticles and Ions on Human Peripheral Leukocytes in Vitro Mutagenesis, 23(5), 377–382.
22. Kim, H. R., Park, Y. J., Shin, D. Y., Oh, S. M., & Chung, K. H. (2013). Appropriate in Vitro Methods for Genotoxicity Testing of Silver Nanoparticles. Environ Health Toxicol 28, 1–8.
23. AshaRani, P. V., Mun, G. L. K., Hande, M. P., & Valiyaveettil, S. (2009). Cytotoxicity and Genotoxicity of Silver Nanoparticles in Human Cells ACS Nano, 3(2), 279–290.
24. Yan, L. I., Chen, D. H., Yan, J., Chen, Y., Mittelstaedt, R. A., Zhang, Y., Biris, A. S., Heflich, R. H., Chen, T. (2012). Genotoxicity of Silver Nanoparticles Evaluated Using the Ames test and In Vitro Micronucleus Assay. Mut Res, 745(1–2), 4–10.
25. Chiu, L. C., Kong, C. K., & Ooi, V. E. (2003). Anti-Proliferative Effect of Chlorophyllin derived from a Traditional Chinese Medicine Bombyxmori Excreta on Human Breast Cancer MCF-7 Cells. International Journal of Oncology, 23(3), 729–735.
26. Chiu, L. C., Kong, C. K., & Ooi, V. E. (2005). The Chlorophyllin-Induced Cell Cycle Arrest and Apoptosis in Human Breast Cancer MCF-7 Cells are Associated with ERK Deactivation and Cyclin D1 Depletion. Int. J. Mol. Med, 16(4), 735–740.

CHAPTER 12

THE TRANSFER VARIANTS OF METAL/ CARBON NANOCOMPOSITES INFLUENCE ON LIQUID MEDIA: A RESEARCH NOTE

V. I. KODOLOV[1, 2], and V. V. TRINEEVA[1, 3]

[1]Basic Research – High Educational Centre of Chemical Physics & Mesoscopy, Udmurt Scientific Centre, Ural Division; Russian Academy of Sciences

[2]M.T. Kalashnikov Izhevsk State Technical University

[3]CInstitute of Mechanics, Ural Division, Russian Academy of Sciences

CONTENTS

Abstract 156
12.1   Introduction 156
12.2   The Analysis of Transfer Variants of the Metal/Carbon Nanocomposites Super Small Quantities Influence on Liquid Media 156
12.3   The Changes in IR Spectra at the Metal/Carbon Nanocomposites Super Small Quantities Addition 158
12.4   Conclusion 159
Keywords 160
References 160

ABSTRACT

The transfer variants of metal/carbon nanocomposites influence on liquid media and also on polymeric compositions are represented. The changes in IR spectra lines intensity of different liquids at the super small quantities addition of metal/carbon nanocomposites are considered. The features of electromagnetic wave motion arise in different media are discussed.

12.1   INTRODUCTION

In the previous papers [1–13] the hypothesis of correspondent nanostructures super small quantities influence on media through nanostructures vibrations transfer on media molecules when these vibrations near to ultrasonic vibrations is proposed. Further this hypothesis is confirmed in some measure by the media IR spectra intensities increasing at the addition of metal/carbon nanocomposites in these media. In this case the self-organization of media molecules and the changes of corresponding media properties are found.

In liquid media the self-organization effect disappeared after some time in depend on media viscosity and their polarity. At the same time the line widening in IR spectra of some media equally to the intensity increasing as well as the C1 s lines widening in X-ray photoelectron spectra is discovered. This fact can be explained by the media electronic structure changes, and also by the coordination interaction of nanocomposites with media molecules.

The observed experimental results demand an answer with point of view concerning fundamentals of polymeric materials modification by super small quantities of metal/carbon nanocomposites (M/C nc).

12.2   THE ANALYSIS OF TRANSFER VARIANTS OF THE METAL/ CARBON NANOCOMPOSITES SUPER SMALL QUANTITIES INFLUENCE ON LIQUID MEDIA

According to the analysis of metal/carbon nanocomposites energetic characteristics the following variants of their influence transfer on liquid media can be proposed:
- through the transference of vibrations with ultrasonic frequency on surroundings molecules;
- trough the electromagnetic radiation from nanocomposites – vibrators;
- action transfer of metal/carbon nanocomposite electromagnetic field for self-organization of media molecules;
- by means of the media molecules activation under the nanocomposite super molecules action.

In the first case the transmission of vibration in medium is stipulated the velocity of vibration expansion and also the vibration transference depends on the medium density. The vibration expansion velocity can be presented by volume in which the vibration wave spreads on the distance l during τ time. The energy of widespread vibrations (W) in medium can be expressed by the following equation –

$$W = \rho V \cdot 2\pi^2 \cdot v_{vib}^2 \cdot \lambda_{vib}^2 = m_{av} \cdot 2\pi^2 \cdot v_{vib}^2 \cdot \lambda_{vib}^2 \tag{1}$$

where ρ – the medium density; V – the medium volume in which the nanocomposite vibrations are expanded, v and λ – the frequency and amplitude of (M/C nc) vibrations, m_{av} – the average mass of medium volume in which the vibration transference takes place.

In the second case it is necessary to note that the super molecule of nanocomposite, having the great dipole and magnetic moments, vibrates with ultrasonic frequency in the medium molecules field. Besides the every molecule have own electrical and magnetic fields. In this notion metal/carbon nanocomposite is considered as vibrator, which radiate electromagnetic waves.

The equations, which describe mechanical and electromagnetic vibrations near in form:

For mechanical vibration,

$$\Delta = A_m \bullet \sin 2\pi \left(\frac{\tau}{P} - \frac{1}{\lambda} \right) \tag{2a}$$

For electromagnetic vibration,

$$\Delta\varphi = \varphi_m \bullet \sin 2\pi \left(\frac{\tau}{P} - \frac{1}{\lambda} \right) \tag{2b}$$

where parameters in these equations mean: Δ – the chemical particles shift to certain point of medium which vibration wave is over; A_m – the amplitude of nanocomposite vibration; τ – the time of certain point reaching by vibration wave in the medium; P – the vibration period; l – the distance from nanocomposite to the certain point in medium which the vibration wave pass; λ – the vibration wave length; $\Delta\varphi$ – the potential change in the medium point at the electromagnetic vibration transference in medium from nanocomposite; φ_m – the maximum potential of electromagnetic radiation of nanocomposite.

It is possible the application of following equation for the electromagnetic wave transference –

$$U_{trans} = \sqrt{\frac{1}{\varepsilon_{ab} \bullet \mu_{ab}}} \tag{3}$$

where U_{trans} – the electromagnetic wave transmission velocity, ε_{ab} – the absolute dielectric constant for medium, μ_{ab} – the absolute magnetic penetrability of medium.

In third case the nanocomposite electromagnetic field action on the medium molecules leads to the medium molecule self-organization and to the formation of nanostructured fragments (fractals) in the material composition. These changes can be accompanied by the processes of active particles appearance.

Therefore, in the last case the processes of nanocomposites functionalization are possible. Besides the interaction of active particles (ions and molecules) can be described by the following equations –

$$E_{c-d} = \frac{q \cdot \mu \cdot \cos Q}{r^2} \qquad (4a)$$

$$E_{d-d} = \frac{\mu_1 \cdot \mu_2 \cdot \cos Q}{r^3} \qquad (4b)$$

where E_{c-d} and E_{d-d} – the interaction energies of charge with dipole and dipole with dipole; q – the charge of ion formed; μ_1 and μ_2 – the dipole moments of nanocomposite super molecule (1) and medium molecule (2); θ – the angle between correspondent vectors of active particles interaction with the participating of super molecules nanocomposite; r – the distance between active particles and molecules, including super molecules metal/carbon nanocomposites.

12.3   THE CHANGES IN IR SPECTRA AT THE METAL/CARBON NANOCOMPOSITES SUPER SMALL QUANTITIES ADDITION

The metal/carbon nanocomposite vibration energy transference through electromagnetic waves on the medium molecules is confirmed by the increasing of intensity in IR spectra for the fine dispersed suspensions, which contain the super small quantities of M/C nanocomposite. The stability of correspondent suspensions changes in dependence on the medium nature. The essential changes of lines intensity in IR spectra are for the polar liquids. In Fig. 12.1, the IR spectrum of iso methyl tetra hydro phthalic anhydrate, containing 0.0001% copper/carbon nanocomposite, is presented.

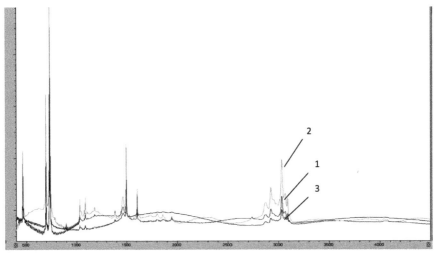

FIGURE 12.1   Changes in IR spectrum of copper/carbon nanocomposite fine suspension based on iso methyl tetra hydro phthalic anhydride (CA) with time (1 – the moment of nanocomposite addition into CA, 2 – IR spectrum of fine suspension later day, 3 – IR spectrum of the same suspension after two days).

It is noted that the iso methyl tetra hydro phthalic anhydrate is curing agent for the cross-linking of epoxy resins. After the addition of nanocomposite in this dispersed medium the lines intensity increasing in IR spectra is observed. According to IR spectra the intensity of some lines is increased more than two times later two days. However after three days the decreasing of these lines intensity takes place.

At the same times this suspension can be activated again by means of ultrasonic machining.

The facts observed are correlated with grow of nanocomposites activities in the self-organization processes during the cross-linking of epoxy resins that leads to improvement of material properties.

12.4   CONCLUSION

In this chapter, the influence of super small quantities of metal/carbon nanocomposites on the different liquid media, including liquid polymeric compositions, is discussed.

It is shown that the addition of super small quantities of nanocomposites in the liquid media leads to the growth of lines intensities in IR spectra of these liquids.

This fact is explained by the vibration increasing of correspondent chemical groups, which are activated by the electromagnetic waves formed at the vibration of metal/carbon nanocomposites particles.

The transference variants of metal/carbon nanocomposites influence on liquid media and also on polymeric compositions are represented.

The changes in IR spectra lines intensity of different liquids at the super small quantities addition of metal/carbon nanocomposites are considered.

The features of electromagnetic wave motion arise in different media are discussed.

KEYWORDS

- **curing agents**
- **fine dispersed suspension**
- **i-methyl tetra hydro phthalic anhydrate**
- **IR spectroscopy**
- **metal/carbon nanocomposites**
- **peptizers**
- **polyethylene polyamine**
- **polymeric compositions solvents**
- **super small quantities**

REFERENCES

1. Kodolov, V. I., & Khokhrikov, N. V. (2009). Chemical Physics of Formation and Transformation Processes of Nanostructures and Nanosystems. Izhevsk: publisher IzhSACA *1*. 361p *2*. 4e15p.
2. Shabanova, I. N., Kodolov, V. I., Terebova, N. S., & Trineeva, V. V. (2012). X-Ray Photoelectron Spectroscopy in Investigations of Metal Carbon Nanosystems and Nanostructured Materials. M. Izhevsk: Publisher "Udmurt University", 252p.
3. Chashkin, M. A., Kodolov, V. I., Zakharov, A. I. et al. (2011). Metal Carbon Nanocomposites for Epoxy Compositions: Quantum-Chemical Investigations and Experimental Modeling. Polymer Research Journal, *5(N1)*, 5–19.
4. Yu, V. Pershin, & Kodolov, V. I. (2011). Polycarbonate modified with Cu C Nanocomposite. Polymer Research Journal, *5(N2)*, 197–202.
5. Chashkin, M. A., Lyakhovitch, A. M., & Kodolov, V. I. (2012). Investigations of Cold Hardening Epoxy Compositions Structural Peculiarities at their Modification by Copper Carbon Nanocomposite, Nanotechnics, *30(N2)*, 19–23.

6. Zagrebin, L. D., Kodolov, V. I., Akhmetshina, L. F. et al. (2012). Thermal Peculiarities of Heat Capacity and Thermal Conductivity of Silicate Compositions Modified by Iron or Nickel Containing Nanostructures. Chemical Physics and Mesoscopy, *14(N2)*, 231–236.

7. Akhmetshina, L. F., Lebedeva, G. A., & Kodolov, V. I. (2012). Phosphorus Containing Metal Carbon Nanocomposites and their Application for the Modification of Intumescent Fireproof Coatings. JCDNM, *4(N4)*, 313–324.

8. Kodolov, V. I., & Trineeva, V. V. (2012). Perspectives of Idea Development about Nanosystems Self-Organization in Polymeric Matrixes. "The Problems of Nanochemistry for the Creation of New Materials" Torun. Poland: IEPMD, 75–100.

9. Kodolov, V. I., & Trineeva, V. V. (2013). Theory of Modification of Polymeric Materials by Super Small Quantities of Metal Carbon Nanocomposites. Chemical Physics and Mesoscopy, *15(N3),* 351–363.

10. Shabanova, I. N., Kodolov, V. I., Terebova, N. S. et al. (2013). X-Ray Photoelectron Spectroscopy Investigation of Super Small Quantities Additives Influence of Metal Carbon Nanocomposites on Polycarbonate Modification Degree. Chemical Physics and Mesoscopy, *15(N4)*, 570–575.

11. Kodolov, V. I., & Trineeva, V. V. (2013). Fundamental Definitions for Domain of Nanostructures and Metal Carbon Nanocomposites. "Nanostructure, Nanosystems and Nanostructured Materials: Theory, Production and Development." Toronto, New Jersey, Apple Academic Press, 1–42.

12. Shabanova, I. N., & Terebova, N. S. (2012). Dependence of the Value of the Atomic Magnetic Moment of D Metals on the Chemical Structure of Nanoforms. "The Problems of Nano Chemistry for the Creation of New Materials" Torun, Poland: IEPMD, 123–131.

13. Kodolov, V. I., Trineeva, V. V., Blagodatskikh, I. I., Yu, M. Vasil'chenko, Vakhrushina, M. A., & Bondar, A. Yu. (2013). The Nanostructures Obtaining and the Synthesis of Metal/Carbon Nanocomposites in Nanoreactors "Nanostructure, Nanosystems and Nanostructured Materials: Theory, Production and Development." Toronto, New Jersey: Apple Academic Press, 101–145.

CHAPTER 13

ANALYSIS OF THE METAL/CARBON NANOCOMPOSITES SURFACE ENERGY: A RESEARCH NOTE

V. I. KODOLOV[1,2], and V. V. TRINEEVA[1,3]

[1]Basic Research – High Educational Centre of Chemical Physics & Mesoscopy, Udmurt Scientific Centre, Ural Branch of Russian Academy of Sciences

[2]M.T. Kalashnikov Izhevsk State Technical University

[3]Institute of Mechanics, Ural Branch of Russian Academy of Sciences

CONTENTS

Abstract 164
13.1   Introduction 164
13.2   The Distribution Variants on the Surface Energy Portions 164
13.3   The Energetic Properties of Metal/Carbon Nanocomposites 166
13.4   Conclusion 167
Keywords 168
References 168

ABSTRACT

The analysis of the metal/carbon nanocomposites surface energy as well as the comparison of their energetic characteristics is carried out. It is shown that the metal/carbon nanocomposites average vibration energy, which is correspondent portion of surface energy, depend on the nanocomposite mass and vibration velocity.

13.1   INTRODUCTION

In Refs. [1–10], some nanocomposites properties and their application for polymeric compositions modification are considered. However, it is necessary to explain some effects of nanocomposites influence on media, especially the nanocomposite super small quantities influence on polymeric compositions. These explanations connect with energetic characteristics of metal/carbon nanocomposites obtained by the redox synthesis into nanoreactors of polymeric matrixes. Therefore, the energetic properties properly metal/carbon nanocomposites are the wide field for the discussion concerning to nature and activity of these nanocomposites.

13.2   THE DISTRIBUTION VARIANTS ON THE SURFACE ENERGY PORTIONS

If the metal/carbon nanocomposites are considered as super molecules, their surface energies analogously energy of usual molecules consists portions of energy which correspond to progressive, rotation and vibration motions and also electronic motion:

$$\varepsilon_s^{NC} = \varepsilon_{prog} + \varepsilon_{rot} + \varepsilon_{vib} + \varepsilon_{elm},$$

where ε_s^{NC} – the surface energy of nanocomposite; ε_{prog} – the nanocomposite surface energy portion which corresponds to the progressive motion; ε_{rot} – the nanocomposite surface energy portion which corresponds to the rotation motion; ε_{vib} – the nanocomposite surface energy portion which corresponds to the vibration motion, ε_{elm} – the nanocomposite surface energy portion which corresponds to electron motion for the interactions of nanocomposites with surroundings molecules.

When the nanocomposite progressive motion portion increases the nanoparticles diffusion processes importance grows in media that leads to their coagulation with the decreasing of the surface energy aggregate obtained. The coagulation of nanoparticles increases at the increasing of their quantities in medium, which has little viscosity.

The decreasing of medium viscosity is arrived at the addition of nanoparticles with the great portion of surface energy, which corresponds to the rotation motion. If the relatively small interaction of nanoparticles with medium molecules takes place, the nanocomposite rotation increasing leads to the local formation of free volumes as well as viscosity decreasing and coagulation of nanoparticles.

The decreasing of active nanoparticles quantities in media with sufficiently biggest viscosity and polarity the energetic portions of progressive motion and rotation can be considerably decreases, and in the same time the surface energy portion of vibration can be essentially increased.

From the metal/carbon nanocomposites Raman and IR spectra analysis it follows that the skeleton vibration of them on the vibrations frequencies corresponds to ultrasonic vibrations. The nanocomposite vibrations energy values are determined by the corresponding nanoparticles sizes and masses. The mass of nanoparticle (metal/carbon nanocomposite) depends on their content and types of metal containing phase clusters.

Usually metal/carbon nanocomposites have the great dipole moment. Therefore, it is possible the proposition that nanocomposite is vibrator which radiates electromagnetic waves. The nanocomposite vibration emission in medium is determined by their dielectric characteristics and the corresponding functional groups presence in medium.

At the metal/carbon nanocomposites obtaining the interaction of polymeric matrix with metal containing phase leads to formation of metal clusters covered by carbon shells accompanied by metal electron structure changes. In some cases the medium characteristics influence on nanocomposites throw into the increasing of nanocomposite surface energy portion, which concerns with changes of their electron structure and equally with electron structure of medium.

In these cases the growth of metal atomic magnetic moment is observed that corresponds to the unpaired electron number increasing (Table 13.1). The considerable changes of metal atomic magnetic moments in nanocomposites proceed when the phosphorus atoms include to carbon shells of nanocomposites.

TABLE 13.1 Atomic Magnetic Moments of Some 3d Metals in Metal/Carbon Nanocomposites and Their Functionalized Analogs [6, 13]

Metal in nanocomposite	Cu	Ni	Co	Fe
Metal atomic magnetic moment for nanocomposite	0.6	1.6	1.7	2.3
Metal atomic magnetic moment for phosphorus containing nanocomposite	2.0	3.0	2.5	2.5

The appearance or increasing of paramagnetic properties is linked with the possibility of chemical bonds formation, and also the growth of metal/carbon nanocomposites reactivity into polar media.

Vibration portion of surface energy of metal/carbon nanocomposites can be presented as:

$$\varepsilon_{vib} = \frac{mu_{vib}^2}{2} \qquad (1)$$

where m – the mass of metal/carbon nanocomposite, which includes summary mass of carbon or carbon, polymeric shell and cluster of metal containing phase, u_{vib} – the vibration velocity correspondent to product of vibration amplitude on vibration frequency.

13.3 THE ENERGETIC PROPERTIES OF METAL/CARBON NANOCOMPOSITES

Usually the fine dispersed suspensions or sols of nanostructures are used for the modification of polymeric compositions (matrixes) to distribute nanoparticles into polymeric matrix. The liquids, which are used for polymeric materials production, are represented as dispersion medium. For the obtaining of fine dispersed suspensions small and super small quantities of metal/carbon nanocomposites are introduced into dispersed media at the mechanical and further the ultrasonic mixing.

The nanocomposite vibration amplitude is the product of wave number on velocity of light. Therefore, the Eq. (1) can be written as

$$\varepsilon_{vib} = \frac{(m_{cl} + m_{sh})dvc}{2} \qquad (2)$$

where m_{cl} – the mass of metal containing phase; m_{sh} – the mass of carbon or carbon polymeric shell; d – the average linear size of metal/carbon nanocomposite nanoparticles; v – the wave number of nanocomposite skeleton vibration; c – the velocity of light.

Under the influence by electromagnetic radiation arising at the ultrasonic vibration of nanocomposite the self-organization of polymeric compositions macromolecules is possible. According to the equation 2 the nanocomposite vibration energy is determined by the average mass and size of nanocomposite (Table 13.2).

TABLE 13.2 Energetic Characteristics of Metal/Carbon Nanocomposites [11–15]

Metal in metal/carbon nanocomposites	Cu	Ni	Co	Fe
Relation of metal to carbon content in metal/carbon nanocomposites (%)	50/50	60/40	65/35	70/30
Density of metal/carbon nanocomposite, g/cm³	1.71	2.17	1.6	2.1
Summary mass of metal/carbon nanocomposite (au)	36.75	40.01	42.50	42.69

Average size of metal/carbon nanocomposite, d, nm	25	11	15	17
Specific surface of metal/carbon nanocomposite, S, m²/g	160	251	209	163
Skeleton vibration frequency of metal/carbon nanocomposite, v_{vib}, s^{-1}	4×10^{11}	4.2×10^{11}	4.1×10^{11}	4×10^{11}
Average vibration velocity of metal/carbon nanocomposites, u_{vib}, cm/s	8.5×10^5	4.6×10^5	6.1×10^5	6.9×10^5
Average vibration energy of metal/carbon nanocomposites, ε_{vib}, erg	1.6×10^{13}	4.3×10^{12}	7.6×10^{12}	9.9×10^{12}

The vibration motion portion energy essentially determines by the vibration amplitude because the nanocomposite skeleton vibration frequencies practically equal (the proximity of metal nature).

If the vibration amplitudes and the nanocomposite sizes are correlated, the nanocomposites sizes show the great influence on the vibration energy and also on the vibration velocity. Therefore, according to the increasing of vibration energy as well as the increasing of metal/carbon nanocomposites linear sizes, the correspondent nanocomposites (M/C nc) are displaced in the following order:

$$\varepsilon_{vib} \ (Cu/C \ nc) > \varepsilon_{vib} \ (Fe/C \ nc) > \varepsilon_{vib} \ (Co/C \ nc) > \varepsilon_{vib} \ (Ni/C \ nc)$$

This disposition of nanocomposites on the vibration energy values is possible, when the vibration motion portion is greatly bigger on the comparison with other portions of surface energy. However, the common surface energy of nanocomposites obtained differs from previous row and corresponds to values of specific surface.

$$S_{sp} \ (Ni/C \ nc) > S_{sp} \ (Co/C \ nc) > S_{sp} \ (Fe/C \ nc) > S_{sp} \ (Cu/C \ nc)$$

Certainly, in dependence on methods and conditions of the metal/carbon nanocomposites synthesis the changes of forms and contents for nanocomposites are possible. At the same time the fundamentals of the metal/carbon nanocomposite activity open the new perspectives for their interaction prognosis in the different media.

13.4 CONCLUSION

Thus, according to the analysis of metal/carbon nanocomposites characteristics, which are determined by their sizes and content, their activities are stipulated the correspondent dipole moments and vibration energies.

The electromagnetic waves are formed at the ultrasonic vibration of metal/carbon nanocomposites. These waves stimulate the changes of electron structure and the growth of magnetic moments of metal containing clusters within nanocomposites.

The introduction of phosphorus in nanocomposites leads to the growth of dipole and magnetic moments of metal/carbon nanocomposites. It is shown that the nanocomposites vibration energies depend on their average masses.

However, the specific surface of metal/carbon nanocomposites particles changes in dependence on the nature of nanocomposite in other order than the correspondent order of the vibration energies.

Therefore, the energetic characteristics of metal/carbon nanocomposites are more important for the activity determination in comparison with their size characteristics.

KEYWORDS

- **energetic characteristics**
- **IR spectroscopy**
- **metal/carbon nanocomposites**
- **refractometry**
- **super molecules**
- **super small quantities**
- **x-ray photoelectron spectroscopy**

REFERENCES

1. Kodolov, V. I., & Khokhrikov, N. V. (2009). Chemical Physics of Formation and Transformation Processes of Nanostructures and Nanosystems. Izhevsk: publisher IzhSACA, 1, 361p 2. 4e15p.
2. Shabanova, I. N., Kodolov, V. I., Terebova, N. S., & Trineeva, V. V. (2012). X-Ray Photoelectron Spectroscopy in Investigations of Metal Carbon Nanosystems and Nanostructured Materials. M. Izhevsk: publisher "Udmurt University", 252p.
3. Kodolov, V. I., Khokhriakov, N. V., Trineeva, V. V., & Blagodatskikh, I. I. (2008). Nanostructure Activity and its Display in Nanoreactors of Polymeric Matrixes and in Active Media. Chemical Physics & Mesoscopy, 10(N 4). 448–460.
4. Kodolov, V. I. (2009). The addition to previous paper (in CP&M, 2008, 10(N4). 448–460.) Chemical Physics & Mesoscopy, 11(N1), 134–136.
5. Trineeva, V. V., Vakhrushina, M. A., Bulatov, D. I., & Kodolov, V. I. (2012). The Obtaining of Metal Carbon Nanocomposites and Investigation of their Structure Phenomena. Nanotechnics, N4. 50–55.

6. Shabanova, I. N., & Terebova, N. S. (2012). Dependence of the Value of the Atomic Magnetic Moment of D Metals on the Chemical Structure of Nanoforms. In book "The Problems of Nano Chemistry for the Creation of New Materials" Torun, Poland: IEPMD, 123–131.
7. Kodolov, V. I., & Trineeva, V. V. (2012). Perspectives of Idea Development about Nanosystems Self-Organization in Polymeric Matrixes, "The Problems of Nanochemistry for the Creation of New Materials" Torun, Poland, IEPMD, 75–100.
8. Kodolov, V. I., & Trineeva, V. V. (2013). Fundamental Definitions for Domain of Nanostructures and Metal Carbon Nanocomposites. "Nanostructure, Nanosystems and Nanostructured Materials: Theory, Production and Development." Toronto New Jersey: Apple Academic Press, 1–42.
9. Khokhriakov, N. V., Kodolov, V. I., Korablev, G. A., Trineeva, V. V., & Zaikov, C. E. Prognostic Investigations of Metal Carbon Nanocomposites and Nanostructures Synthesis Processes Characterization ibid. 43–100.
10. Kodolov, V. I., Trineeva, V. V., Blagodatskikh, I. I., Vasil'chenko, Yu. M., Vakhrushina, M. A., & Bondar, A. Yu. The Nanostructures Obtaining and the Synthesis of Metal Carbon Nanocomposites in Nanoreactors. ibid. 101–145.
11. Kodolov, V. I., Akhmetshina, L. F., Chashkin, M. A., & Trineeva, V. V. The Functionalization of Metal Carbon Nanocomposites or the Introduction of Functional Groups in Metal Carbon Nanocomposites. ibid. 147–176.
12. Shabanova, I. N., Terebova, N. S., Kodolov, V. I., Blagodatskikh, I. I., Trineeva, V. V., Akhmetshina, L. F., & Chashkin, M. A. The Investigation of Metal Carbon Nanocomposites Electron Structure by X-Ray Photoelectron Spectroscopy. ibid. 177–230.
13. Kodolov, V. I., & Trineeva, V. V. (2013). Theory of Modification of Polymeric Materials by Super Small Quantities of Metal Carbon Nanocomposites. Chemical Physics & Mesoscopy, 15(N3), 351–363.
14. Shabanova, I. N., Kodolov, V. I., Terebova, N. S. et al. (2013). X-Ray Photoelectron Spectroscopy Investigation of Super Small Quantities Additives Influence of Metal Carbon Nanocomposites on Polycarbonate Modification Degree. Chemical Physics & Mesoscopy, 15(N4), 570–575.
15. Zagrebin, L. D., Kodolov, V. I., Akhmetshina, L. F. et al (2012). Thermal Peculiarities of Heat Capacity and Thermal Conductivity of Silicate Compositions Modified by Iron or Nickel Containing Nanostructures. Chemical Physics & Mesoscopy, 14(N2), 231–236.

CHAPTER 14

NANOPOLYMER FIBERS: A VERY COMPREHENSIVE REVIEW IN AXIAL AND COAXIAL ELECTROSPINNING PROCESS

SAEEDEH RAFIEI and A. K. HAGHI

University of Guilan, Rasht, Iran

CONTENTS

Abstract 172
14.1   An Introduction to Nanotechnology 173
14.2   Nanostructured Materials..174
14.3   Nanofiber Technology...182
14.4   Design Multifunctional Product by Nanostructures 190
14.5   Introduction to Theoretical Study of Electrospinning Process 213
14.6   Study of Electrospinning Jet Path 214
14.7   Electrospinning Draw Backs...217
14.8   Modeling of the Electrospinning Process 218
14.9   Electrospinning Simulation...251
14.10 Electrospinning Simulation Example 252
14.11 Applied Numerical Methods for Electrospinning....................255
14.12 Concluding Remarks of Electrospinning Modeling 268
14.13 Numerical Study of Coaxial Electrospinning 269
14.14 Application of Coaxial Electrospun Nanofibers 314
14.15 Advantages and Disadvantages of Co-Electrospinning Process...............323
14.16 General Assumptions in Co-Electrospinning Process..............324
14.17 Conclusions and Future Perspective 324
Appendixes 326
Keywords 333
References...333

ABSTRACT

The nanostructure materials productions are most challenging and innovative processes, introducing, in the manufacturing, a new approaches such as self-assembly and self-replication. The fast growing of nanotechnology with modern computational/experimental methods give the possibility to design multifunctional materials and products in human surroundings. Smart clothing, portable fuel cells, medical devices are some of them. Research in nanotechnology began with applications outside of everyday life and is based on discoveries in physics and chemistry. The reason for that is needed to understand the physical and chemical properties of molecules and nanostructures in order to control them.

A new approach in nanostructured materials is a computational-based material development. It is based on multiscale material and process modeling spanning, on a large spectrum of time as well as on length scales. The cost of designing and producing novel multifunctional nanomaterials can be high and the risk of investment to be significant. Computational nanomaterials research that relies on multiscale modeling has the potential to significantly reduce development costs of new nanostructured materials for demanding applications by bringing physical and microstructural information into the realm of the design engineer.

One of the most significant types of these one-dimensional nanomaterials is nanofibers, which can be produced widely through electrospinning procedure. A drawback of this method, however, is the unstable behavior of the liquid jet, which causes the fibers to be collected randomly. So a critical concern in this process is to achieve desirable control. Studying the dynamics of electrospinning jet would be easier and faster if it can be modeled and simulated, rather than doing experiments. By replacing the single capillary with a coaxial spinneret in electrospinning set up, it is possible to generate core-sheath and hollow nanofibers made of polymers, ceramics or composites. These nanofibers, as a kind of one dimensional nanostructure, have attracted special attention in several research groups, because these structures could further enhance material property profiles and these unique core–sheath structures offer potential in a number of applications including nanoelectronics, microfluidics, photonics, and energy storage. This chapter focuses on modeling and then simulating of electrospinning and coaxial electrospinning process in various views. In order to study the applicability of the electrospinning modeling equations, which discussed in detail in earlier parts of this approach, an existing mathematical model in which the jet was considered as a mechanical system, was interconnected with viscoelastic elements and used to build a numeric method. The simulation features the possibility of predicting essential parameters of electrospinning and coaxial electrospinning process and the results have good agreement with other numeric studies of electrospinning, which modeled this process based on axial direction.

14.1 AN INTRODUCTION TO NANOTECHNOLOGY

Understanding the nanoworld makes up one of the frontiers of modern science. One reason for this is that technology based on nanostructures promises to be hugely important economically [1–3]. Nanotechnology literally means any technology on a nanoscale that has applications in the real world. It includes the production and application of physical, chemical, and biological systems at scales ranging from individual atoms or molecules to submicron dimensions, as well as the integration of the resulting nanostructures into larger systems. Nanotechnology is likely to have a profound impact on our economy and society in the early twenty-first century, comparable to that of semiconductor technology, information technology, or cellular and molecular biology. Science and technology research in nanotechnology promises breakthroughs in areas such as materials and manufacturing [4], nanoelectronics [5], medicine and healthcare [6], energy [7], biotechnology [8], information technology [9], and national security [10]. It is widely felt that nanotechnology will be the next Industrial Revolution [9].

As far as "nanostructures" are concerned, one can view this as objects or structures whereby at least one of its dimensions is within nano-scale. A "nanoparticle" can be considered as a zero dimensional nano-element, which is the simplest form of nanostructure. It follows that a "nanotube" or a nanorod" is a one-dimensional nano-element from which slightly more complex nanostructure can be constructed of [11–12].

Following this fact, a "nanoplatelet" or a "nanodisk" is a two-dimensional element, which, along with its one-dimensional counterpart, is useful in the construction of nanodevices. The difference between a nanostructure and a nanodevice can be viewed upon as the analogy between a building and a machine (whether mechanical, electrical or both) [1]. It is important to know that as far as nanoscale is concerned; these nano-elements should not consider only as an element that form a structure while they can be used as a significant part of a device. For example, the use of carbon nanotube as the tip of an Atomic Force Microscope (AFM) would have it classified as a nanostructure. The same nanotube, however, can be used as a single molecule circuit, or as part of a miniaturized electronic component, thereby appearing as a nanodevice. Hence, the function, along with the structure, is essential in classifying which nanotechnology subarea it belongs to. This classification will be discussed in detail in further sections [11, 13].

As long as nanostructures clearly define the solids' overall dimensions, the same cannot be said so for nanomaterials. In some instances a nanomaterial refers to a nano-sized material while in other instances a nanomaterial is a bulk material with nano-scaled structures. Nanocrystals are other groups of nanostructured materials. It is understood that a crystal is highly structured and that the repetitive unit is indeed small enough. Hence a nanocrystal refers to the size of the entire crystal itself being nano-sized, but not of the repetitive unit [14].

Nanomagnetics are the other type of nanostructured materials, which are known as highly miniaturized magnetic data storage materials with very high memory. This can be attained by taking advantage of the electron spin for memory storage, hence the term "spin-electronics," which has since been more popularly and more conveniently known as "spintronics" [1, 9, 15]. In nanobioengineering, the novel properties of nano-scale are taken advantage of for bioengineering applications. The many naturally occurring nanofibrous and nanoporous structure in the human body further adds to the impetus for research and development in this subarea. Closely related to this is molecular functionalization whereby the surface of an object is modified by attaching certain molecules to enable desired functions to be carried out such as for sensing or filtering chemicals based on molecular affinity [16, 17].

With the rapid growth of nanotechnology, nanomechanics are no longer the narrow field, which it used to be [13]. This field can be broadly categorized into the molecular mechanics and the continuum mechanics approaches which view objects as consisting of discrete many-body system and continuous media, respectively. As long as the former inherently includes the size effect, it is a requirement for the latter to factor in the influence of increasing surface-to-volume ratio, molecular reorientation and other novelties as the size shrinks. As with many other fields, nanotechnology includes nanoprocessing novel materials processing techniques by which nano-scale structures and devices are designed and constructed [18, 19].

Depending on the final size and shape, a nanostructure or nanodevice can be created from the top-down or the bottom-up approach. The former refers to the act of removal or cutting down a bulk to the desired size, while the latter takes on the philosophy of using the fundamental building blocks – such as atoms and molecules, to build up nanostructures in the same manner. It is obvious that the top-down and the bottom-up nanoprocessing methodologies are suitable for the larger and two smaller ends respectively in the spectrum of nano-scale construction. The effort of nanopatterning – or patterning at the nanoscale would hence fall into nanoprocessing [1, 12, 18].

14.2   NANOSTRUCTURED MATERIALS

Strictly speaking, a nanostructure is any structure with one or more dimensions measuring in the nanometer (10^{-9} m) range. Various definitions refine this further, stating that a nanostructure should have a characteristic dimension lying between 1 nm and 100 nm, putting nanostructures as intermediate in size between a molecule and a bacterium. Nanostructures are typically probed either optically (spectroscopy, photoluminescence ...) or in transport experiments. This field of investigation is often given the name mesoscopic transport, and the following considerations give an idea of the significance of this term [1, 2, 12, 20, 21].

What makes nanostructured materials very interesting and award them with their unique properties is that their size is smaller than critical lengths that characterize many physical phenomena. Generally, physical properties of materials can be characterized by some critical length, a thermal diffusion length, or a scattering length, for example. The electrical conductivity of a metal is strongly determined by the distance that the electrons travel between collisions with the vibrating atoms or impurities of the solid. This distance is called the mean free path or the scattering length. If the sizes of the particles are less than these characteristic lengths, it is possible that new physics or chemistry may occur [1, 9, 17].

Several computational techniques have been employed to simulate and model nanomaterials. Since the relaxation times can vary anywhere from picoseconds to hours, it becomes necessary to employ Langevin dynamics besides molecular dynamics in the calculations. Simulation of nanodevices through the optimization of various components and functions provides challenging and useful task [20, 22]. There are many examples where simulation and modeling have yielded impressive results, such as nanoscale lubrication [23]. Simulation of the molecular dynamics of DNA has been successful to some extent [24]. Quantum dots and nanotubes have been modeled satisfactorily [25, 26]. First principles calculations of nanomaterials can be problematic if the clusters are too large to be treated by Hartree–Fock methods and too small for density functional theory [1]. In the next section various classifications of these kinds of materials are considered in detail.

14.2.1 CLASSIFICATION OF NANOSTRUCTURED MATERIALS

Nanostructure materials as a subject of nanotechnology are low dimensional materials comprising of building units of a submicron or nanoscale size at least in one direction and exhibiting size effects. The first classification idea of NSMs was given by Gleiter in 1995 [3]. A modified classification scheme for these materials, in which 0D, 1D, 2D and 3D dimensions are included suggested in later researches [21]. These classifications are as follows:

14.2.1.1 0D NANOPARTICLES

A major feature that distinguishes various types of nanostructures is their dimensionality. In the past 10 years, significant progress has been made in the field of 0 dimension nanostructure materials. A rich variety of physical and chemical methods have been developed for fabricating these materials with well-controlled dimensions [3, 18]. Recently, 0D nanostructured materials such as uniform particles arrays (quantum dots), heterogeneous particles arrays, core-shell quantum dots, onions, hollow spheres and nanolenses have been synthesized by several research groups [21]. They have been extensively studied in light emitting diodes (LEDs), solar cells, single-electron transistors, and lasers.

14.2.1.2   1D NANOPARTICLES

In the last decade, 1D nanostructured materials have focused an increasing interest due to their importance in research and developments and have a wide range
of potential applications [27]. It is generally accepted that these materials are ideal
systems for exploring a large number of novel phenomena at the nanoscale and
investigating the size and dimensionality dependence of functional properties. They
are also expected to play an important role as both interconnects and the key units
in fabricating electronic, optoelectronic, and EEDs with nanoscale dimensions. The
most important types of this group are nanowires, nanorods, nanotubes, nanobelts,
nanoribbons, hierarchical nanostructures and nanofibers [1, 18, 28].

14.2.1.3   2D NANOPARTICLES

2D nanostructures have two dimensions outside of the nanometric size range. In
recent years, synthesis of 2D nanomaterial has become a focal area in materials
research, owing to their many low dimensional characteristics different from the
bulk properties. Considerable research attention has been focused over the past few
years on the development of them. 2D nanostructured materials with certain geometries exhibit unique shape-dependent characteristics and subsequent utilization
as building blocks for the key components of nanodevices [21]. In addition, these
materials are particularly interesting not only for basic understanding of the mechanism of nanostructure growth, but also for investigation and developing novel applications in sensors, photocatalysts, nanocontainers, nanoreactors, and templates
for 2D structures of other materials. Some of the 3 dimension nanoparticles are junc-
tions (continuous islands), branched structures, nanoprisms, nanoplates, nanosheets,
nanowalls, and nanodisks [1].

14.2.1.4   3D NANOPARTICLES

Owing to the large specific surface area and other superior properties over their
bulk counterparts arising from quantum size effect, they have attracted considerable research interest and many of them have been synthesized in the past 10 years
[1, 12]. It is well known that the behaviors of NSMs strongly depend on the sizes,
shapes, dimensionality and morphologies, which are thus the key factors to their
ultimate performance and applications. Therefore, it is of great interest to synthesize
3D NSMs with a controlled structure and morphology. In addition, 3D nanostructures are an important material due to its wide range of applications in the area of
catalysis, magnetic material and electrode material for batteries [2]. Moreover, the
3D NSMs have recently attracted intensive research interests because the nanostructures have higher surface area and supply enough absorption sites for all involved
molecules in a small space [58]. On the other hand, such materials with porosity

in three dimensions could lead to a better transport of the molecules. Nanoballs (dendritic structures), nanocoils, nanocones, nanopillers and nanoflowers are in this group [1, 2, 18, 29].

14.2.2 SYNTHESIS METHODS OF NANOMATERIALS

The synthesis of nanomaterials includes control of size, shape, and structure. Assembling the nanostructures into ordered arrays often becomes necessary for rendering them functional and operational. In the last decade, nanoparticles (powders) of ceramic materials have been produced in large scales by employing both physical and chemical methods. There has been considerable progress in the preparation of nanocrystals of metals, semiconductors, and magnetic materials by employing colloid chemical methods [18, 30].

The construction of ordered arrays of nanostructures by employing techniques of organic self-assembly provides alternative strategies for nanodevices. Two and three-dimensional arrays of nanocrystals of semiconductors, metals, and magnetic materials have been assembled by using suitable organic reagents [1, 31]. Strain directed assembly of nanoparticle arrays (e.g., of semiconductors) provides the means to introduce functionality into the substrate that is coupled to that on the surface [32].

Preparation of nanoparticles is an important branch of the materials science and engineering. The study of nanoparticles relates various scientific fields, e.g., chemistry, physics, optics, electronics, magnetism and mechanism of materials. Some nanoparticles have already reached practical stage. In order to meet the nanotechnology and nano-materials development in the next century, it is necessary to review the preparation techniques of nanoparticles.

All particle synthesis techniques fall into one of the three categories: vapor-phase, solution precipitation, and solid-state processes. Although vapor-phase processes have been common during the early days of nanoparticles development, the last of the three processes mentioned above is the most widely used in the industry for production of micron-sized particles, predominantly due to cost considerations [18, 33].

Methods for preparation of nanoparticles can be divided into physical and chemical methods based on whether there exist chemical reactions [33]. On the other hand, in general, these methods can be classified into the gas phase, liquid phase and solid phase methods based on the state of the reaction system. The gas phase method includes gas-phase evaporation method (resistance heating, high frequency induction heating, plasma heating, electron beam heating, laser heating, electric heating evaporation method, vacuum deposition on the surface of flowing oil and exploding wire method), chemical vapor reaction (heating heat pipe gas reaction, laser induced chemical vapor reaction, plasma enhanced chemical vapor reaction), chemical vapor condensation and sputtering method. Liquid phase method for synthesizing nanoparticles mainly includes precipitation, hydrolysis, spray, solvent thermal method (high temperature and high pressure), solvent evaporation pyrolysis, oxidation reduction (room pressure), emulsion, radiation chemical synthesis and sol-gel processing. The solid phase method includes thermal decomposition, solid-state reaction, spark discharge, stripping and milling method [30, 33].

In other classification, there are two general approaches to the synthesis of nanomaterials and the fabrication of nanostructures, bottom-up and Top-down approach. The first one includes the miniaturization of material components (up to atomic level) with further self-assembly process leading to the formation assembly of nanostructures. During self-assembly the physical forces operating at nanoscale are used to combine basic units into larger stable structures. Typical examples are quantum dot formation during epitaxial growth and formation of nanoparticles from colloidal dispersion. The latter uses larger (macroscopic) initial structures, which can be externally controlled in the processing of nanostructures. Typical examples are etching through the mask, ball milling, and application of severe plastic deformation [3, 13].

Some of the most common methods are described in following subsections.

14.2.2.1   PLASMA BASED METHODS

Metallic, semiconductive and ceramic nanomaterials are widely synthesized by hot and cold plasma methods. A plasma is sometimes referred to as being "hot" if it is nearly fully ionized, or "cold" if only a small fraction, (for instance 1%), of the gas molecules are ionized, but other definitions of the terms "hot plasma" and "cold plasma" are common. Even in cold plasma, the electron temperature is still typically several thousand degrees centigrade. Generally the related equipment consists of an arc melting chamber and a collecting system. The thin films of alloys were prepared from highly pure metals by arc melting in an inert gas atmosphere. Each arc-melted ingot was flipped over and remelted three times. Then, the thin films of alloy were produced by arc melting a piece of bulk materials in a mixing gas atmosphere at a low pressure. Before the ultrafine particles were taken out from the arc-melting chamber, they were passivized with a mixture of inert gas and air to prevent the particles from burning up [34, 35].

Cold plasma method is used for producing nanowires in large scale and bulk quantity. The general equipment of this method consists of a conventional horizontal quartz tube furnace and an inductively coupled coil driven by a 13.56 MHz radio-frequency (RF or radio-frequency) power supply. This method often is called as an RF plasma method. During RF plasma method, the starting metal is contained in a pestle in an evacuated chamber. The metal is heated above its evaporation point using high voltage RF coils wrapped around the evacuated system in the vicinity of the pestle. Helium gas is then allowed to enter the system, forming a high temperature plasma in the region of the coils. The metal vapor nucleates on the He gas atoms and diffuses up to a colder collector rod where nanoparticles are formed. The particles are generally passivized by the introduction of some gas such as oxygen. In the case of aluminum nanoparticles the oxygen forms a layer of aluminum oxide about the particle [1, 36].

14.2.2.2   CHEMICAL METHODS

Chemical methods have played a major role in developing materials imparting technologically important properties through structuring the materials on the nanoscale. However, the primary advantage of chemical processing is its versatility in designing and synthesizing new materials that can be refined into the final end products. The secondary most advantage that the chemical processes offer over physical methods is a good chemical homogeneity, as a chemical method offers mixing at the molecular level. On the other hand, chemical methods frequently involve toxic reagents and solvents for the synthesis of nanostructured materials. Another disadvantage of the chemical methods is the unavoidable introduction of byproducts, which require subsequent purification steps after the synthesis in other words, this process is time consuming. In spite of these facts, probably the most useful methods of synthesis in terms of their potential to be scaled up are chemical methods [33, 37]. There are a number of different chemical methods that can be used to make nanoparticles of metals, and we will give some examples. Several types of reducing agents can be used to produce nanoparticles such as $NaBEt_3H$, $LiBEt_3H$, and $NaBH_4$ where Et denotes the ethyl $(-C_2H_5)$ radical. For example, nanoparticles of molybdenum (Mo) can be reduced in toluene solution with $NaBEt_3H$ at room temperature, providing a high yield of Mo nanoparticles having dimensions of 1–5 nm [30].

14.2.2.3   THERMOLYSIS AND PYROLYSIS

Nanoparticles can be made by decomposing solids at high temperature having metal cations, and molecular anions or metal organic compounds. The process is called thermolysis. For example, small lithium particles can be made by decomposing lithium oxide, LiN_3. The material is placed in an evacuated quartz tube and heated to 400°C in the apparatus. At about 370°C the LiN_3 decomposes, releasing N_2 gas, which is observed by an increase in the pressure on the vacuum gauge. In a few minutes the pressure drops back to its original low value, indicating that all the N_2 has been removed. The remaining lithium atoms coalesce to form small colloidal metal particles. Particles less than 5 nm can be made by this method. Passivation can be achieved by introducing an appropriate gas [1].

Pyrolysis is commonly a solution process in which nanoparticles are directly deposited by spraying a solution on a heated substrate surface, where the constituent react to form a chemical compound. The chemical reactants are selected such that the products other than the desired compound are volatile at the temperature of deposition. This method represents a very simple and relatively cost-effective processing method (particularly in regard to equipment costs) as compared to many other film deposition techniques [30].

The other pyrolysis based method that can be applied in nanostructures production is a laser pyrolysis technique, which requires the presence in the reaction medium of a molecule absorbing the CO_2 laser radiation [38–39]. In most cases, the atoms of a molecule are rapidly heated via vibrational excitation and are dissociated. But in some cases, a sensitizer gas such as SF_6 can be directly used. The heated gas molecules transfer their energy to the reaction medium by collisions leading to dissociation of the reactive medium without, in the ideal case, dissociation of this molecule. Rapid thermalization occurs after dissociation of the reactants due to transfer collision. Nucleation and growth of NSMs can take place in the as-formed supersaturated vapor. The nucleation and growth period is very short time (0.1–10 ms). Therefore, the growth is rapidly stopped as soon as the particles leave the reaction zone. The flame-excited luminescence is observed in the reaction region where the laser beam intersects the reactant gas stream. Since there is no interaction with any walls, the purity of the desired products is limited by the purity of the reactants. However, because of the very limited size of the reaction zone with a faster cooling rate, the powders obtained in this wellness reactor present a low degree of agglomeration. The particle size is small (~5–50 nm range) with a narrow size distribution. Moreover, the average size can be manipulated by optimizing the flow rate, and, therefore, the residence time in the reaction zone [39, 40].

14.2.2.4   LASER BASED METHODS

The most important laser based techniques in the synthesis of nanoparticles is pulsed laser ablation. As a physical gas-phase method for preparing nanosized particles, pulsed laser ablation has become a popular method to prepare high-purity and ultra-fine nanomaterials of any composition [41, 42]. In this method, the material is evaporated using pulsed laser in a chamber filled with a known amount of a reagent gas and by controlling condensation of nanoparticles onto the support. It's possible to prepare nanoparticles of mixed molecular composition such as mixed oxides/nitrides and carbides/nitrides or mixtures of oxides of various metals by this method. This method is capable of a high rate of production of 2–3 g/min [40].

Laser chemical vapor deposition method is the next laser-based technique in which photoinduced processes are used to initiate the chemical reaction. During this method, three kinds of activation should be considered. First, if the thermalization of the laser energy is faster than the chemical reaction, pyrolytic and/or photothermal activation is responsible for the activation. Secondly, if the first chemical reaction step is faster than the thermalization, photolytic (nonthermal) processes are responsible for the excitation energy. Thirdly, combinations of the different types of activation are often encountered. During this technique a high intensity laser beam is incident on a metal rod, causing evaporation of atoms from the surface of the metal. The atoms are then swept away by a burst of helium and passed through an orifice into a vacuum where the expansion of the gas causes cooling and formation of clusters of the metal atoms. These clusters are then ionized by UV radiation and passed into a mass spectrometer that measures their mass: charge ratio [1, 41–43].

Laser-produced nanoparticles have found many applications in medicine, biophotonics, in the development of sensors, new materials and solar cells. Laser interactions provide a possibility of chemical clean synthesis, which is difficult to achieve under more conventional NP production conditions [42]. Moreover, a careful optimization of the experimental conditions can allow a control over size distributions of the produced nanoclusters. Therefore, many studies were focused on the investigation the laser nanofabrication. In particular, many experiments were performed to demonstrate nanoparticles formation in vacuum, in the presence of a gas or a liquid. Nevertheless, it is still difficult to control the properties of the produced particles. It is believed that numerical calculations can help to explain experimental results and to better understand the mechanisms involved [43].

Despite rapid development in laser physics, one of the fundamental questions still concern the definition of proper ablation mechanisms and the processes leading to the nanoparticles formation. Apparently, the progress in laser systems implies several important changes in these mechanisms, which depend on both laser parameters and material properties. Among the more studied ablation mechanisms there are thermal, photochemical and photomechanical ablation processes. Frequently, however, the mechanisms are mixed, so that the existing analytical equations are hardly applicable. Therefore, numerical simulation is needed to better understand and to optimize the ablation process [44].

So far, thermal models are commonly used to describe nanosecond (and longer) laser ablation. In these models, the laser-irradiated material experiences heating, melting, boiling and evaporation. In this way, three numerical approaches were used [29, 45]:

- Atomistic approach based on such methods as molecular dynamics (MD) and Direct Monte Carlo Simulation (DSMC). Typical calculation results provide detailed information about atomic positions, velocities, kinetic and potential energy;
- Macroscopic approach based hydrodynamic models. These models allow the investigations of the role of the laser-induced pressure gradient, which is particularly important for ultra-short laser pulses. The models are based on a one fluid two-temperature approximation and a set of additional models (equation of state) that determines thermal properties of the target;
- Multi-scale approach based on the combination of two approaches cited above was developed by several groups and was shown to be particularly suitable for laser applications.

14.3   NANOFIBER TECHNOLOGY

Nano fiber consists of two terms "Nano" and "fiber," as the latter term is looking more familiar. Anatomists observed fibers as any of the filament constituting the extracellular matrix of connective tissue, or any elongated cells or thread like structures, muscle fiber or nerve fiber. According to textile industry fiber is a natural or synthetic filament, such as cotton or nylon, capable of being spun into simply as materials made of such filaments. Physiologists and biochemists use the term fiber for indigestible plant matter consisting of polysaccharides such as cellulose, that when eaten stimulates intestinal peristalsis. Historically, the term "fiber" in US English or "Fibre" in British English comes from Latin "fibra." Fiber is a slender, elongated thread like structure. Nano is originated from Greek word "nanos" or "nannos" refer to "little old man" or "dwarf." The prefixes "nannos" or "nano" as nannoplanktons or nanoplanktons used for very small planktons measuring 2 to 20 micrometers. In modern "nano" is used for describing various physical quantities within the scale of a billionth as nanometer (length), nanosecond (time), nanogram (weight) and nanofarad (charge) [1, 4, 9, 46]. As it was mentioned before, nanotechnology refers to the science and engineering concerning materials, structures and devices which has at least one dimension is 100 nm or less. This term also refers for a fabrication technology, where molecules, specification and individual atoms, which have at least one dimension in nanometers or less, is used to design or built objects. Nano fiber, as the name suggests the fiber having a diameter range in nanometer. Fibrous structure having at least one dimension in nanometer or less is defined as Nano fiber according to National Science Foundation (NSC). The term Nano describes the diameter of the fibrous shape at anything below one micron or 1000 nm [4, 18].

Nanofiber technology is a branch of nanotechnology whose primary objective is to create materials in the form of nanoscale fibers in order to achieve superior functions [1, 2, 4]. The unique combination of high specific surface area, flexibility, and superior directional strength makes such fibers a preferred material form for many applications ranging from clothing to reinforcements for aerospace structures. Indeed, while the primary classification of nanofibers is that of nanostructure or nanomaterial, other aspects of nanofibers such as its characteristics, modeling, application and processing would enable nanofibers to penetrate into many subfields of nanotechnology [4, 46, 47].

It is obvious that nanofibers would geometrically fall into the category of one dimensional nano-scale elements that includes nanotubes and nanorods. However, the flexible nature of nanofibers would align it along with other highly flexible nanoelements such as globular molecules (assumed as zero dimensional soft matter), as well as solid and liquid films of nanothickness (two dimensional). A nanofiber is a nanomaterial in view of its diameter, and can be considered a nanostructured material if filled with nanoparticles to form composite nanofibers [1, 48].

The study of the nanofiber mechanical properties as a result of manufacturing techniques, constituent materials, processing parameters and other factors would fall into the category of nanomechanics. Indeed, while the primary classification of nanofibers is that of nanostructure or nanomaterial, other aspects of nanofibers such as its characteristics, modeling, application and processing would enable nanofibers to penetrate into many subfields of nanotechnology [1, 18].

Although the effect of fiber diameter on the performance and processibility of fibrous structures has long been recognized, the practical generation of fibers at the nanometer scale was not realized until the rediscovery and popularization of the electrospinning technology by Professor Darrell Reneker almost a decade ago [49, 50]. The ability to create nanoscale fibers from a broad range of polymeric materials in a relatively simple manner using the electrospinning process, coupled with the rapid growth of nanotechnology in recent years have greatly accelerated the growth of nanofiber technology. Although there are several alternative methods for generating fibers in a nanometer scale, none matches the popularity of the electrospinning technology due largely to the simplicity of the electrospinning process [18]. These methods will be discussed in following sections.

14.3.1   VARIOUS NANOFIBERS PRODUCTION METHODS

As it was discussed in detail, nanofiber is defined as the fiber having at least one dimension in nanometer range which can be used for a wide range of medical applications for drug delivery systems, scaffold formation, wound healing and widely used in tissue engineering, skeletal tissue, bone tissue, cartilage tissue, ligament tissue, blood vessel tissue, neural tissue etc. It is also used in dental and orthopedic implants [4, 51, 52]. Nano fiber can be formed using different techniques including: drawing, template synthesis, phases separation, Self-assembly and electrospinning.

14.3.1.1   DRAWING

In 1998, nanofibers were fabricated with citrate molecules through the process of drawing for the first time [53]. During drawing process, the fibers are fabricated by contacting a previously deposited polymer solution droplet with a sharp tip and drawing it as a liquid fiber, which is then solidified by rapid evaporation of the solvent due to the high surface area. The drawn fiber can be connected to another previously deposited polymer solution droplet thus forming a suspended fiber. Here, the predeposition of droplets significantly limits the ability to extend this technique, especially in three-dimensional configurations and hard to access spatial geometries. Furthermore, there is a specific time in which the fibers can be pulled. The viscosity of the droplet continuously increases with time due to solvent evaporation from the deposited droplet. The continual shrinkage in the volume of the polymer solution droplet affects the diameter of the fiber drawn and limits the continuous drawing of fibers [54].

To overcome the above-mentioned limitation is appropriate to use hollow glass micropipettes with a continuous polymer dosage. It provides greater flexibility in drawing continuous fibers in any configuration. Moreover, this method offers increased flexibility in the control of key parameters of drawing such as waiting time before drawing (due to the required viscosity of the polymer edge drops), the drawing speed or viscosity, thus enabling repeatability and control on the dimensions of the fabricated fibers. Thus, drawing process requires a viscoelastic material that can undergo strong deformations while being cohesive enough to support the stresses developed during pulling [54, 55].

14.3.1.2   TEMPLATE SYNTHESIS

Template synthesis implies the use of a template or mold to obtain a desired material or structure. Hence the casting method and DNA replication can be considered as template-based synthesis. In the case of nanofiber creation by [56], the template refers to a metal oxide membrane with through-thickness pores of nano-scale diameter. Under the application of water pressure on one side and restrain from the porous membrane causes extrusion of the polymer which, upon coming into contact with a solidifying solution, gives rise to nanofibers whose diameters are determined by the pores [1, 57].

This method is an effective route to synthesize nanofibrils and nanotubes of various polymers. The advantage of the template synthesis method is that the length and diameter of the polymer fibers and tubes can be controlled by the selected porous membrane, which results in more regular nanostructures. General feature of the conventional template method is that the membrane should be soluble so that it can be removed after synthesis in order to obtain single fibers or tubes. This restricts practical application of this method and gives rise to a need for other techniques [1, 56, 57].

14.3.1.3   PHASE SEPARATION METHOD

This method consists of five basic steps: polymer dissolution, gelation, solvent extraction, freezing and freeze-drying. In this process, it is observed that gelatin is the most difficult step to control the porous morphology of nanofiber. Duration of gelation varied with polymer concentration and gelation temperature. At low gelation temperature, nano-scale fiber network is formed, whereas, high gelation temperature led to the formation of platelet-like structure. Uniform nanofiber can be produced as the cooling rate is increased, polymer concentration affects the properties of nanofiber, as polymer concentration is increased porosity of fiber decreased and mechanical properties of fiber are increased [1, 58].

14.3.1.4   SELF-ASSEMBLY

Self–assembly refers to the build-up of nanoscale fibers using smaller molecules. In this technique, a small molecule is arranged in a concentric manner so that they can form bonds among the concentrically arranged small molecules, which, upon extension in the plane's normal gives the longitudinal axis of a nanofiber. The main mechanism for a generic self-assembly is the intramolecular forces that bring the smaller unit together. A hydrophobic core of alkyl residues and a hydrophilic exterior lined by peptide residues was found in obtained fiber. It is observed that the nanofibers produced with this technique have diameter range 5–8 mm approximately and several microns in length [1, 59].

Although there are a number of techniques used for the synthesis of nanofiber but Electrospinning represents an attractive technique to fabricate polymeric biomaterial into nanofibers. Electrospinning is one of the most commonly used methods for the production of nanofiber. It has a wide advantage over the previously available fiber formation techniques because here electrostatic force is used instead of conventionally used mechanical force for the formation of fibers. This method will be debated comprehensively in following.

14.3.1.5   ELECTROSPINNING OF NANOFIBERS

Electrospinning is a straightforward and cost-effective method to produce novel fibers with diameters in the range of from less than 3 nm to over 1 mm, which overlaps contemporary textile fiber technology. During this process, an electrostatic force is applied to a polymeric solution to produce nanofiber [60–61] with diameter ranging from 50 nm to 1000 nm or greater [49, 62–63]; Due to surface tension the solution is held at the tip of syringe. Polymer solution is charged due to applied electric force. In the polymer solution, a force is induced due to mutual charge repulsion that is directly opposite to the surface tension of the polymer solution. Further increases in the electrical potential led to the elongation of the hemispherical surface of the solution at the tip of the syringe to form a conical shape known as "Taylor cone" [50, 64]. The electric potential is increased to overcome the surface tension forces to cause the formation of a jet, ejects from the tip of the Taylor cone. Due to elongation and solvent evaporation, charged jet instable and gradually thins in air primarily [62, 65–67]. The charged jet forms randomly oriented nanofibers that can be collected on a stationary or rotating grounded metallic collector [50]. Electrospinning provides a good method and a practical way to produce polymer fibers with diameters ranging from 40–2000 nm [49, 50].

14.3.1.5.1   THE HISTORY OF ELECTROSPINNING METHODOLOGY

William Gilbert discovered the first record of the electrostatic attraction of a liquid in 1600 [68]. The first electrospinning patent was submitted by John Francis Cooley in 1900 [69]. After that in 1914 John Zeleny studied on the behavior of fluid droplets at the end of metal capillaries, which caused the beginning of the mathematical model the behavior of fluids under electrostatic forces [65]. Between 1931 and 1944 Anton Formhals took out at least 22 patents on electrospinning [69]. In 1938, N.D. Rozenblum and I.V. Petryanov-Sokolov generated electrospun fibers, which they developed into filter materials [70]. Between 1964 and 1969 Sir Geoffrey Ingram Taylor produced the beginnings of a theoretical foundation of electrospinning by mathematically modeling the shape of the (Taylor) cone formed by the fluid droplet under the effect of an electric field [71, 72]. In the early 1990s several research groups (such as Reneker) demonstrated electrospun nano-fibers. Since 1995, the number of publications about electrospinning has been increasing exponentially every year [69].

14.3.1.5.2   ELECTROSPINNING PROCESS

Electrospinning process can be explained in five significant steps including [48, 73–75]:

a)   **Charging of the Polymer Fluid**

The syringe is filled with a polymer solution, the polymer solution is charged with a very high potential around, that is, 10–30 kV. The nature of the fluid and polarity of the applied potential free electrons, ions or ion-pairs are generated as the charge carriers form an electrical double layer. This charging induction is suitable for conducting fluid, but for nonconducting fluid charge directly injected into the fluid by the application of electrostatic field.

b)   **Formation of the Cone Jet (Taylor Cone)**

The polarity of the fluid depends upon the voltage generator. The repulsion between the similar charges at the free electrical double layer works against the surface tension and fluid elasticity in the polymer solution to deform the droplet into a conical shaped structure, that is, known as a Taylor cone. Beyond a critical charge density Taylor-cone becomes unstable and a jet of fluid is ejected from the tip of the cone.

c)   **Thinning of the Jet in the Presence of an Electric Field**

The jet travels a path to the ground; this fluid jet form a slender continuous liquid filament. The charged fluid is accelerated in the presence of an electrical field. This region of fluid is generally linear and thin.

d)   **Instability of the Jet**

Fluid elements accelerated under electric field and thus stretched and succumbed to one or more fluid instabilities, which distort as they grow following many spiral and distort the path before collected on the collector electrode. This region of instability is also known as whipping region.

e) Collection of the Jet

Charged electro spun fibers travel downfield until its impact with a lower potential collector plate. The orientation of the collector affects the alignment of the fibers. Different type of collector also affects the morphology and the properties of producing nanofiber. Different type of collectors are used: Rotating drum collector, moving belt collector, rotating wheel with beveled edge, multifilament thread, parallel bars, simple mesh collector etc.

14.3.1.5.3   ELECTROSPINNING SET UPS

Electrospinning is conducted at room temperature with atmospheric conditions. The typical set up of electrospinning apparatus is shown in Fig. 14.1. Basically, an electrospinning system consists of three major components: a high voltage power supply, a spinneret (such as a pipette tip) and a grounded collecting plate (usually a metal screen, plate, or rotating mandrel) and uses a high voltage source to inject charge of a certain polarity into a polymer solution or melt, which is then accelerated towards a collector of opposite polarity [73, 76, 77]. Most of the polymers are dissolved in some solvents before electrospinning, and when it completely dissolves, forms polymer solution. The polymer fluid is then introduced into the capillary tube for electrospinning. However, some polymers may emit unpleasant or even harmful smells, so the processes should be conducted within chambers having a ventilation system. In the electrospinning process, a polymer solution held by its surface tension at the end of a capillary tube is subjected to an electric field and an electric charge is induced on the liquid surface due to this electric field. When the electric field applied reaches a critical value, the repulsive electrical forces overcome the surface tension forces. Eventually, a charged jet of the solution is ejected from the tip of the Taylor cone and an unstable and a rapid whipping of the jet occurs in the space between the capillary tip and collector which leads to evaporation of the solvent, leaving a polymer behind. The jet is only stable at the tip of the spinneret and after that instability starts. Thus, the electrospinning process offers a simplified technique for fiber formation [50, 73, 78, 79].

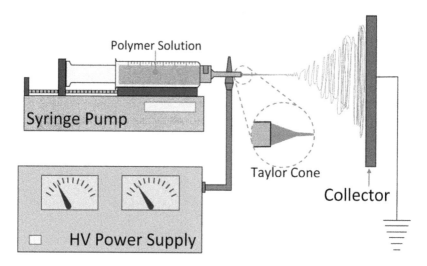

FIGURE 14.1   Scheme of a conventional electrospinning set-up.

14.3.1.5.4   THE EFFECTIVE PARAMETERS ON ELECTROSPINNING

The electrospinning process is generally governed by many parameters, which can be classified broadly into solution parameters, process parameters, and ambient parameters. Each of these parameters significantly affects the fiber morphology obtained as a result of electrospinning, and by proper manipulation of these parameters we can get nanofibers of desired morphology and diameters. These effective parameters are sorted as below [63, 67, 73, 76]:

   a)   **Polymer solution parameters**, which includes molecular weight and solution viscosity, surface tension, solution conductivity and dielectric effect of solvent.

   b)   **Processing parameters**, which include voltage, feed rate, temperature, effect of collector, the diameter of the orifice of the needle.

a) Polymer Solution Parameters

   1)   Molecular Weight and Solution Viscosity

Higher the molecular weight of the polymer, increases molecular entanglement in the solution, hence there is an increase in viscosity. The electro spun jet eject with high viscosity during it is stretched to a collector electrode leading to formation of continuous fiber with higher diameter, but very high viscosity makes difficult to pump the solution and also lead to the drying of the solution at the needle tip. As a very low viscosity lead in bead formation in the resultant electro spun fiber, so the molecular weight and viscosity should be acceptable to form nanofiber [48, 80].

2) Surface Tension

Lower viscosity leads to decrease in surface tension resulting bead formation along the fiber length because the surface area is decreased, but at the higher viscosity effect of surface tension is nullified because of the uniform distribution of the polymer solution over the entangled polymer molecules. So, lower surface tension is required to obtain smooth fiber and lower surface tension can be achieved by adding of surfactants in polymer solution [80, 81].

3) Solution Conductivity

Higher conductivity of the solution followed a higher charge distribution on the electrospinning jet, which leads to increase in stretching of the solution during fiber formation. Increased conductivity of the polymer solution lowers the critical voltage for the electro spinning. Increased charge leads to the higher bending instability leading to the higher deposition area of the fiber being formed, as a result jet path is increased and finer fiber is formed. Solution conductivity can be increased by the addition of salt or polyelectrolyte or increased by the addition of drugs and proteins, which dissociate into ions when dissolved in the solvent formation of smaller diameter fiber [67, 80].

4) Dielectric Effect of Solvent

Higher the dielectric property of the solution lesser is the chance of bead formation and smaller is the diameter of electro-spun fiber. As the dielectric property is increased, there is increase in the bending instability of the jet and the deposition area of the fiber is increased. As jet path length is increased fine fiber deposit on the collector [67, 80].

b) Processing Condition Parameters

1) Voltage

Taylor cone stability depends on the applied voltage; at the higher voltage greater amount of charge causes the jet to accelerate faster leading to smaller and unstable Taylor cone. Higher voltage leads to greater stretching of the solution due to fiber with small diameter formed. At lower voltage the flight time of the fiber to a collector plate increases that led to the formation of fine fibers. There is greater tendency to bead formation at high voltage because of increased instability of the Taylor cone, and theses beads join to form thick diameter fibers. It is observed that the better crystallinity in the fiber obtained at higher voltage, because with very high voltage acceleration of fiber increased that reduced flight time and polymer molecules do not have much time to align them and fiber with less crystallinity formed. Instead of DC if AC voltage is provided for electro spinning it forms thicker fibers [48, 80].

2) Feed Rate

As the feed rate is increased, there is an increase in the fiber diameter because greater volume of solution is drawn from the needle tip [80].

3) Temperature

At high temperature, the viscosity of the solution is decreased and there is increase in higher evaporation rate, which allows greater stretching of the solution, and a uniform fiber is formed [82].

4) Effect of Collector

In electro spinning, collector material should be conductive. The collector is grounded to create stable potential difference between needle and collector. A non-conducting material collector reduces the amount of fiber being deposited with lower packing density. But in case of conducting collector there is accumulation of closely packed fibers with higher packing density. Porous collector yields fibers with lower packing density as compared to nonporous collector plate. In porous collector plate the surface area is increased so residual solvent molecules gets evaporated fast as compared to nonporous. Rotating collector is useful in getting dry fibers as it provides more time to the solvents to evaporate. It also increases fiber morphology[83]. The specific hat target with proper parameters has a uniform surface electric field distribution, the target can collect the fiber mats of uniform thickness and thinner diameters with even size distribution [80].

5) Diameter of Pipette Orifice

Orifice with small diameter reduces the clogging effect due to less exposure of the solution to the atmosphere and leads to the formation of fibers with smaller diameter. However, very small orifice has the disadvantage that it creates problems in extruding droplets of solution from the tip of the orifice [80].

14.4   DESIGN MULTIFUNCTIONAL PRODUCTBY NANOSTRUCTURES

The largest variety of efficient and elegant multifunctional materials is seen in naturalbiological systems, which occur sometimes in the simple geometrical forms in man-madematerials. The multifunctionality of a material could be achieved by designing the materialfrom the micro to macroscales (bottom up design approach), mimicking the structural formations created by nature [84]. Biological materials present around us have a largenumber of ingenious solutions and serve as a source of inspiration. There are different waysof producing multifunctional materials that depend largely on whether these materials arestructural composites, smart materials, or nanostructured materials. Thenanostructurematerials are most challenging and innovative processes, introducing, in the manufacturing,a new approaches such as self-assembly and self-replication. For bio-materials involved insurface-interface related processes, common geometries involve capillaries, dendrites, hair,or fin-like attachments supported on larger substrates. It may be useful to incorporatesimilar hierarchical structures in the design and fabrication of multifunctional syntheticproducts that include surface sensitive functions such as sensing, reactivity, charge storage,transport property or stress transfer. Significant effort is being directed in order to fabricateand understand materials involving multiple length scales and functionalities. Porousfibrous structures can behave like lightweight solids providing significantly higher surfacearea compared to compact ones. Depending on what is attached on their surfaces, or what matrix is infiltrated in them, these core structures can be envisioned in a wide variety ofsurface-active components or net-shape composites. If nanoelements can be attached in thepores, the surface area within the given space can be increased by severalorders ofmagnitude, thereby increasing the potency of any desired surface functionality. Recentdevelopments in electrospinning have made these possible, thanks to a co-electrospinningpolymer suspension [85]. This opens up the possibility of taking a functional material of anyshape and size, and attaching nanoelements on them for added surface functionality. Thefast growing nanotechnology with modern computational/experimental methods gives thepossibility to design multifunctional materials and products in human surroundings. Smartclothing, portable fuel cells, medical devices are some of them. Research in nanotechnologybegan with applications outside of everyday life and is based on discoveries in physics andchemistry. The reason for that is need to understand the physical and chemical properties ofmolecules and nanostructures in order to control them. For example, nanoscalemanipulation results in new functionalities for textile structures, including self-cleaning,sensing, actuating, and communicating. Development of precisely controlled orprogrammable medical nanomachines and nanorobots is great promise for nanomedicine.Once nanomachines are available, the ultimate dream of every medical man becomes reality.The miniaturization of instruments on micro and nano-dimensions promises to make ourfuture lives safer with more humanity. A new approach in material synthesis is acomputational-based material development. It is based on multiscale material and processmodeling spanning, on a large spectrum of time as well as on length scales. Multi-scalematerials design means to design materials from a molecular scale up to a macro scale. Theability to manipulate at atomic and molecular level is also creating materials and structuresthat have unique functionalities and characteristics. Therefore, it will be and revolutionizingnext-generation technology ranging from structural materials to nano-electro-mechanicalsystems (NEMs), for medicine and bioengineering applications. Recent researchdevelopment in nanomaterials has been progressing at a tremendous speed for it can totallychange the ways in which materials can be made with unusual properties. Such researchincludes the synthetic of nanomaterials, manufacturing processes, in terms of the controls oftheir nano-structural and geometrical properties, moldability and mixability with othermatrix for nanocomposites. The cost of designing and producing a novel multifunctionalmaterial can be high and the risk of investment to be significant [12, 22].

Computational materialsresearch that relies on multiscale modeling has the potential to significantly reducedevelopment costs of new nanostructured materials for demanding applications by bringingphysical and microstructural information into the realm of the design engineer. As there arevarious potential applications of nanotechnology in design multifunctional product, onlysome of the well-known properties come from by nano-treatment are critically highlighted [12, 22, 30].This section review current research in nanotechnology application of the electrospinningnano-fiber, from fibber production and development to end uses as multifunctionalnano-structure device and product. Theelectrospinning phenomena are described fromex-perimental point of view to it simulation as multiscale problem.

14.4.1   THE MULTIFUNCTIONAL MATERIALS AND PRODUCTS

14.4.1.1   RESPONSIVE NANOPARTICLES

There are several directions in the research and development of the responsive nanoparticle (RNP) applications. Development of particles that respond by chang-ing stability of colloidal dispersions is the first directions. Stimuli-responsive emul-sions and foams could be very attractive for various technologies in coating indus-tries, cosmetic, and personal care. The RNPs compete with surfactants and, hence, the costs for the particle production will play a key role. The main challenge is the development of robust and simple methods for the synthesis of RNPs from inexpen-sive colloidal particles and suspensions. That is indeed not a simple job since most of commercially available NPs are more expensive than surfactants. Another impor-tant application of RNPs for tunable colloidal stability of the particle suspensions is a very broad area of biosensors [86, 87].

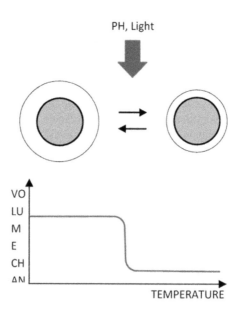

FIGURE 14.2   Stimuli-responsive nanoparticles.

The second direction is stimuli-responsive capsules that can release the cargo upon externalstimuli (*see*Fig. 14.2). The capsules are interesting for biomedical applications (drugs deliveryagents) and for composite materials (release of chemicals for self-healing). The mostchallenging task in many cases is to engineering systems capable to work with demandedstimuli. It is not a simple job for many biomedical applications wheresignalingbiomolecules are present in very small concentrations and a range of changes of manyproperties is limited by physiological conditions. A well-known challenge is related to theacceptable size production of capsules. Many medical applications need capsules less than50 nm in diameter. Fabrication of capsules with a narrow pore size distribution and tunablesizes could dramatically improve the mass transport control [86, 88].

A hierarchically organizedmulticompartment RNPs are in the focus. These particles could respond to weak signals, tomultiple signals, and could demonstrate a multiple response. They can perform logicaloperations with multiple signals, store energy, absorb and consume chemicals, andsynthesize and release chemicals. In other words, they could operate as an autonomousintelligent minidevice. The development of such RNPs can be considered as a part of biomimetics inspired by living cells or logic extension of the bottom up approach innanotechnology. The development of the intelligent RNPs faces numerous challengesrelated to the coupling of many functional building blocks in a single hierarchicallystructured RNP. These particles could find applications for intelligent drug delivery, removalof toxic substances, diagnostics in medicine, intelligent catalysis, microreactors for chemicalsynthesis and biotechnology, new generation of smart products for personal use, and others [88, 89].

14.4.1.2   NANOCOATINGS

In general, the coating's thickness is at least an order of magnitude lower than the size of the geometry to be coated. The coating's thickness less than 10 nm is called nanocoating. Nanocoatingsare materials that are produced by shrinking the material at the molecular level toform a denser product. Nanostructure coatings have an excellent toughness, good corrosionresistance, wear and adhesion properties. These coatings can be used to repair componentparts instead of replacing them, resulting in significant reductions in maintenance costs.Additionally, the nanostructure coatings will extend the service life of the component due tothe improved properties over conventional coatings [90, 91] (Fig. 14.3).

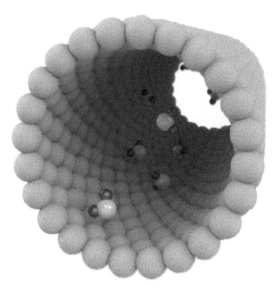

FIGURE 14.3   Nanocoatings.

14.4.1.3   *FIBROUS NANOSTRUCTURE*

The nanofibers are basic building block for plants and animals. From the structural point ofview, a uniaxial structure is able to transmit forces along its length and reducing requiredmass of materials. Nanofibers serves as another platform for multifunctional hierarchicalexample. The successful design concepts of nature, the nanofiber become an attractive basicbuilding component in the construction of hierarchically organized nanostructures. Tofollow nature's design, a process that is able to fabricate nanofiber from a variety materialsand mixtures is a prerequisite [92, 93].

Control of the nanofibers arrangement is also necessary tooptimize structural requirements. Finally, incorporation of other components into thenanofibers is required to form a complex, hierarchically organized composite. A nanofiber-fabrication technique known as electrospinning process has the potential to play a vital rolein the construction of a multilevels nanostructure [94]. In this chapter, we will introduce electrospinning as a potential technology for use as aplatform for multifunctional, hierarchically organized nanostructures. Electrospinning is a method of producing superfine fibers with diameters ranging from 10 nm to 100 nm.Electrospinning occurs when the electrical forces at the surface of a polymer solutionovercome the surface tension and cause an electrically charged jet of polymer solution to beejected. A schematic drawing of the electrospinning process is shown in Fig.14.4. Theelectrically charged jet undergoes a series of electrically induced instabilities during itspassage to the collection surface, which results in complicated stretching and looping of the jet [50, 60]. This stretching process is accompanied by the rapid evaporation of the solventmolecules, further reducing the jet diameter. Dry fibers are accumulated on the surface ofthe collector, resulting in a nonwoven mesh of nanofibers.

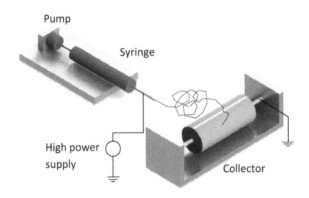

FIGURE 14.4   The electrospinning process.

Basically, an electrospinning system consists of three major components: a high voltage power supply, an emitter (e.g., a syringe) and a grounded collecting plate (usually a metal screen,plate, or rotating mandrel). There are a wide range of polymers that used in electrospinningand are able to form fine nanofibers within the submicron range and used for variedapplications. Electrospun nanofibers have been reported as being from various syntheticpolymers, natural polymers or a blend of both including proteins, nucleic acids [74]. The electrospinning process is solely governed by many parameters, classified broadly intorheological, processing, and ambient parameters. Rheological parameters include viscosity,conductivity, molecular weight, and surface tension and process parameters include appliedelectric field, tip to collector distance and flow rate. Each of these parameters significantlyaffect the fibers morphology obtained as a result of electrospinning, and by propermanipulation of these parameters we can get nanofibers fabrics of desired structure andproperties on multiple material scale.

Among these variables, ambient parametersencompass the humidity and temperature of the surroundings, whichplay a significant rolein determining the morphology and topology of electrospun fabrics. Nanofibrousassemblies such as nonwoven fibrous sheet, aligned fibrous fabric, continuous yarn and 3Dstructure have been fabricated using electrospinning [51]. Physical characteristics of theelectrospun nanofibers can also be manipulated by selecting the electrospinning conditionsand solution. Structure organization on a few hierarchical levels (*see*Fig. 14.5) has beendeveloped using electrospinning. Such hierarchy and multifunctionality potential will bedescribed in the following sections. Finally, we will describe how electrospunmultifunctional, hierarchically organized nanostructure can be used in applications such ashealthcare, defense and security, and environmental.

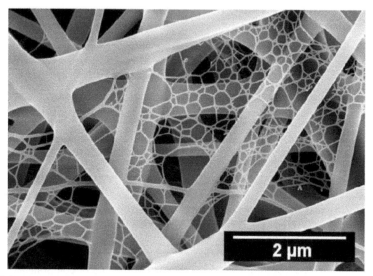

FIGURE 14.5   Multiscale electrospun fabric.

The slender-body approximation is widely used in electrospinning analysis of commonfluids [51]. The presence of nanoelements (nanoparticles, carbon nanotube, clay) insuspension jet complicate replacement 3D axisymmetric with 1D equivalent jet problem undersolid-fluid interaction force on nanolevel domain. The applied electric field induced dipolemoment, while torque on the dipole rotate and align the nanoelement with electric field. Thetheories developed to describe the behavior of the suspension jet fall into two levelsmacroscopic and microscopic. The macroscopic governing equations of the electrospinningare equation of continuity, conservation of the charge, balance of momentum and electricfield equation. Conservation of mass for the jet requires that [61, 95] (Fig. 14.6).

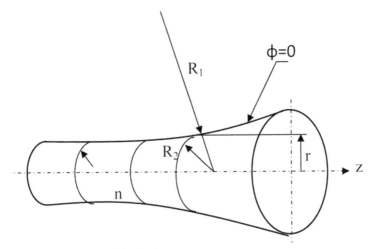

FIGURE 14.6   Geometry of the jet flow.

For polymer suspension stress tensor τ_{ij} come from polymeric $\hat{\tau}_{ij}$ andsolvent contribution tensor via constitutive equation:

$$\tau_v = \hat{\tau}_v + n_s \cdot \mathring{y}_v \tag{1}$$

where n_s is solvent viscosity, and γ_{ij} strain rate tensor. The polymer contribution tensor $\hat{\tau}_{ij}$ depend on microscopic models of the suspension. Microscopic approach represents themicrostructural features of material by means of a large number of micromechanical elements (beads, platelet, rods) obeying stochastic differential equations. The evolution equations of the microelements arise from a balance of momentum on the elementary level. For example, rheological behavior of the dilute suspension of the carbon nanotube (CNTs) in polymer matrix can be described as FENE dumbbell model [96].

$$\lambda \langle Q.Q \rangle \overset{\triangledown}{=} \delta_v - \frac{c \langle Q.Q \rangle}{1 - tr \langle Q.Q \rangle / b_{mAx}} \tag{2}$$

where $\langle Q.Q \rangle$ is the suspension configuration tensor (see Fig.14.7), c is a spring constant, andmax b is maximum CNT extensibility. Subscript $\triangledown$ represent the upper convected derivative, and λ denote a relaxation time. The polymeric stress can be obtained from the following relation:

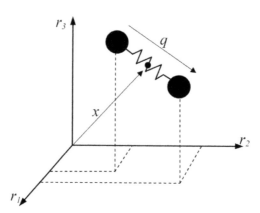

FIGURE 14.7   FENE dumbbell model.

$$\frac{\hat{\tau}_v}{nkT} = \delta_v - \frac{c(Q.Q)}{1-tr(Q.Q)/b_{mAx}}$$

(3)

where k is Boltzmann's constant, T is temperature, and n is dumbbells density. Orientation probability distribution function ψ of the dumbbell vector Q can be described by the Fokker-Planck equation, neglecting rotary diffusivity.

$$\frac{\partial \psi}{\partial t} + \frac{\partial}{\partial Q}(\psi.Q) = 0$$

(4)

Solution Eqs.(3) and (4) with supposition that flow in orifice is Hamel flow [97], give value orientation probability distribution function ψ along streamline of the jet. Rotation motion of a nanoelement (CNTs for example) in a Newtonian flow can be described as short fiber suspension model as another rheological model [8].

$$\frac{dp}{dt} = \frac{1}{2}\omega_v P_i + \frac{1}{2}\Theta\left[\frac{d\gamma_v}{dt}P_j - \frac{d\gamma_{ki}}{dt}P_k P_t P_l\right] - D_r \frac{1}{\psi}\frac{\partial \psi}{\partial t}$$

(5)

where p is a unit vector in nanoelement axis direction, ω_{ij} is the rotation rate tensor, γ_{ij} is the deformation tensor, D_r is the rotary diffusivity and θ is shape factor. Microscopic models for evolution of suspension microstructure can be coupled to macroscopic transport equations of mass and momentum to yield micro–macro multiscale flow models. The presence of the CNTs in the solution contributes to new form of instability with influences on the formation of the electrospun mat. The high strain rate on the nanoscale with complicated microstructure requires innovative research approach from the computational modeling point of view [98].

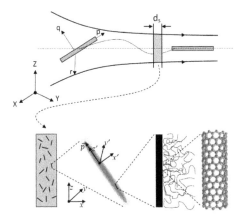

FIGURE 14.8   The CNTs alignment in jet flow.

Figure 14.8 illustrates the multiscale treatment the CNTs suspension in the jet, first time as short flexible cylinder in solution (microscale), and second time as coarse grain system with polymer chain particles and CNT(nanoscale level).

14.4.2   MULTIFUNCTIONAL NANOFIBER-BASED STRUCTURE

The variety of materials and fibrous structures that can be electrospun allow for the incorporation and optimization of various functions to the nanofiber, either during spinning or through postspinning modifications. A schematic of the multilevel organization of an electrospun fiber based composite is shown in Fig. 14.9.

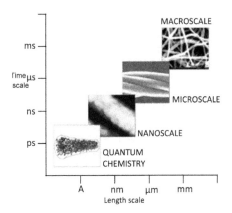

FIGURE 14.9   Multiscale electrospun fabrics.

Based on current technology, at least four different levels of organization can be put together to form a nanofiber based hierarchically organized structure. At the first level, nanoparticles or a second polymer can be mixed into the primary polymer solution and electrospun to form composite nanofiber. Using a dual-orifice spinneret design, a second layer of material can be coated over an inner core material during electrospinning to give rise to the second level organization. Two solution reservoirs, one leading to the inner orifice and the other to the outer orifice will extrude the solutions simultaneously. Otherwise, other conditions for electrospinning remain the same. Rapid evaporation of the solvents during the spinning process reduces mixing of the two solutions therefore forming core-shell nanofiber. At the same level, various surface coating or functionalization techniques may be used to introduce additional property to the fabricated nanofiber surface. Chemical functionality is a vital component in advance multifunctional composite material to detect and respond to changes in its environment. Thus various surface modifications techniques have been used to construct the preferred arrangement of chemically active molecules on the surface with the nanofiber as a supporting base. The third level organization will see the fibers oriented and organized to optimize its performance. A multilayered nanofiber membrane or mixed materials nanofibers membrane can be fabricated *in situ*through selective spinning or using a multiple orifice spinneret design, respectively. Finally, the nanofibrous assembly may be embedded within a matrix to give the fourth-level organization. The resultant structure will have various properties and functionality due its hierarchical organization. Nanofiber structure at various levels have been constructed and tested for various applications and will be covered in the following sections. To follow surface functionality and modification, jet flow must be solved on multiple scale level. All above scale (nanoscale) can be solved by use particle method together with coarse grain method on supramolecular level [50–51].

14.4.3 NANOFIBER EFFECTIVE PROPERTIES

The effective properties of the nanofiber can be determined by homogenization procedure using representative volume element (RVE). There is need for incorporating more physical information on microscale in order to precise determine material behavior model. For electrospun suspension with nanoelements (CNTs, …), a concentric composite cylinder embedded with a caped carbon nanotube represents RVE as shown by Fig.14.10. A carbon nanotube with a length 2l, radii 2a is embedded at the center of matrix materials with a radii R and length 2L.

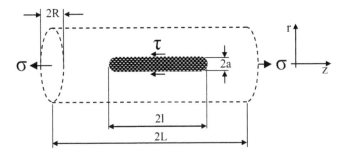

FIGURE 14.10   The nanofiber representative volume element.

The discrete atomic nanotube structure replaced the effective (solid) fiber having the same length and outer diameter as a discrete nanotube with effective Young's nanotube modulus determined from atomic structure. The stress and strain distribution in RVE was determined using modified shear-lag model [99]. For the known stress and strain distribution under RVE we can calculate elastic effective properties quantificators. The effective axial module E_{33}, and the transverse module$E_{11} = E_{22}$, can be calculated as follow:

$$E_{33} = \frac{\langle \sigma_{zz} \rangle}{\langle \varepsilon_{zz} \rangle}$$

$$E_{11} = \frac{\langle \sigma_{xx} \rangle}{\langle \varepsilon_{xx} \rangle} \tag{6}$$

where denotes a volume average under volume V as defined by:

$$\langle \Xi \rangle = \frac{1}{V} \int_V \Xi(x, y, z) \, dV. \tag{7}$$

The three-phase concentric cylindrical shell model has been proposed to predict effective modulus of nanotube-reinforced nanofibers. The modulus of nanofiber depends strongly upon the thickness of the interphase and CNTs diameter [12].

14.4.4   NETWORK MACROSCOPIC PROPERTIES

Macroscopic properties of the multifunctional structure determine final value of the any engineering product. The major objective in the determination of macroscopic properties is the link between atomic and continuum types of modeling and simulation approaches. The multiscale method such as quasi-continuum, bridge method, coarse-grain method, and dissipative particle dynamics are some popular methods of solution [98, 100]. The main advantage of the mesoscopic model is its higher computational efficiency than the molecular modeling without a loss of detailed properties at molecular level. Peridynamic modeling of fibrous network is another promising method, which allows damage, fracture and long-range forces to be treated as natural components of the deformation of materials [101]. In the first stage, effective fiber properties are determined by homogenization procedure, while in the second stage the point-bonded stochastic fibrous network at mesoscale is replaced by continuum plane stress model. Effective mechanical properties of nanofiber sheets at the macro scale level can be determined using the 2D Timoshenko beam-network. The critical parameters are the mean number of crossings per nanofiber, total nanofiber crossing in sheet and mean segment length [102]. Let as first consider a general planar fiber network characterized by fiber concentration n and fiber angular and length distribution ψ (φ,1), where φ and l are fiber orientation angle and fiber length, respectively. The fiber radius r is considered uniform and the fiber concentration n is defined as the number of fiber per unite area (Fig. 14.11).

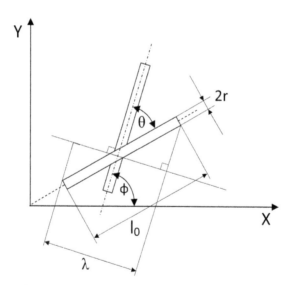

FIGURE 14.11   The fiber contact analysis.

The Poisson probability distribution can be used to describe the fiber segment length distribution for electrospun fabrics, a portion of the fiber between two neighboring contacts:

$$f(\ell) = \frac{1}{\ell}\exp\left(-\ell/\bar{\ell}\right)$$

(8)

where ℓ is the mean segment length. The total number fibersegments $N^{\hat{}}$ in the rectangular region $b*h$:

$$\overline{N} = \{n.\ell_0(\langle\lambda\rangle + 2r) - 1\}.n.b.h \tag{9}$$

With

$$\langle\lambda\rangle = \int_0^\Phi \int_0^\infty \psi(9.\ell).\lambda(9).d\ell.d9 \tag{10}$$

where the dangled segments at fiber ends have been excluded. The fiber network will be deformed in several ways. The strain energy in fiber segments come from bending, stretching and shearing modes of deformation (see Fig. 14.12)

$$U = N.\ell_0.b.h\frac{1}{2}\iint \frac{E.A}{\ell}\,\varepsilon_{XX}^2.\psi(\varphi,\ell).\ell.d\ell.d\varphi$$

$$+n.\ell_0.\{ \quad (\langle\lambda\rangle + 2r) - 1 \quad \}.b.h.\frac{1}{2}\left\{\iint \frac{G.A}{\ell}..\gamma_{xy}^2.\psi(\varphi,\ell).\ell.d\ell.d\varphi\right.$$

$$\left. + \iint \frac{3.E.I}{\ell^3}\,\gamma_{xy}^2.\psi(\varphi,\ell).\ell.d\ell.d\varphi \right\} \tag{11}$$

where A and I are beam cross-section area and moment of inertia, respectively. The first term on right side is stretching mode, while second and last term are shear bending modes, respectively.

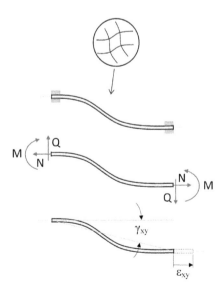

FIGURE 14.12   Fiber network 2D model.

The effective material constants for fiber network can be determined using homogenization procedure concept for fiber network. The strain energy fiber network for representative volume element is equal to strain energy continuum element with effective material constant. The strain energy of the representative volume element under plane stress conditions are:

$$U = \frac{1}{2} \cdot \langle \varepsilon_v \rangle . C_{vkl} \langle \varepsilon_{kl} \rangle . V \qquad (12)$$

where $V.b.h.2.r$ is the representative volume element, C_{ijkl} are effective elasticity tensor. The square bracket $\langle \rangle$ means macroscopic strain value. Microscopic deformation tensor was assume of a fiber segments ε_{ij} is compatible with effective macroscopic strain $\langle \varepsilon_{ij} \rangle$ of effective continuum (affine transformation). This is bridge relations between fiber segment microstrain ε_{ij} and macroscopic strain $\langle \varepsilon_{ij} \rangle$ in the effective medium. Properties of this nanofibrous structure on the macro scale depend on the 3D joint morphology. The joints can be modeled as contact torsional elements with spring and dashpot [102]. The elastic energy of the whole random fiber network can be calculated numerically, from the local deformation state of the each segment by finite element method [103]. The elastic energy of the network is then the sum of the elastic energies of all segments. We consider here tensile stress, and the fibers are rigidly bonded to each other at every fiber–fiber crossing points. To mimic the microstructure of electrospun mats, we generated fibrous structures with fibers positioned in horizontal planes, and stacked the planes on top of one another to form a 2D or 3D structure. The representative volume element dimensions are considered to be an input parameter that can be used among other parameters to control the solid volume fraction of the structure, density number of fiber in the simulations. The number of intersections/unit area and mean lengths are obtained from image analysis of electrospun sheets. For the random point field the stochastic fiber network was generated. Using polar coordinates and having the centerline equation of each fiber, the relevant parameters confined in the simulation box is obtained. The procedure is repeated until reaching the desired parameters is achieved [22, 95, 104]. The nonload bearing fiber segments were removed and trimmed to keep dimensions $b*h$ of the representative window (see Fig. 14.13).

A line representative network model is replaced by finite element beam mesh. The finite element analyzes were performed in a network of 100 fibers, for some CNTs volume fractions values. Nanofibers were modeled as equivalent cylindrical beam as mentioned above. Effective mechanical properties of nanofiber sheets at the macro scale level can be determined using the 2D Timoshenko beam-network.

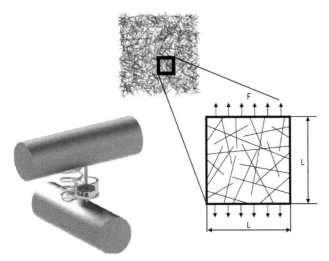

FIGURE 14.13   Representative volume element of the network.

For a displacement-based form of beam element, the principle of virtual work is assumed valid. For a beam system, a necessary and sufficient condition for equilibrium is that the virtual work done by sum of the external forces and internal forces vanish for any virtual displacement $\delta W = 0$, where W is the virtual work, which the work is done by imaginary or virtual displacements.

$$\delta W = \iiint_v v \delta \varepsilon_v + \iiint_v F_j \delta u_s dV + \oiint_A T \delta u_l dA \qquad (13)$$

where, ε is the strain, σ is the stress, F is the body force, δu is the virtual displacement, and T is the traction on surface A. The symbol δ is the variational operator designating the virtual quantity. Finite element interpolation for displacement field [15]:

$$\{u\} = [N]\{\hat{u}\} \qquad (14)$$

where is $\{u\}$ u displacement vector of arbitrary point and $\{u^\wedge\}$ is nodal displacement-point's vector. $[N]$ is shape function matrix. After FEM procedure the problem is reduced to the solution of the linear system of equations

$$[K_e].\{u\} = \{f\} \qquad (15)$$

where are $\{u\}$ is global displacement vector, $\{f\}$ global nodal force vector, and $[K_e]$ global stiffness matrix. Finite element analyzes were performed for computer generated network of 100 fibers. The comparison of calculated data with experimental data [99] for nanotube sheet shows some discrepancies (Fig. 14.14). A rough morphological network model for the sheets can explain this on the one hand and simple joint morphology on the other hand [103].

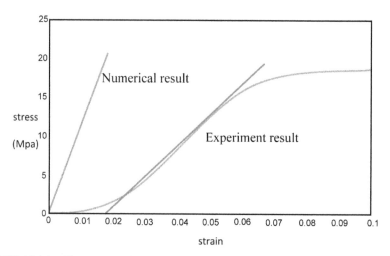

FIGURE 14.14 The stress-strain curve.

14.4.5 FLOW IN FIBER NETWORK

Electrospun nanofiber materials are becoming an integral part of many recent applicationsand products. Such materials are currently being used in tissue engineering, air and liquidfiltration, protective clothing's, drug delivery, and many others. Permeability of fibrousmedia is important in many applications, therefore during the past few decades, there havebeen many original studies dedicated to this subject. Depending on the fiber diameter andthe air thermal conditions, there are four different regimes of flow around a fiber:

a) Continuum regime ($K_N \leq 10^{-3}$),
b) Slip-flow regime ($10^{-3} \leq K_N \leq 0.25$),
c) Transient regime ($0.25 \leq K_N \leq 10$).
d) Free molecule regime ($N_k \geq 10$),

Here, $K_N = 2\lambda/d$ is the fiber Knudson number, where$\lambda = RT//\sqrt{2}N^{\wedge}\pi. d^2p$ is themean free path of gas molecules, d is fiber diameter, $N^{\wedge}$ is Avogadro number. Air flowaround most electrospun nanofibers is typically in the slip or transition flow regimes. In thecontext of air filtration, the 2D analytical work of Kuwabara [105] has long been used forpredicting thepermeability across fibrous filters. The analytical expression has beenmodified by Brown [106] todevelop an expression for predicting the permeability acrossfilter media operating in the slip flow regime. The ratio of the slip to no-slip pressure dropsobtained from the simplified 2D models may be used to modify the more realistic, and somore accurate, existing 3D permeability models in such a way that they could be used topredict the permeability of nanofiber structure. To test this supposition, for abovedeveloped 3D virtual nanofibrous structure, the Stokes flow equations solved numericallyinside these virtual structures with an appropriate slip boundary condition that isdeveloped for accounting the gas slip at fiber surface.

14.4.5.1 FLOW FIELD CALCULATION

A steady state, laminar, incompressible model has been adopted for the flow regime insideour virtual media. Implemented in the Fluent code is used to solve continuity andconservation of linear momentum in the absence of inertial effects [107]:

$$\nabla . v = 0 \tag{16}$$

$$\nabla P = \mu . \Delta^2 v \tag{17}$$

The grid size required to mesh the gap between two fibers around their crossover point isoften too small. The computational grid used for computational fluid dynamics (CFD)simulations needs to be fine enough to resolve the flow field in the narrow gaps, and at thesame time coarse enough to cover the whole domain without requiring infinitecomputational power. Permeability of a fibrous material is often presented as a function offiber radius, r, and solid volume fraction α, of the medium. Here, we use the continuumregime analyticalexpressions of Jackson and James [108], developed for 3D isotropic fibrousstructures given as

$$\frac{k}{r^2} = \frac{3r^2}{20a}[-\ell n(a) - 0.931] \tag{18}$$

Brown [106] has proposed an expression for the pressure drop across a fibrous mediumbased on the 2D cell model of Kuwabara [105] with the slip boundary condition:

$$\Delta P_{SLIP} = \frac{4\mu a.hV.(1 + 1.996K_N}{r^2[\tilde{K} + 1.996K_N(-0.5.\ell na - 0.25 + 0.25a^2]}$$

$$\tilde{K} = -0.5.\ell na - 0.25 + 0.25a^2 \tag{19}$$

where $K^{\wedge} = -0.5.lna - 0.75 + a - 0.25a^2$,Kuwabara hydrodynamic factor, h is fabric thickness, and V is velocity. As discussed in the some reference [*], permeability (orpressure drop) models obtained using ordered 2D fiber arrangements are known forunder predictingthe permeability of a fibrous medium. In order to overcome this problem, if acorrection factor can be derived based on the above 2D expression, and used with therealistic expressions developed for realistic 3D fibrous structures. From Eq.(19), wehave for the case of no-slip boundarycondition ($K_N = 0$):

$$\Delta P_{NONSLIP} = \frac{4\mu a.hV}{r^2\hat{K}} \tag{20}$$

The correction factor is defined as:

$$\Xi = \frac{\Delta P_{NONSLIP}}{\Delta P_{SLIP}} \tag{21}$$

This correction factor is to be used in modifying the original permeability expressions of Jackson andJames [109], and/or any other expressionbased on the no-slip boundary condition, in order to incorporate the slip effect. For instance,the modified expression of Jackson and James can be presented as:

$$k_z = \frac{3r^2}{20\alpha}[-\ln(\alpha) - 0.931].\Xi \tag{22}$$

Operating pressure has no influence on the pressure drop in the continuum region, while pressure drop in the free molecular region is linearly proportional to the operating pressure. While there are many equations available for predicting the permeability of fibrous materials made up of coarse fibers, there are no accurate "easy-to-use" permeability expressions that can be used for nanofiber media. On Fig.14.15 are drown corrected Jackson and James data (blue line). Points on figure are CFD numerical data.

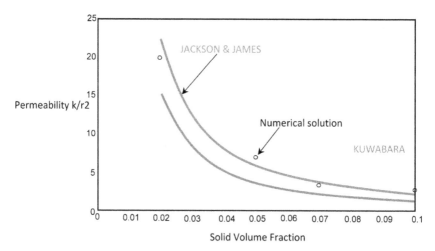

FIGURE 14.15   Permeability k/r^2dependence on solid volume fraction.

14.4.6 SOME ILLUSTRATIVE EXAMPLES

14.4.6.1 FUEL CELL EXAMPLE

Fuel cells are electrochemical devices capable of converting hydrogen or hydrogen-rich fuelsinto electrical current by a metal catalyst. There are many kinds of fuel cells, such as protonexchange mat (PEM) fuel cells, direct methanol fuel cells, alkaline fuel cells and solid oxidefuel cells [110]. PEM fuel cells are the most important one among thembecause of high power density and low operating temperature.Pt nanoparticle catalyst is a main component in fuel cells. The price of Pt has driven up thecell cost and limited the commercialization. Electrospun materials have been prepared asalternative catalyst with high catalytic efficiency, good durability and affordable cost. BinaryPtRh and PtRu nanowires were synthesized by electrospinning, and they had better catalyticperformance than commercial nanoparticle catalyst because of the one-dimensional features [111]. Pt nanowires also showed higher catalytic activities in a polymerelectrolyte membrane fuel cell [112]. Instead of direct use as catalyst, catalyst supporting material is another importantapplication area for electrospun nanofibers. Pt clusters were electrodeposited on a carbonnanofiber mat for methanol oxidation, and the catalytic peak current of the compositecatalyst reached 420 mA/mg compared with 185 mA/mg of a commercial Pt catalyst [113]. Pt nanoparticles were immobilized on polyimide-based nanofibers using ahydrolysis process and Pt nanoparticles were also loaded on the carbonnanotube containing polyamic acid nanofibers to achieve high catalytic current with long-term stability [114]. Proton exchange mat is the essential element of PEM fuel cells and normally made of aNafion film for proton conduction. Because pure Nafion is not suitable for electrospinningdue to its low viscosity in solution, it is normally mixed with other polymers to make blendnanofibers. Blend Nafion/PEO nanofibers were embedded in an inert polymer matrix tomake a proton conducting mat [115], and a high proton conductivity of 0.06–0.08S/cm at 15°C in water and low water swelling of 12–23 wt% at 25°C were achieved [116].

14.4.6.2 PROTECTIVE CLOTHING EXAMPLE

The development of smart nanotextiles has the potential to revolutionize the functionality of our clothing and the fabrics in our surroundings. This is made possible by suchdevelopments as new materials, fibbers, and finishing; inherently conducting polymers;carbon nanotubes; an antimicrobial nanocoatings. These additional functionalities havenumerous applications in healthcare, sports, military applications, fashion, etc. Smart textilesbecome a critical part of the emerging area of body sensor networks incorporating sensing,actuation, control and wireless data transmission (Fig. 14.16)[51, 52, 117].

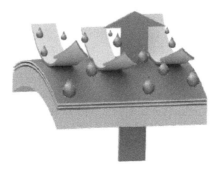

FIGURE 14.16   Ultrathin layer for selective transport.

14.4.6.3   MEDICAL DEVICE

Basic engineered nanomaterial and biotechnology products will be enormously useful infuture medical applications. We know nanomedicine as the monitoring, repair, constructionand control of biological systems at the nanoscale level, using engineered nanodevices andnanostructures. The upper portion of the dress contains cotton coated with silvernanoparticles. Silver possesses natural antibacterial qualities that are strengthened at thenanoscale, thus giving the ability to deactivate many harmful bacteria and viruses (Fig. 14.17). Thesilver infusion also reduces the need to wash the garment, since it destroys bacteria, and thesmall size of the particles prevents soiling and stains [16, 118].

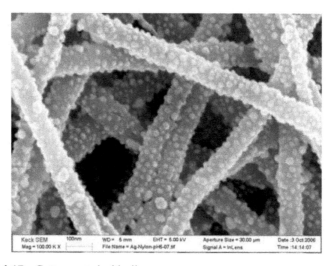

FIGURE 14.17   Cotton coated with silver.

14.4.6.3.1 DRUG DELIVERY AND RELEASE CONTROL

Controlled release is an efficient process of delivering drugs in medical therapy. It canbalance the delivery kinetics, immunize the toxicity and side effects, and improve patientconvenience [119]. In a controlled release system, the active substance isloaded into a carrier or device first, and then released at a predictable rate in vivowhenadministered by an injected or noninjected route.As a potential drug delivery carrier, electrospun nanofibers have exhibited manyadvantages. The drug loading is very easy to implement via electrospinning process, andthe high-applied voltage used in electrospinning process had little influence on the drugactivity. The high specific surface area and short diffusion passage length give the nanofiberdrug system higher overall release rate than the bulk material (e.g., film). The release profilecan be finely controlled by modulation of nanofiber morphology, porosity and composition.

Nanofibers for drug release systems mainly come from biodegradable polymers andhydrophilic polymers. Model drugs that have been studied include water soluble [120], poor-water soluble [121] and water insoluble drugs[122]. The release of macromolecules, such as DNA [123] and bioactive proteinsfrom nanofibers was also investigated.In most cases, water soluble drugs, including DNA and proteins, exhibited an early stageburst [124]. For some applications, preventing postsurgery induced adhesionfor instance, and such an early burst release will be an ideal profile because most infectionsoccur within the first few hours after surgery. A recent study also found that when a poorlywater-soluble drug was loaded into PVP nanofibers [125], 84.9% of the drug canbe released in the first 20 seconds when the drug-to-PVP ratio was kept as 1:4, which can be used for fast drug delivery systems. However, for a long-lasting release process, it would beessential to maintain the release at an even and stable pace, and any early burst releaseshould be avoided. For a water insoluble drug, the drug release from hydrophobicnanofibers into buffer solution is difficult. However, when an enzyme capable of degradingnanofibers exists in the buffer solution, the drug can be released at a constant rate because of the degradation of nanofibers [122]. For example, when rifampin wasencapsulated in PLA nanofibers, no drug release was detected from the nanofibers.However, when the buffer solution contained proteinase K, the drug release took placenearly in zero-order kinetics, and no early burst release happened. Similarly, initial burstrelease did not occur for poor-water soluble drugs, but the release from a nonbiodegradablenanofiber could follow different kinetics [126]. In another example,blending a hydrophilic but water-insoluble polymer (PEG-g-CHN) with PLGA could assistin the release of a poor-water soluble drug Iburprofen [127]. However, when awater-soluble polymer was used, the poorly soluble drug was released accompanied withdissolving of the nanofibers, leading to a low burst release [128]. The early burst release can be reduced when the drug is encapsulated within the nanofibermatrix. When an amphiphilic block copolymer, PEG-b-PLA was added into Mefoxin/PLGAnanofibers, the cumulative amount of the released drug at earlier time points was reducedand the drug release was prolonged [129]. The reason for the reduced burst release was attributed to the encapsulation of some drug molecules within the hydrophilicblock of the PEG-b-PLA. Amphiphilic block copolymer also assisted the dispersion andencapsulation of water-soluble drug into nanofibers when the polymer solution used anoleophilic solvent, such as chloroform, during electrospinning [130]. In this case, awater-in-oil emulation can be electrospun into uniform nanofibers, and drug molecules aretrapped by hydrophilic chains. The swelling of the hydrophilic chains during releasingassists the diffusion of drug from nanofibers to the buffer.Coating nanofibers with a shell could be an effective way to control the release profile.

When a thin layer of hydrophobic polymer, such as poly (p-xylylene)(PPX), was coated onPVA nanofibers loaded with bovine serum albumin (BSA)/luciferase, the early burst releaseof the enzyme was prevented [131]. Fluorination treatment [132] onPVA nanofibers introduced functional C-F groups and made the fiber surface hydrophobic,which dramatically decreased the initial drug burst and prolonged the total release time.The polymer shell can also be directly applied via a coaxial co-electrospinning process, andthe nanofibers produced are normally named "core-sheath" bicomponent nanofibers. In thiscase, even a pure drug can be entrapped into nanofiber as the core, and the release profilewas less dependent on the solubility of drug released [133].A research has compared the release behavior of two drug-loaded PLLA nanofibersprepared using blend and coaxial electrospinning[134]. It was found that theblend fibers still showed an early burst release, while the threads made of core-sheath fibersprovided a stable release of growth factor and other therapeutic drugs. In addition, the earlyburst release can also be lowered via encapsulating drugs into nanomaterial, followed byincorporating the drug-loaded nanomaterials into nanofibers. For example, halloysitenanotubes loaded with tetracycline hydrochloride were incorporated into PLGA nanofibersand showed greatly reduced initial burst release [135].

14.4.7   CONCLUDING REMARKS OF MULTIFUNCTIONAL NANOSTRUCTURES DESIGN

Electrospinning is a simple, versatile, and cost-effective technology, which generates nonwoven fibers with high surface area to volume ratio, porosity and tunable porosity. Becauseof these properties this process seems to be a promising candidate for various applicationsespecially nanostructure applications. Electrospun fibers are increasingly being used in avariety of applications such as, tissue engineering scaffolds, wound healing, drug delivery,immobilization of enzymes, as membrane in biosensors, protective clothing, cosmetics,affinity membranes, filtration applications etc. In summary, Mother Nature has always usedhierarchical structures such as capillaries and dendrites to increase multifunctional of livingorgans. Material scientists are at beginning to use this concept and create multiscalestructures where nanotubes, nanofillers can be attached to larger surfaces and subsequentlyfunctionalized. In principle, many more applications can be envisioned and created. Despiteof several advantages and success of electrospinning there are some critical limitations inthis process such as small pore size inside the fibers. Several attempts in these directions arebeing made to improve the design through multilayering, inclusion of nanoelements andblending with polymers with different degradation behavior. As new architecturesdevelop, a new wave of surface-sensitive devices related to sensing, catalysis, photo-voltaic,cell scaffolding, and gas storage applications is bound to follow.

14.5 INTRODUCTION TO THEORETICAL STUDY OF ELECTROSPINNING PROCESS

Electrospinning is a procedure in which an electrical charge to draw very fine (typically on the micro or nanoscale) fibers from polymer solution or molten. Electrospinning shares characteristics of both electrospraying and conventional solution dry spinning of fibers. The process does not require the use of coagulation chemistry or high temperatures to produce solid threads from solution. This makes the process more efficient to produce the fibers using large and complex molecules. Recently, various polymers havebeen successfully electrospun into ultrafine fibers mostly in solvent solution and some in melt form [79, 136]. Optimization of the alignment and morphology of the fibers is produced by fitting the composition of the solution and the configuration of the electrospinning apparatus such as voltage, flow rate, etc. As a result, the efficiency of this method can be improved [137]. Mathematical and theoretical modeling and simulating procedure will assist to offer an in-depth insight into the physical understanding of complex phenomenaduring electrospinningand might be very useful to manage contributing factors toward increasing production rate [75, 138].

Despite the simplicity of the electrospinning technology, industrial applications of it are still relatively rare, mainly due to the notable problems with very low fiber production rate and difficulties in controlling the process [67].

Modeling and simulation (M&S) give information about how something will act without actually testing it in real. The model is a representation of a real object or system of objects for purposes of visualizing its appearance or analyzing its behavior. Simulation is a transition from a mathematical or computational model for description of the system behavior based on sets of input parameters [104, 139]. Simulation is often the only means for accurately predicting the performance of the modeled system [140]. Using simulation is generally cheaper and safer than conducting experiments with a prototype of the final product. Also simulation can often be even more realistic than traditional experiments, as they allow the free configuration of environmental and operational parameters and can often be run faster than in real time. In a situation with different alternatives analysis, simulation can improve the efficiency, in particular when the necessary data to initialize can easily be obtained from operational data. Applying simulation adds decision support systems to the tool box of traditional decision support systems [141].

Simulation permits set up a coherent synthetic environment that allows for integration of systems in the early analysis phase for a virtual test environment in the final system. If managed correctly, the environment can be migrated from the development and test domain to the training and education domain in real systems under realistic constraints [142].

A collection of experimental data and their confrontation with simple physical models appears as an effective approach towards the development of practical tools for controlling and optimizing the electrospinning process. On the other hand, it is necessary to develop theoretical and numerical models of electrospinning because of demanding a different optimization procedure for each material[143]. Utilizing a model to express the effect of electrospinning parameters will assist researchers to make an easy and systematic way of presenting the influence of variables and by means of that, the process can be controlled. Additionally, it causes to predict the results under a new combination of parameters. Therefore, without conducting any experiments, one can easily estimate features of the product under unknown conditions [95].

14.6   STUDY OF ELECTROSPINNING JET PATH

To yield individual fibers, most, if not all of the solvents must be evaporated by the time the electrospinning jet reaches the collection plate. As a result, volatile solvents are often used to dissolve the polymer. However, clogging of the polymer may occur when the solvent evaporates before the formation of the Taylor cone during the extrusion ofthe solution from several needles. In order to maintain a stable jet while still using a volatile solvent, an effective method is to use a gas jacket around the Taylor cone through two coaxial capillary tubes. The outer tube, which surrounds the inner tube, will provide a controlled flow of inert gas, which is saturated with the solvent used to dissolve the polymer. The inner tube is then used to deliver the polymer solution. For 10 wt% poly (L-lactic acid)(PLLA) solution in dichloromethane, electrospinning was not possible due to clogging of the needle. However, when N2 gas was used to create a flowing gas jacket, a stable Taylor cone was formed and electrospinning was carried out smoothly.

14.6.1   THE THINNING JET (JET STABILITY)

The conical meniscus eventually gives rise to a slender jet that emerges from the apex of the meniscus and propagates downstream. Hohman et al. [60] first reported this approach for the relatively simple case of Newtonian fluids. This suggests that the shape of the thinning jet depends significantly on the evolution of the surface charge density and the local electric field. As the jet thins down and the charges relax to the surface of the jet, the charge density and local field quickly pass through a maximum, and the current due to advection of surface charge begins to dominate over that due to bulk conduction.

The crossover occurs on the length scale given by [6]:

$$L_N = \left(K^4 Q^7 \rho^3 (\ln X)^2 / 8\pi^2 E_\infty I^5 \varepsilon^{-2} \right)^{1/5} \tag{23}$$

This length scale defines the 'nozzle regime' over which the transition from the meniscus to the steady jet occurs. Sufficiently far from the nozzle regime, the jet thins less rapidly and finally enters the asymptotic regime, where all forces except inertial and electrostatic forces ceases to influence the jet. In this regime, the radius of the jet decreases as follows:

$$h = \left(\frac{Q^3 \rho}{2\pi^2 E_\infty I} \right)^{1/4} z^{-1/4} \tag{24}$$

Here z is the distance along the centerline of the jet. Between the 'nozzle regime' and the 'asymptotic regime,' the evolution of the diameter of the thinning jet can be affected by the viscous response of the fluid. Indeed by balancing the viscous and the electrostatic terms in the force balance equation it can be shown that the diameter of the jet decreases as:

$$h = \left(\frac{6\mu Q^2}{\pi E_\infty I} \right)^{1/2} z^{-1} \tag{25}$$

In fact, the straight jet section has been studied extensively to understand the influence of viscoelastic behavior on the axisymmetric instabilities [93] and crystallization [60] and has even been used to extract extensional viscosity of polymeric fluids at very high strain rates.

For highly strain-hardening fluids, Yu et al. [144]demonstrated that the diameter of the jet decreased with a power-law exponent of $-1/2$, rather than $-1/4$ or -1, as discussed earlier for Newtonian fluids. This $-1/2$ power-law scaling for jet thinning in viscoelastic fluids has been explained in terms of a balance between electromechanical stresses acting on the surface of the jet and the viscoelastic stress associated with extensional strain hardening of the fluid. Additionally, theoretical studies of viscoelastic fluids predict a change in the shape of the jet due tonon-Newtonian fluid behavior. Both Yu et al.[144] and Han et al.[145] have demonstrated that substantial elastic stresses can be accumulated in the fluid as a result of the high strain rate in the transition from the meniscus into thejetting region. This elastic stress stabilizes the jet against external perturbations. Further downstream the rate of stretching slows down, and the longitudinal stresses relax through viscoelastic processes. The relaxation of stresses following an extensional deformation, such as those encountered in electrospinning, has been studied in isolation for viscoelastic fluids [146]. Interestingly, Yu et al.[144] also observed that, elastic behavior notwithstanding, the straight jet transitions into the whipping region when the jet diameter becomes of the order of 10 mm.

14.6.2   THE WHIPPING JET (JET INSTABILITY)

While it is in principle possible to draw out the fibers of small diameter by electro-spinning in the cone-jet mode alone, the jet does not typically solidify enough en route to the collector and succumbs to the effect of force imbalances that lead to one or more types of instability. These instabilities distort the jet as they grow. A family of these instabilities exists, and can be analyzed in the context of various symmetries (axisymmetric or nonaxisymmetric) of the growing perturbation to the jet.

Some of the lower modes of this instability observed in electrospinning have been discussed in a separate review [81]. The 'whipping instability' occurs when the jet becomes convectively unstable and its centerline bends. In this region, small perturbations grow exponentially, and the jet is stretched out laterally. Shin et al. [62] and Fridrikh et al. [63] have demonstrated how the whipping instability can be largely responsible for the formation of solid fiber in electrospinning. This is signifi-cant, since as recently as the late 1990 s the bifurcation of the jet into two more or less equal parts (also called 'splitting' or 'splaying') were thought to be the mecha-nism through which the diameter of the jet is reduced, leading to the fine fibers formed in electrospinning. In contrast to 'splitting' or 'splaying, ' the appearance of secondary, smaller jets from the primary jet have been observed more frequently and in situ[64, 147]. These secondary jets occur when the conditions on the surface of the jet are such that perturbations in the local field, for example, due to the onset of the slight bending of the jet, is enough to overcome the surface tension forces and invoke a local jetting phenomenon.

The conditions necessary for the transition of the straight jet to the whipping jet has been discussed in the works of Ganan-Calvo [148], Yarin et al.[64], Reneker et al.[66] and Hohman et al.[60].

During this whipping instability, the surface charge repulsion, surface tension, and inertia were considered to have more influence on the jet path than Maxwell's stress, which arises due to the electric field and finite conductivity of the fluid. Us-ing the equations reported by Hohman et al.[60] and Fridrikh et al.[63] obtained an equation for the lateral growth of the jet excursions arising from the whipping in-stability far from the onset and deep into the nonlinear regime. These developments have been summarized in the review article of Rutledge and Fridrikh.

The whipping instability is postulated to impose the stretch necessary to draw out the jet into fine fibers. As discussed previously, the stretch imposed can make an elastic response in the fluid, especially if the fluid is polymeric in nature. An empiri-cal rheological model was used to explore the consequences of nonlinear behavior of the fluid on the growth of the amplitude of the whipping instability in numerical calculations [63, 79]. There it was observed that the elasticity of the fluid significantly reduces the amplitude of oscillation of the whipping jet. The elastic response also stabilizes the jet against the effect of surface tension. In the absence of any elasticity, the jet eventually breaks up and forms an aerosol. However, the presence of a polymer in the fluid can stop this breakup if:

$$\tau / \left(\frac{\rho h^3}{\gamma} \right)^{1/2} \geq 1 \tag{26}$$

Where τ is the relaxation time of the polymer, ρ is the density of the fluid, h is a characteristic radius, and γ is the surface tension of the fluid.

14.7 ELECTROSPINNING DRAW BACKS

Electrospinning has attracted much attention both to academic research and industry application because electrospinning (1) can fabricate continuous fibers with diameters down to a few nanometers, (2) is applicable to a wide range of materials such as synthetic and natural polymers, metals as well as ceramics and composite systems, (3) can prepare nanofibers with low cost and high yielding [47].

Despite the simplicity of the electrospinning technology, industrial applications of it are still relatively rare, mainly due to the notable problems of very low fiber production rate and difficulties in controlling the process [50, 67]. The usual feed rate for electrospinning is about 1.5 mL/h. Given a solution concentration of 0.2 g/mL, the mass of nanofiber collected from a single needle after an hour is only 0.3 g. In order for electrospinning to be commercially viable, it is necessary to increase the production rate of the nanofibers. To do so, multiple-spinning setup is necessary to increase the yield while at the same time maintaining the uniformity of the nanofiber mesh [48].

Optimization of the alignment and morphology of the fibers which is produced by fitting the composition of the solution and the configuration of the electrospinning apparatus such as voltage, flow rate, etc., can be useful to improve the efficiency of this method [137]. Mathematical and theoretical modeling and simulating procedure will assist to offer an in-depth insight into the physical understanding of complex phenomenaduring electrospinningand might be very useful to manage contributing factors toward increasing production rate [75, 138].

Presently, nanofibers have attracted the attention of researchers due to their remarkable micro and nano structural characteristics, high surface area, small pore size, and the possibility of their producing three dimensional structure that enable the development of advanced materials with sophisticated applications [73].

Controlling the property, geometry, and mass production of the nanofibers, is essential to comprehend quantitatively how the electrospinning process transforms the fluid solution through a millimeter diameter capillary tube into solid fibers which are four to five orders smaller in diameter [74].

As mentioned above, the electrospinning gives us the impression of being a very simple and easily controlled technique for the production of nanofibers. But, actually the process is very intricate. Thus, electrospinning is usually described as the interaction of several physical instability processes. The bending and stretching of the jet are mainly caused by the rapidly whipping which is an essential element of the process induced by these instabilities. Until now, little is known about the detailed mechanisms of the instabilities and the splaying process of the primary jet into multiple filaments. It is thought to be responsible that the electrostatic forces overcome surface tensions of the droplet during undergoing the electrostatic field and the deposition of jets formed nanofibers [47].

Though electrospinning has been become an indispensable technique for generating functional nanostructures, many technical issues still need to be resolved. For example, it is not easy to prepare nanofibers with a same scale in diameters by electrospinning; it is still necessary to investigate systematically the correlation between the secondary structure of nanofiber and the processing parameters; the mechanical properties, photoelectric properties and other property of single fiber should be systematically studied and optimized; the production of nanofiber is still in laboratory level, and it is extremely important to make efforts for scaled-up commercialization; nanofiber from electrospinning has a the low production rate and low mechanical strength which hindered it's commercialization; in addition, another more important issue should be resolved is how to treat the solvents volatilized in the process.

Until now, lots of efforts are putted on the improvement of electrospinning installation, such as the shape of collectors, modified spinnerets and so on. The application of multijets electrospinning provides a possibility to produce nanofibers in industrial scale. The development of equipments, which can collect the poisonous solvents, and the application of melt electrospinning, which would largely reduce the environment problem, create a possibility of the industrialization of electrospinning. The application of water as the solvent for electrospinning provide another approach to reduce environmental pollution, which is the main fact hindered the commercialization of electrospinning. In summary, electrospinning is an attractive and promising approach for the preparation of functional nanofibers due to its wide applicability to materials, low cost and high production rate [47].

14.8   MODELING OF THE ELECTROSPINNING PROCESS

The electrospinning process is a fluid dynamics related problem. Controlling the property, geometry, and mass production of the nanofibers, is essential to comprehend quantitatively how the electrospinning process transforms the fluid solution through a millimeter diameter capillary tube into solid fibers which are four to five orders smaller in diameter [74]. Although information on the effect of various processing parameters and constituent material properties can be obtained experimentally, theoretical models offer in-depth scientific understanding which can be useful to clarify the affecting factors that cannot be exactly measured experimentally. Results from modeling also explained how processing parameters and fluid behavior lead to the nanofiber of appropriate properties. The term "properties" refers to basic properties (such as fiber diameter, surface roughness, fiber connectivity, etc.), physical properties (such as stiffness, toughness, thermal conductivity, electrical resistivity, thermal expansion coefficient, density, etc.) and specialized properties (such as biocompatibility, degradation curve, etc. for biomedical applications)[48, 73].

For example, the developed models can be used for the analysis of mechanisms of jet deposition and alignment on various collecting devices in arbitrary electric fields [149].

The various method formulated by researchers are prompted by several applications of nanofibers. It would be sufficient to briefly describe some of these methods to observed similarities and disadvantages of these approaches. An abbreviated literature review of these models will be discussed in the following sections.

14.8.1 MODELING ASSUMPTIONS

Just as in any other process modeling, a set of assumptions are required for the following reasons:

   a)   To furnish industry – based applications whereby speed of calculation, but not accuracy, is critical,
   b)   To simplify – hence enabling checkpoints to be made before more detailed models can proceed.
   c)   For enabling the formulations to be practically traceable.

The first assumption to be considered as far as electrospinning is concerned is conceptualizing the jet itself. Even though the most appropriate view of a jet flow is that of a liquid continuum, the use of nodes connected in series by certain elements that constitute rheological properties has proven successful [64, 66]. The second assumption is the fluid constitutive properties. In the discrete node model [66], the nodes are connected in series by a Maxwell unit, that is, a spring and dashpot in series, for quantifying the viscoelastic properties.

In analyzing viscoelastic models, we apply two types of elements: the dashpot element, which describes the force as being in proportion to the velocity (recall friction), and the spring element, which describes the force as being in proportion to elongation. One can then develop viscoelastic models using combinations of these elements. Among all possible viscoelastic models, the Maxwell model was selected by Ref. [66]due to its suitability for liquid jet as well as its simplicity. Other models are either unsuitable for liquid jet or too detailed.

In the continuum models a power law can be used for describing the liquid behavior under shear flow for describing the jet flow [150]. At this juncture, we note that the power law is characterized from a shear flow, while the jet flow in electrospinning undergoes elongational flow. This assumption will be discussed in detail in following sections.

The other assumption, which should be applied in electrospinning modeling, is about the coordinate system. The method for coordinate system selection in electrospinning process is similar to other process modeling, the system that best bring out the results by (i) allowing the computation to be performed in the most convenient manner and, more importantly, (ii) enabling the computation results to be accurate. In view of the linear jet portion during the initial first stage of the jet, the spherical coordinate system is eliminated. Assuming the second stage of the jet to be perfectly spiraling, due to bending of the jet, the cylindrical coordinate system would be ideal. However, experimental results have shown that the bending instability portion of the jet is not perfectly expanding spiral. Hence the Cartesian coordinate system, which is the most common of all coordinate system, is adopted.

Depending on the processing parameters (such as applied voltage, volumeflow rate, etc.) and the fluid properties (such as surface tension, viscosity,etc.) as many as 10 modes of electrohydrodynamically driven liquid jethave been identified [151]. The scope of jetmodes is highlyabbreviated in this chapter because most electrospinningprocesses that lead to nanofibers consist of only two modes, the straightjet portion and the spiraling (or whipping) jet portion. Insofar aselectrospinning process modeling is involved, the followingclassification indicates the considered modes or portion of theelectrospinning jet.

1. Modeling the criteria for jet initiation from the droplet [64, 152].
2. Modeling the straight jet portion [150, 153–155].
3. Modeling the entire jet [60, 61, 66, 156].

14.8.2 CONSERVATION RELATIONS

Balance of the producing accumulation is, particularly, a basic source of quantitative models of phenomena or processes. Differential balance equations are formulated for momentum, mass and energy through the contribution of local rates of transport expressed by the principle of Newton's, Fick's and Fourier laws. For a description of more complex systems like electrospinning that involved strong turbulence of the fluid flow, characterization of the product property is necessary and various balances are required [157].

The basic principle used in modeling of chemical engineering process is a concept of balance of momentum, mass and energy, which can be expressed in a general form as:

$$A = I + G - O - C \tag{27}$$

whereA –accumulation built up within the system; I –input entering through the system surface;G –generation produced in system volume; O –output leaving through system boundary;C – consumption used in system volume. The form of expression depends on the level of the process phenomenon description [157, 158].

According to the electrospinning models, the jet dynamics are governed by a set of three equations representing mass, energy and momentum conservation for the electrically charge jet [159].

In electrospinning modeling for simplification of describing the process, researchers consider an element of the jet and the jet variation versus time is neglected.

14.8.2.1   MASS CONSERVATION

The concept of mass conservation is widely used in many fields such as chemistry, mechanics, and fluid dynamics. Historically, mass conservation was discovered in chemical reactions by Antoine Lavoisier in the late eighteenth century, and was of decisive importance in the progress from alchemy to the modern natural science of chemistry. The concept of matter conservation is useful and sufficiently accurate for most chemical calculations, even in modern practice [160].

The equations for the jet follow from Newton's Law and the conservation laws obey, namely, conservation of mass and conservation of charge [60].

According to the conservation of mass equation,

$$\pi R^2 v = Q \tag{28}$$

$$\frac{\partial}{\partial t}\left(\pi R^2\right) + \frac{\partial}{\partial z}\left(\pi R^2 v\right) = 0 \tag{29}$$

For incompressible jets, by increasing the velocity the radius of the jet decreases. At the maximum level of the velocity, the radius of the jet reduces. The macromolecules of the polymers are compacted together closer while the jet becomes thinner as it shown in Fig. 14.18. When the radius of the jet reaches the minimum value and its speed becomes maximum to keep the conservation of mass equation, the jet dilates by decreasing its density, which called electrospinning dilation [161, 162].

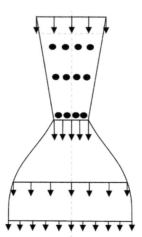

FIGURE 14.18 Macromolecular chains are compacted during the electrospinning.

14.8.2.2 ELECTRIC CHARGE CONSERVATION

An electric current is a flow of electric charge. Electric charge flows when there is voltage present across a conductor. In physics, charge conservation is the principle that electric charge can neither be created nor destroyed. The net quantity of electric charge, the amount of positive charge minus the amount of negative charge in the universe, is always conserved. The first written statement of the principle was by American scientist and statesman Benjamin Franklin in 1747 [163]. Charge conservation is a physical law, which states that the change in the amount of electric charge in any volume of space is exactly equal to the amount of charge in a region and the flow of charge into and out of that region [164].

During the electrospinning process, the electrostatic repulsion between excess charges in the solution stretches the jet. This stretching also decreases the jet diameter that this leads to the law of charge conservation as the second governing equation [165].

In electrospinning process, the electric current, which induced by electric field included two parts, conduction and convection.

The conventional symbol for current is I:

$$I = I_{conduction} + I_{convection} \tag{30}$$

Electrical conduction is the movement of electrically charged particles through a transmission medium. The movement can form an electric current in response to an electric field. The underlying mechanism for this movement depends on the material.

$$I_{conduction} = J_{cond} \times S = KE \times \pi R^2 \tag{31}$$

$$J = \frac{I}{A(s)} \tag{32}$$

$$I = J \times S \tag{33}$$

Convection current is the flow of current with the absence of an electric field.

$$I_{conduction} = J_{cond} \times S = KE \times \pi R^2 \tag{34}$$

$$J_{conv} = \sigma v \tag{35}$$

So, the total current can be calculated as:

$$\pi R^2 KE + 2\pi R v \sigma = I \tag{36}$$

$$\frac{\partial}{\partial t}(2\pi R\sigma) + \frac{\partial}{\partial z}\left(\pi R^2 KE + 2\pi R v\sigma\right) = 0 \tag{37}$$

14.8.3 MOMENTUM BALANCE

In classical mechanics, linear momentum or translational momentum is the product of the mass and velocity of an object. Like velocity, linear momentum is a vector quantity, possessing a direction as well as a magnitude:

$$P = mv \tag{38}$$

Linear momentum is also a conserved quantity, meaning that if a closed system (one that does not exchange any matter with the outside and is not acted on by outside forces) is not affected by external forces, its total linear momentum cannot change. In classical mechanics, conservation of linear momentum is implied by Newton's laws of motion; but it also holds in special relativity (with a modified formula) and, with appropriate definitions, a (generalized) linear momentum conservation law holds in electrodynamics, quantum mechanics, quantum field theory, and general relativity [163]. For example, according to the third law, the forces between two particles are equal and opposite. If the particles are numbered 1 and 2, the second law states:

$$F_1 = \frac{dP_1}{dt} \tag{38}$$

$$\frac{dP_1}{dt} = -\frac{dP_2}{dt} \tag{39}$$

Therefore,

$$\frac{dP_1}{dt} = -\frac{dP_2}{dt} \tag{40}$$

$$\frac{d}{dt}(P_1 + P_2) = 0 \tag{41}$$

If the velocities of the particles are υ_{11} and υ_{12} before the interaction, and afterwards they are υ_{21} and υ_{22}, then

$$m_1 \upsilon_{11} + m_2 \upsilon_{12} = m_1 \upsilon_{21} + m_2 \upsilon_{22} \tag{42}$$

This law holds no matter how complicated the force is between the particles. Similarly, if there are several particles, the momentum exchanged between each pair of particles adds up to zero, so the total change in momentum is zero. This conservation law applies to all interactions, including collisions and separations caused by explosive forces. It can also be generalized to situations where Newton's laws do not hold, for example in the theory of relativity and in electrodynamics [153, 166]. The momentum equation for the fluid can be derived as follow:

$$\rho(\frac{d\upsilon}{dt} + \upsilon\frac{d\upsilon}{dz}) = \rho g + \frac{d}{dz}[\tau_{zz} - \tau_{rr}] + \frac{\gamma}{R^2}.\frac{dr}{dz} + \frac{\sigma}{\varepsilon_0}\frac{d\sigma}{dz} + (\varepsilon - \varepsilon_0)(E\frac{dE}{dz}) + \frac{2\sigma E}{r} \tag{43}$$

But commonly the momentum equation for electrospinning modeling is formulated by considering the forces on a short segment of the jet [153, 166].

$$\frac{d}{dz}(\pi R^2 \rho \upsilon^2) = \pi R^2 \rho g + \frac{d}{dz}\left[\pi R^2(-p + \tau_{zz})\right] + \frac{\gamma}{R}.2\pi RR' + 2\pi R(t_t^e - t_n^e R') \tag{44}$$

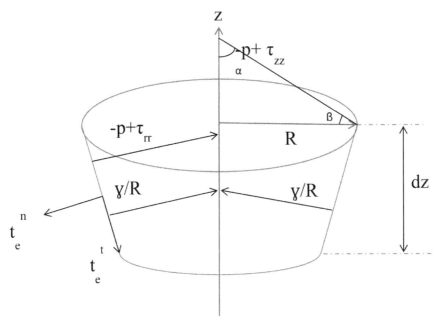

FIGURE 14.19 Momentum balance on a short section of the jet.

As it is shown in the Fig. 14.19, the element's angels could be defined as α and β. According to the mathematical relationships, it is obvious that:

$$\alpha + \beta = \frac{\pi}{2} \tag{45}$$

$$\sin \alpha = \tan \alpha$$
$$\cos \alpha = 1 \tag{46}$$

Due to the figure, relationships between these electrical forces are as below:

$$t_n^e \sin \alpha \cong t_n^e \tan \alpha \cong -t_n^e \tan \beta \cong -\frac{dR}{dz} t_n^e = -R't_n^e \tag{47}$$

$$t_t^e \cos \alpha \cong t_t^e \tag{48}$$

So the effect of the electric forces in the momentum balance equation can be presented as:

$$2\pi RL(t_t^e - R't_n^e)dz \tag{49}$$

(Notation: In the main momentum equation, final formula is obtained by dividing into dz)

Generally, the normal electric force is defined as:

$$t_n^e \cong \frac{1}{2}\bar{\varepsilon}E_n^2 = \frac{1}{2}\bar{\varepsilon}(\frac{\sigma}{\bar{\varepsilon}})^2 = \frac{\sigma^2}{2\bar{\varepsilon}} \tag{50}$$

A little amount of electric forces is perished in the vicinity of the air.

$$E_n = \frac{\sigma}{\bar{\varepsilon}} \tag{51}$$

The electric force can be presented by:

$$F = \frac{\Delta We}{\Delta l} = \frac{1}{2}(\varepsilon - \bar{\varepsilon})E^2 \times \Delta S \tag{52}$$

The force per surface unit is:

$$\frac{F}{\Delta S} = \frac{1}{2}(\varepsilon - \bar{\varepsilon})E^2 \tag{53}$$

Generally, the electric potential energy is obtained by:

$$Ue = -We = -\int F.ds \tag{54}$$

$$\Delta We = \frac{1}{2}(\varepsilon - \bar{\varepsilon})E^2 \times \Delta V = \frac{1}{2}(\varepsilon - \bar{\varepsilon})E^2 \times \Delta S.\Delta l \tag{55}$$

So, finally it could be resulted:

$$t_n^e = \frac{\sigma^2}{2\bar{\varepsilon}} - \frac{1}{2}(\varepsilon - \bar{\varepsilon})E^2 \tag{56}$$

$$t_t^e = \sigma E \tag{57}$$

14.8.4 COULOMB'S LAW

Coulomb's law is a mathematical description of the electric force between charged objects, which is formulated by the eighteenth-century French physicist Charles-Augustin de Coulomb. It is analogous to Isaac Newton's law of gravity. Both gravitational and electric forces decrease with the square of the distance between the objects, andboth forces act along a line between them [167]. In Coulomb's law, the magnitude and sign of the electric force are determined by the electric charge, more than the mass of an object. Thus, a charge, which is a basic property matter, determines how electromagnetism affects the motion of charged targets [163].

Coulomb force is thought to be the main cause for the instability of the jet in the electrospinning process [168]. This statement is based on the Earnshaw's theorem, named after Samuel Earnshaw [169], which claims that "A charged body placed in an electric field of force cannot rest in stable equilibrium under the influence of the electric forces alone." This theorem can be notably adapted to the electrospinning process [168]. The instability of charged jet influences on jet deposition and as a consequence on nanofiber formation. Therefore, some researchers applied developed models to the analysis of mechanisms of jet deposition and alignment on various collecting devices in arbitrary electric fields [66].

The equation for the potential along the centerline of the jet can be derived from Coulomb's law. Polarized charge density is obtained:

$$\rho_{p'} = -\vec{\nabla}.\vec{P}' \tag{58}$$

where P' is polarization:

$$\vec{P}' = (\varepsilon - \bar{\varepsilon})\vec{E} \tag{59}$$

By substituting P' in the last equation:

$$\rho_{p'} = -(\bar{\varepsilon} - \varepsilon)\frac{dE}{dz'} \tag{60}$$

Beneficial charge per surface unit can be calculated as below:

$$\rho_{p'} = \frac{Q_b}{\pi R^2} \tag{61}$$

$$Q_b = \rho_b.\pi R^2 = -(\bar{\varepsilon} - \varepsilon)\pi R^2 \frac{dE}{dz'} \tag{62}$$

$$Q_b = -(\bar{\varepsilon} - \varepsilon)\pi \frac{d(ER^2)}{dz'} \tag{63}$$

$$\rho_{sb} = Q_b.dz' = -(\bar{\varepsilon} - \varepsilon)\pi \frac{d}{dz'}(ER^2)dz' \tag{64}$$

The main equation of Coulomb's law:

$$F = \frac{1}{4\pi\varepsilon_0}\frac{qq_0}{r^2}$$

(65)

The electric field is:

$$E = \frac{1}{4\pi\varepsilon_0}\frac{q}{r^2}$$

(66)

The electric potential can be measured:

$$\Delta V = -\int E.dL$$

(67)

$$V = \frac{1}{4\pi\varepsilon_0}\frac{Q_b}{r}$$

(68)

According to the beneficial charge equation, the electric potential could be rewritten as:

$$\Delta V = Q(z) - Q_\infty(z) = \frac{1}{4\pi\bar{\varepsilon}}\int\frac{(q - Q_b)}{r}dz'$$

(69)

$$Q(z) = Q_\infty(z) + \frac{1}{4\pi\bar{\varepsilon}}\int\frac{q}{r}dz' - \frac{1}{4\pi\bar{\varepsilon}}\int\frac{Q_b}{r}dz'$$

(70)

$$Q_b = -(\bar{\varepsilon} - \varepsilon)\pi\frac{d(ER^2)}{dz'}$$

(71)

The surface charge density's equation is:

$$q = \sigma.2\pi RL$$

(72)

$$r^2 = R^2 + (z - z')^2$$

(73)

$$r = \sqrt{R^2 + (z - z')^2}$$

(74)

The final equation, which obtained by substituting the above-mentioned equations is:

$$Q(z) = Q_\infty(z) + \frac{1}{4\pi\bar{\varepsilon}}\int\frac{\sigma.2\pi R}{\sqrt{(z - z')^2 + R^2}}dz' - \frac{1}{4\pi\bar{\varepsilon}}\int\frac{(\bar{\varepsilon} - \varepsilon)\pi}{\sqrt{(z - z')^2 + R^2}}\frac{d(ER^2)}{dz'}$$

(75)

It is assumed that β is defined:

$$\beta = \frac{\varepsilon}{\bar{\varepsilon}} - 1 = -\frac{(\bar{\varepsilon} - \varepsilon)}{\bar{\varepsilon}}$$

(76)

So, the potential equation becomes:

$$Q(z) = Q_{\infty}(z) + \frac{1}{2\bar{\varepsilon}} \int \frac{\sigma.R}{\sqrt{(z-z')^2 + R^2}} dz' - \frac{\beta}{4} \int \frac{1}{\sqrt{(z-z')^2 + R^2}} \frac{d(ER^2)}{dz'}$$

(77)

The asymptotic approximation of χ is used to evaluate the integrals mentioned above:

$$\chi = \left(-z + \xi + \sqrt{z^2 - 2z\xi + \xi^2 + R^2} \right)$$

(78)

where χ is "aspect ratio" of the jet (L= length, R_0= Initial radius) leads to the final relation to the axial electric fiel

$$E(z) = E_{\infty}(z) - \ln \chi \left(\frac{1}{\bar{\varepsilon}} \frac{d(\sigma R)}{dz} - \frac{\beta}{2} \frac{d^2(ER^2)}{dz^2} \right)$$

(79)

14.8.5 FORCES CONSERVATION

There exists a force, as a result of charge build-up, acting upon thedroplet coming out of the syringe needle pointing toward thecollecting plate, which can be either grounded or oppositely charged.Furthermore, similar charges within the droplet promote jet initiation dueto their repulsive forces. Nevertheless, surface tension and otherhydrostatic forces inhibit the jet initiation because the total energy of adroplet is lower than that of a thin jet of equal volume uponconsideration of surface energy. When the forces that aid jet initiation(such as electric field and Coulombic) over-come the opposing forces(such as surface tension and gravitational), the droplet ac-celerates towardthe collecting plate. This forms a jet of very small diameter. Other thaninitiating jet flow, the electric field and Coulombic forces tend to stretchthe jet, thereby contributing towards the thinning effect of the resultingnanofibers.

In the flow path modeling, we recall the Newton's Second Law of motion

$$m\frac{d^2 P}{dt^2} = \Sigma f$$

(80)

where, m (equivalent mass) and the various forces are summed as

$$\Sigma f = f_C + f_E + f_V + f_S + f_A + f_G + \dots$$

(81)

In which subscripts C, E, V, S, A and G correspond to the Coulombic, electric field, viscoelastic, surface tension, air drag and gravitational forces respectively. A description of each of these forces based on the literature [66] is summarized in Table 14.1, where, V_0= applied voltage;h = distance from pendent drop to ground collector; σ_V= viscoelastic stress;v = kinematic viscosity.

TABLE 14.1   Description of Itemized Forces or Terms Related to Them

Forces	Equations
Coulombic	$f_C = \dfrac{q^2}{l^2}$
Electric field	$f_E = -\dfrac{qV_0}{h}$
Viscoelastic	$f_V = \dfrac{d\sigma_V}{dt} = \dfrac{G}{l}\dfrac{dl}{dt} - \dfrac{G}{\eta}\sigma_V$
Surface Tension	$f_S = \dfrac{\alpha\pi R^2 k}{\sqrt{x_i^2 + y_i^2}}\left[i|x|Sin(x) + i|y|Sin(y)\right]$
Air drag	$f_A = 0.65\pi R\rho_{air}v^2\left(\dfrac{2vR}{v_{air}}\right)^{-0.81}$
Gravitational	$f_G = \rho g\pi R^2$

14.8.6   *CONSTITUTIVE EQUATIONS*

In modern condensed matter physics, the constitutive equation plays a major role. In physics and engineering, a constitutive equation or relation is a relation between two physical quantities that is specific to a material or substance, and approximates the response of that material to external stimulus, usually as applied fields or forces [170]. There are a sort of mechanical equation of state, and describe how the material is constituted mechanically. With these constitutive relations, the vital role of the material is reasserted [171]. There are two groups of constitutive equations: Linear and nonlinear constitutive equations [172]. These equations are combined with other governing physical laws to solve problems; for example, in fluid mechanics the flow of a fluid in a pipe, in solid state physics the response of a crystal to an electric field, or in structural analysis, the connection between applied stresses or forces to strains or deformations [170].

The first constitutive equation (constitutive law) was developed by Robert Hooke and is known as Hooke's law. It deals with the case of linear elastic materials. Following this discovery, this type of equation, often called a "stress-strain relation" in this example, but also called a "constitutive assumption" or an "equation of state" was commonly used [173]. Walter Noll advanced the use of constitutive equations, clarifying their classification and the role of invariance requirements, constraints, and definitions of terms like "material," "isotropic," "aeolotropic," etc. The class of "constitutive relations" of the form stress rate = f (velocity gradient, stress, density) was the subject of Walter Noll's dissertation in 1954 under Clifford Truesdell [170]. There are several kinds of constitutive equations, which are applied commonly in electrospinning. Some of these applicable equations are discussed as following:

14.8.6.1   OSTWALD–DE WAELE POWER LAW

Rheological behavior of many polymer fluids can be described by power law constitutive equations [172]. The equations that describe the dynamics in electrospinning constitute, at a minimum, those describing the conservation of mass, momentum and charge, and the electric field equation. Additionally a constitutive equation for the fluid behavior is also required [76]. A Power-law fluid, or the Ostwald–de Waele relationship, is a type of generalized Newtonian fluid for which the shear stress, τ, is given by:

$$\tau = K' \left(\frac{\partial v}{\partial y} \right)^m \tag{82}$$

where $\partial v / \partial y$ is the shear rate or the velocity gradient perpendicular to the plane of shear. The power law is only a good description of fluid behavior across the range of shear rates to which the coefficients are fitted. There are a number of other models that better describe the entire flow behavior of shear-dependent fluids, but they do so at the expense of simplicity, so the power law is still used to describe fluid behavior, permit mathematical predictions, and correlate experimental data [166, 174].

Nonlinear rheological constitutive equations applicable for polymer fluids (Ostwald–de Waele power law) were applied to the electrospinning process by Spivak and Dzenis [77, 150, 175].

$$\hat{\tau}^c = \mu \left[r \left(\dot{\hat{\gamma}}^2 \right) \right]^{(m-1)/2} \dot{\hat{\gamma}} \tag{83}$$

$$\mu = K\left(\frac{\partial v}{\partial y}\right)^{m-1}$$

(84)

Viscous Newtonian fluids are described by a special case of equation above with the flow index $m=1$. Pseudoplastic (shear thinning) fluids are described by flow indices $0 \leq m \leq 1$. Dilatant (shear thickening) fluids are described by the flow indices $m>1$ [150].

14.8.6.2   GIESEKUS EQUATION

In 1966, Giesekus established the concept of anisotropic forces and motions into polymer kinetic theory. With particular choices for the tensors describing the anisotropy, one can obtained Giesekus constitutive equation from elastic dumbbell kinetic theory [176, 177]. The Giesekus equation is known to predict, both qualitatively and quantitatively, material functions for steady and nonsteady shear and elongational flows. However, the equation sustains two drawbacks: it predicts that the viscosity is inversely proportional to the shear rate in the limit of infinite shear rate and it is unable to predict any decrease in the elongational viscosity with increasing elongation rates in uniaxial elongational flow. The first one is not serious because of the retardation time, which is included in the constitutive equation, but the second one is more critical because the elongational viscosity of some polymers decreases with increasing of elongation rate [178, 179].

In the main Giesekus equation, the tensor of excess stresses depending on the motion of polymer units relative to their surroundings was connected to a sequence of tensors characterizing the configurational state of the different kinds of network structures present in the concentrated solution or melt. The respective set of constitutive equations indicates [180, 181]:

$$S_k + \eta \frac{\partial C_k}{\partial t} = 0$$

(85)

The equation below indicates the upper convective time derivative (Oldroyd derivative):

$$\frac{\partial C_k}{\partial t} = \frac{DC_k}{Dt} - \left[C_k \nabla v + (\nabla v)^T C_k \right]$$

(86)

(Note: The upper convective derivative is the rate of change of any tensor property of a small parcel of fluid that is written in the coordinate system rotating and stretching with the fluid.)

C_k also can be measured as follows:

$$C_k = 1 + 2E_k \qquad (87)$$

According to the concept of "recoverable strain" S_k may be understood as a function of E_k and vice versa. If linear relations corresponding to Hooke's law are adopted.

$$S_k = 2\mu_k E_k \qquad (88)$$

So:

$$S_k = \mu_k (C_k - 1) \qquad (89)$$

The Eq. (85) becomes:

$$S_k + \lambda_k \frac{\partial S_k}{\partial t} = 2\eta D \qquad (90)$$

$$\lambda_k = \frac{\eta}{\mu_k} \qquad (91)$$

As a second step in order to rid the model of the shortcomings is the scalar mobility constants B_k, which are contained in the constants η. This mobility constant can be represented as:

$$\tfrac{1}{2}(\beta_k S_k + S_k \beta_k) + \tilde{\eta} \frac{\partial C_k}{\partial t} = 0 \qquad (92)$$

The two parts of Eq.(92) reduces to the single constitutive equation:

$$\beta_k + \tilde{\eta} \frac{\partial C_k}{\partial t} = 0 \qquad (93)$$

The excess tension tensor in the deformed network structure where the well-known constitutive equation of a so-called Neo-Hookean material is proposed [180, 182]:

Neo-Hookean equation:

$$\left\{ \begin{array}{l} S_k = 2\mu_k E_k = \mu_k (C_k - 1) \qquad (94) \\ \\ \\ \mu_k = NKT \end{array} \right.$$

$$\beta_k = 1 + \alpha(C_k - 1) = (1 - \alpha) + \alpha C_k \qquad (95)$$

Where K is Boltzmann's constant.

By substitution Eqs.(94) and (95) in the Eq.(89), it can obtained where the condition $0 \leq \alpha \leq 1$ must be fulfilled, the limiting case $\alpha=0$ corresponds to an isotropic mobility [183].

$$\begin{cases} 0 \leq \alpha \leq 1 \quad [1 + \alpha(C_k - 1)](C_k - 1) + \lambda_k \dfrac{\partial C_k}{\partial t} = 0 & (96) \\[3mm] \alpha = 1 \quad 0 \leq \alpha \leq 1 & (97) \\[3mm] 0 \leq \alpha \leq 1 \quad C_k = \dfrac{S_k}{\mu_k} + 1 & (98) \end{cases}$$

By substituting equations above in Eq.(89), it becomes:

$$\left[1 + \frac{\alpha S_k}{\mu_k}\right] \frac{S_k}{\mu_k} + \lambda_k \frac{\partial C_k}{\partial t} = 0 \qquad (99)$$

$$\frac{S_k}{\mu_k} + \frac{\alpha S_k^2}{\mu_k^2} + \lambda_k \frac{\partial \left(S_k / \mu_k + 1 \right)}{\partial t} = 0 \qquad (100)$$

$$S_k + \frac{\alpha S_k^2}{\mu_k} + \lambda_k \frac{\partial S_k}{\partial t} = 0 \qquad (101)$$

$$S_k + \frac{\alpha S_k^2}{\mu_k} + \lambda_k \frac{\partial S_k}{\partial t} = 0 \qquad (102)$$

D means the rate of strain tensor of the material continuum [180].

$$D = \frac{1}{2}\left[\nabla v + (\nabla v)^T\right] \qquad (103)$$

The equation of the upper convective time derivative for all fluid properties can be calculated as:

$$\frac{\partial \otimes}{\partial t} = \frac{D \otimes}{Dt} - \left[\otimes . \nabla v + (\nabla v)^T . \otimes \right] \tag{104}$$

$$\frac{D \otimes}{Dt} = \frac{\partial \otimes}{\partial t} + \left[(v.\nabla). \otimes \right] \tag{105}$$

By replacing S_k instead of the symbol:

$$\lambda_k \frac{\partial S_k}{\partial t} = \lambda_k \frac{DS_k}{Dt} - \lambda_k \left[S_k \nabla v + (\nabla v)^T S_k \right] = \lambda_k \frac{DS_k}{Dt} - \lambda_k (v.\nabla) S_k \tag{106}$$

By simplification the equation above:

$$S_k + \frac{\alpha S_k^2}{\mu_k} + \lambda_k \frac{DS_k}{Dt} = \lambda_k (v.\nabla) S_k \tag{107}$$

$$S_k = 2\mu_k E_k \tag{108}$$

The assumption of $E_k = 1$ would lead to the next equation:

$$S_k + \frac{\alpha \lambda_k S_k^2}{\eta} + \lambda_k \frac{DS_k}{Dt} = \frac{\eta}{\mu_k} (2\mu_k) D = 2\eta D = \eta \left[\nabla v + (\nabla v)^T \right] \tag{109}$$

In electrospinning modeling articles τ is used commonly instead of S_k [154, 159, 161].

$$S_k \leftrightarrow \tau$$

$$\tau + \frac{\alpha \lambda_k \tau^2}{\eta} + \lambda_k \tau_{(1)} = \eta \left[\nabla v + (\nabla v)^T \right] \tag{110}$$

14.8.6.3. MAXWELL EQUATION

Maxwell's equations are a set of partial differential equations that, together with the Lorentz force law, form the foundation of classical electrodynamics, classical optics, and electric circuits. These fields are the bases of modern electrical and communications technologies. Maxwell's equations describe how electric and magnetic fields are generated and altered by each other and by charges and currents. They are named after the Scottish physicist and mathematician James Clerk Maxwell who published an early form of those equations between 1861 and 1862 [184, 185]. It will be discussed in the next section in detail.

14.8.7 MICROSCOPIC MODELS

One of the aims of computer simulation is to reproduce experiment to elucidate the invisible microscopic details and further explain the experiments. Physical phenomena occurring in complex materials cannot be encapsulated within a single numerical paradigm. In fact, they should be described within hierarchical, multilevel numerical models in which each submodel is responsible for different spatial-temporal behavior and passes out the averaged parameters of the model, which is next in the hierarchy. The understanding of the nonequilibrium properties of complex fluids such as the viscoelastic behavior of polymeric liquids, the rheological properties of ferrofluids and liquid crystals subjected to magnetic fields, based on the architecture of their molecular constituents is useful to get a comprehensive view of the process. The analysis of simple physical particle models for complex fluids has developed from the molecular computation of basic systems (atoms, rigid molecules) to the simulation of macromolecular 'complex' system with a large number of internal degrees of freedom exposed to external forces [186, 187].

The most widely used simulation methods for molecular systems are Monte Carlo, Brownian dynamics and molecular dynamics. The microscopic approach represents the microstructural features of material by means of a large number of micromechanical elements (beads, platelet, rods) obeying stochastic differential equations. The evolution equations of the microelements arise from a balance of momentum at the elementary level. The Monte Carlo method is a stochastic strategy that relies on probabilities. The Monte Carlo sampling technique generates large numbers of configurations or microstates of equilibrated systems by stepping from one microstate to the next in a particular statistical ensemble. Random changes are made to the positions of the species present, together with their orientations and conformations where appropriate. Brownian dynamics are an efficient approach for simulations of large polymer molecules or colloidal particles in a small molecule solvent. Molecular dynamics is the most detailed molecular simulation method, which computes the motions of individual molecules. Molecular dynamics efficiently evaluates different configurational properties and dynamic quantities, which cannot generally be obtained by Monte Carlo [188, 189].

The first computer simulation of liquids was carried out in 1953. The model was an idealized two-dimensional representation of molecules as rigid disks. For macromolecular systems, the coarse-grained approach is widely used as the modeling process is simplified, hence becomes more efficient, and the characteristic topological features of the molecule can still be maintained. The level of detail for a coarse-grained model varies in different cases. The whole molecule can be represented by a single particle in a simulation and interactions between particles incorporate average properties of the whole molecule. With this approach, the number of degrees of freedom is greatly reduced [190].

On the other hand, a segment of a polymer molecule can also be represented by a particle (bead). The first coarse-grained model, called the 'dumbbell' model (Fig. 14.20), was introduced in the 1930s. Molecules are treated as a pair of beads interacting via a harmonic potential. However, by using this model, it is possible to perform kinetic theory derivations and calculations for nonlinear rheological properties and solve some flow problems. The analytical results for the dumbbell models can also be used to check computer simulation procedures in molecular dynamics and Brownian dynamics [191, 192].

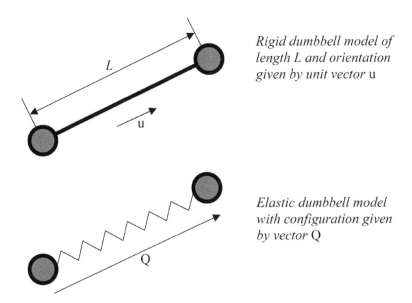

Rigid dumbbell model of length L and orientation given by unit vector u

Elastic dumbbell model with configuration given by vector Q

FIGURE 14.20   The first coarse-grained models – the rigid and elastic dumbbell models.

The bead-rod and bead-spring model (Fig. 14.21) were introduced to model chainlike macromolecules. Beads in the bead-rod model do not represent the atoms of the polymer chain backbone, but some portion of the chain, normally 10 to 20 monomer units. These beads are connected by rigid and massless rods. While in the bead-spring model, a portion of the chain containing several hundreds of backbone atoms are replaced by a "spring" and the masses of the atoms are concentrated on the mass of beads [193].

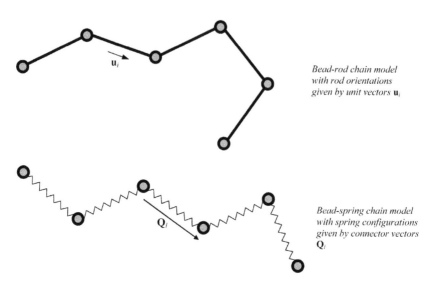

Bead-rod chain model with rod orientations given by unit vectors **u**$_i$

Bead-spring chain model with spring configurations given by connector vectors **Q**$_i$

FIGURE 14.21   The freely jointed bead-rod and bead-spring chain models.

If the springs are taken to be Hookean springs, the bead-spring chain is referred to as a Rouse chain or a Rouse-Zimm chain. This approach has been applied widely as it has a large number of internal degrees of freedom and exhibits orientability and stretchability. However, the disadvantage of this model is that it does not have a constant contour length and can be stretched out to any length. Therefore, in many cases finitely extensible springs with two more parameters, the spring constant and the maximum extensibility of an individual spring, can be included so the contour length of the chain model cannot exceed a certain limit [194, 195].

The understanding of the nonequilibrium properties of complex fluids such as the viscoelastic behavior of polymeric liquids, the rheological properties of ferrofluids and liquid crystals subjected to magnetic fields, based on the architecture of their molecular constituents [186] (Fig. 14.22).

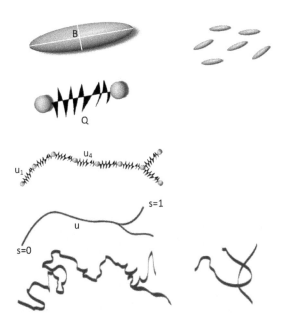

FIGURE 14.22 Simple microscopic models for complex fluids by using dumbbell model.

Dumbbell models are very crude representations of polymer molecules. Too crude to be of much interest to a polymer chemist, since it in no way accounts for the details of the molecular architecture. It certainly does not have enough internal degrees of freedom to describe the very rapid motions that contribute, for example, to the complex viscosity at high frequencies. On the other hand, the elastic dumbbell model is orientable and stretchable, and these two properties are essential for the qualitative description of steady-state rheological properties and those involving slow changes with time. For dumbbell models one can go through the entire program of endeavor—from molecular model for fluid dynamics—for illustrative purposes, in order to point the way towards the task that has ultimately to be performed for more realistic models. According to the researches, dumbbell models must, to some extend then, be regarded as mechanical playthings, somewhat disconnected from the real world of polymers. When used intelligently, however, they can be useful pedagogically and very helpful in developing a qualitative understanding of rheological phenomena [186, 196].

The simplest model of flexible macromolecules in a dilute solution is the elastic dumbbell (or bead-spring) model. This has been widely used for purely mechanical theories of the stress in electrospinning modeling [197] (Fig. 14.23).

A Maxwell constitutive equation was first applied by Reneker et al. in 2000 [66]. Consider an electrified liquid jet in an electric field parallel to its axis. They modeled a segment of the jet by a viscoelastic dumbbell. They used a Gaussian electrostatic system of units. According to this model each particle in the electric field exerts repulsive force on another particle.

He had three main assumptions [66, 198]:

a) the background electric field created by the generator is considered static;
b) the fiber is a perfect insulator;
c) the polymer solution is a viscoelastic medium with constant elastic modulus, viscosity and surface tension.

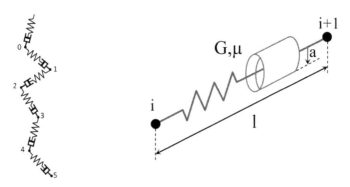

FIGURE 14.23   A schematic of one section of the model.

The researcher considered the governing equations for each bead as [198]:

$$\frac{d}{dt}\left(\pi a^2 l\right) = 0 \tag{111}$$

Therefore, the stress between these particles can be measured by [66]:

$$\frac{d\sigma}{dt} = G\frac{dl}{l\,dt} - \frac{G}{\eta}\sigma \tag{112}$$

The stress can be calculated by a Maxwell viscoelastic constitutive equation [199]:

$$\dot{\tau} = G\left(\varepsilon' - \frac{\tau}{\eta}\right) \tag{113}$$

where ε' is the Lagrangian axial strain Ref. [199]:

$$\varepsilon' \equiv \frac{\partial \dot{x}}{\partial \xi}\hat{\imath}. \tag{114}$$

Equation of motion for beads can be written as [200]:

mass × acceleration = viscous drag + Brownian motion force + force of one bead on another through the connector   (115)

The momentum balance for a bead is [198]:

$$m\frac{dv}{dt} = \underbrace{-\frac{q^2}{l^2}}_{Coulomb\ forces} - \underbrace{qE}_{Electric\ force} + \underbrace{\pi a^2 \sigma}_{Mechanical\ forces}$$ (116)

So the momentum conservation for model charges can be calculated as [201]:

$$m_i\frac{dv_i}{dt} = \underbrace{q_i \sum_{i \neq j} q_j K \frac{r_i - r_j}{\left|r_i - r_j\right|^3}}_{Coulomb\ forces} + \underbrace{q_i E}_{Electric\ force} + \underbrace{\pi a_{i,i+1}^2 \sigma_{i,i+1} \frac{r_{i+1} - r_i}{\left|r_{i+1} - r_i\right|} - \pi a_{i-1,i}^2 \sigma_{i-1,i} \frac{r_i - r_{i-1}}{\left|r_i - r_{i-1}\right|}}_{Mechanical\ forces}$$ (117)

Boundary condition assumptions: A small initial perturbation is added to the position of the first bead, the background electric field is axial and uniform and the first bead is described by a stationary equation. For solving these equations some dimensionless parameters are defined then by simplifying equations, the equations are solved by using boundary conditions [198, 201] (Fig. 14.24).

Now, an example for using this model for the polymer structure is mentioned. For a dumbbell consists of two, which are connected with a nonlinear spring, the spring force law is given by [96]:

$$F = -\frac{HQ}{1 - Q^2/Q_0^2}$$ (118)

Now, if we considered the model for the polymer matrix such as carbon nanotube, the rheological behavior can be obtained as [96, 202]:

$$\tau_{ij} = \tau_p + \tau_s$$ (119)

$$\tau_p = \underbrace{n_a \langle Q_a F_a \rangle}_{aggregated\ dumbbells} + \underbrace{n_f \langle Q_f F_f \rangle}_{free\ dumbbells} - nkT\delta_{ij}$$ (120)

$$\tau_s = \eta \dot{\gamma}$$ (121)

$$\lambda \langle Q.Q \rangle^{\nabla} = \delta_{ij} - \frac{c \langle Q.Q \rangle}{1 - tr \langle Q.Q \rangle / b_{max}}$$ (122)

The polymeric stress can be obtained from the following relation [96]:

$$\frac{\hat{\tau}_{ij}}{n_d kT} = \delta_{ij} - \frac{c\langle Q.Q \rangle}{1 - tr\langle Q.Q \rangle / b_{max}} \tag{123}$$

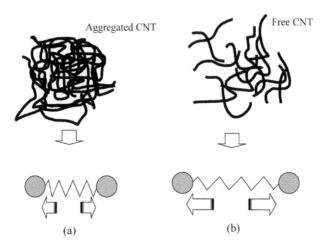

(a)   (b)

FIGURE 14.24 Modeling of two kinds of dumbbell sets, (a) aggregate FENE dumbbell, which has lower mobility and (b) free FENE dumbbell, which has higher mobility.

14.8.8 SCALING

The physical aspect of a phenomenon can use the language of differential equation, which represents the structure of the system by selecting the variables that characterize the state of it, and certain mathematical constraint on the values of those variables can take on. These equations can predict the behavior of the system over a quantity such as time. For an instance, a set of continuous functions of time that describe the way the variables of the system developed over time starting from a given initial state [208]. In general, the renormalization group theory, scaling and fractal geometry, are applied to the understanding of the complex phenomena in physics, economics and medicine [209].

In more recent times, in statistical mechanics, the expression "scaling laws" has referred to the homogeneity of form of the thermodynamic and correlation functions near critical points, and to the resulting relations among the exponents that occur in those functions. From the viewpoint of scaling, electrospinning modeling can be studied in two ways, allometric and dimensionless analysis. Scaling and dimensional analysis actually started with Newton, and allometry exists everywhere in our daily life and scientific activity [209, 210].

14.8.8.1   ALLOMETRIC SCALING

Electrospinning applies electrically generated motion to spin fibers. So, it is difficult to predict the size of the produced fibers, which depends on the applied voltage in principal. Therefore, the relationship between the radius of the jet and the axial distance from the nozzle is always the subject ofinvestigation [211, 212]. It can be described as an allometric equation by using the values of the scaling exponent for the initial steady, instability and terminal stages [213].

The relationship between r and z can be expressed as an allometric equation of the form:

$$r \approx z^b \tag{124}$$

When the power exponent, $b = 1$ the relationship is isometric and when $b \neq 1$ the relationship is allometric [211, 214]. In another view, $b = -1/2$ is considered for the straight jet, $b = -1/4$ for instability jet and $b = 0$ for final stage [172, 212].

Due to high electrical force acting on the jet, it can be illustrated [211]:

$$\frac{d}{dz}\left(\frac{v^2}{2}\right) = \frac{2\sigma E}{\rho r} \tag{125}$$

Equations of mass and charge conservations applied here as mentioned before [211, 214, 215].

From the above equations it can be seen that [161, 211]

$$r \approx z^b,\ \sigma \approx r,\ E \approx r^{-2},\ \frac{dv^2}{dz} \approx r^{-2} \tag{126}$$

So it is obtained for the initial part of jet $r \approx z^{-1/2}$, $r \approx z^{-1/4}$ for the instable stage and $r \approx z^0$ for the final stage.

The charged jet can be considered as a one-dimensional flow as mentioned. If the conservation equations modified, they would change as [211]:

$$2\pi r\sigma^\alpha v + K\pi r^2 E = I \tag{127}$$

$$r \approx z^{-\alpha/(\alpha+1)} \tag{128}$$

where α is a surface charge parameter, the value of α depends on the surface charge in the jet. When $\alpha = 0$ no charge in jet surface, and in $\alpha = 1$ use for full surface charge.

Allometric scaling equations are more widely investigated by different researchers. Some of the most important allometric relationships for electrospinning are presented in Table 14.2.

TABLE 14.2   Investigated Scaling Laws Applied in Electrospinning Model

Parameters	Equation	Ref.
The conductance and polymer concentration	$g \approx c^{\beta}$	[161]
The fiber diameters and the solution viscosity	$d \approx \eta^{\alpha}$	[212]
The mechanical strength and threshold voltage	$\bar{\sigma} \approx E_{threshold}^{-\alpha}$	[216]
The threshold voltage and the solution viscosity	$E_{threshold} \approx \eta^{1/4}$	[216]
The viscosity and the oscillating frequency	$\eta \approx \omega^{-0.4}$	[216]
The volume flow rate and the current	$I \approx Q^{b}$	[215]
The current and the fiber radius	$I \approx r^{2}$	[217]
The surface charge density and the fiber radius	$\sigma \approx r^{3}$	[217]
The induction surface current and the fiber radius	$\phi \approx r^{2}$	[217]
The fiber radius and AC frequency	$r \approx \Omega^{1/4}$	[172]

β, α and b= scaling exponent.

14.8.8.2   DIMENSIONLESS ANALYSIS

One of the simplest, yet most powerful, tools in the physics is dimensional analysis in which there are two kinds of quantities: dimensionless and dimensional.

In physics and all science, dimensional analysis is the analysis of the relationships between different physical quantities by identifying their dimensions. The dimension of any physical quantity is the combination of the basic physical dimensions that compose it, although the definitions of basic physical dimensions may vary. Some fundamental physical dimensions, based on the SI system of units, are length, mass, time, and electric charge. (The SI unit of electric charge is, however, defined in terms of units of length, mass and time, and, for example, the time unit and the length unit are not independent but can be linked by the speed of light c.)

Other physical quantities can be expressed in terms of these fundamental physical dimensions. Dimensional analysis is based on the fact that a physical law must be independent of the units used to measure the physical variables. A straightforward practical consequence is that any meaningful equation (and any inequality and in-equation) must have the same dimensions on the left and right sides. Dimensional analysis is routinely used as a check on the plausibility of derived equations and computations. It is also used to categorize types of physical quantities and units based on their relationship to or dependence on other units.

Dimensionless quantities, which are without associated physical dimensions,are widely used in mathematics, physics, engineering, economics, and in everyday life (such as in counting). Numerous well-known quantities, such as π, e, and φ, are dimensionless. They are "pure" numbers, and as such always have a dimension of 1 [218, 219].

Dimensionless quantities are often defined as products or ratios of quantities that are not dimensionless, but whose dimensions cancel out when their powers are multiplied [220].

The basic principle of dimensional analysis was known to Isaac Newton (1686) who referred to it as the "Great Principle of Similitude." James Clerk Maxwell played a major role in establishing modern use of dimensional analysis by distin-guishing mass, length, and time as fundamental units, while referring to other units as derived. The 19th-century French mathematician Joseph Fourier made important contributions based on the idea that physical laws likeF = ma should be independent of the units employed to measure the physical variables. This led to the conclu-sion that meaningful laws must be homogeneous equations in their various units of measurement, a result that was eventually formalized in the Buckingham π theo-rem. This theorem describes how every physically meaningful equation involving n variables can be equivalently rewritten as an equation of n − m dimensionless parameters, where m is the rank of the dimensional matrix. Furthermore, and most importantly, it provides a method for computing these dimensionless parameters from the given variables.

A dimensional equation can have the dimensions reduced or eliminated through nondimensionalization, which begins with dimensional analysis, and involves scal-ing quantities by characteristic units of a system or natural units of nature. This gives insight into the fundamental properties of the system, as illustrated in the examples below.

In nondimensional scaling, there are two key steps:

(a) identify a set of physically relevant dimensionless groups; and
(b) determine the scaling exponent for each one.

Dimensional analysis will help you with step (a), but it cannot be applicable possibly for step (b).

A good approach to systematically getting to grips with such problems is through the tools of dimensional analysis [221]. The dominant balance of forces controlling the dynamics of any process depends on the relative magnitudes of each underlying physical effect entering the set of governing equations [221]. Now, the most general characteristics parameters, which used in dimensionless analysis in electrospinning are introduced in Table 14.3.

TABLE 14.3   Characteristics Parameters Employed and Their Definitions

Parameter	Definition
Length	R_0
Velocity	$v_0 = \dfrac{Q}{\pi R_0^2 K}$
Electric field	$E_0 = \dfrac{I}{\pi R_0^2 K}$
Surface charge density	$\sigma_0 = \bar{\varepsilon} E_0$
Viscose stress	$\tau_0 = \dfrac{\eta_0 v_0}{R_0}$

For achievement of a simplified form of equations and reduction a number of unknown variables, the parameters should be subdivided into characteristic scales in order to become dimensionless. Electrospinning dimensionless groups are shown in Table 14.4 [222].

TABLE 14.4   Dimensionless Groups Employed and Their Definitions

Name	Definition	Field of application
Froude number	$Fr = \dfrac{v_0^2}{g R_0}$	The ratio of inertial to gravitational forces
Reynolds number	$Re = \dfrac{\rho v_0 R_0}{\eta_0}$	The ratio of the inertia forces of the viscous forces
Weber number	$We = \dfrac{\rho v_0^2 R_0}{\gamma}$	The ratio of the surface tension forces to the inertia forces

Deborah number	$De = \dfrac{\lambda v_0}{R_0}$	The ratio of the fluid relaxation time to the instability growth time
Electric Peclet number	$Pe = \dfrac{2\bar{\varepsilon}v_0}{KR_0}$	The ratio of the characteristic time for flow to that for electrical conduction
Euler number	$Eu = \dfrac{\varepsilon_0 E^2}{\rho v_0^2}$	The ratio of electrostatic forces to inertia forces
Capillary number	$Ca = \dfrac{\eta v_0}{\gamma}$	The ratio of inertia forces of viscous forces
Ohnesorge number	$oh = \dfrac{\eta}{(\rho\gamma R_0)^{1/2}}$	The ratio of viscous force to surface force
Viscosity ratio	$r_\eta = \dfrac{\eta_p}{\eta_0}$	The ratio of the polymer viscosity to total viscosity
Aspect ratio	$\chi = \dfrac{L}{R_0}$	The ratio of the length of the primary radius of jet
Electrostatic force parameter	$\varepsilon = \dfrac{\bar{\varepsilon}E_0^2}{\rho v_0^2}$	The relative importance of the electrostatic and hydrodynamic forces
Dielectric constant ratio	$\beta = \dfrac{\varepsilon}{\bar{\varepsilon}} - 1$	The ratio of the field without the dielectric to the net field with the dielectric

The governing and constitutive equations can be transformed into a dimensionless form using the dimensionless parameters and groups.

14.8.9 SOME OF ELECTROSPINNING MODELS

The most important mathematical models for electrospinning process are classified in the Table 14.5. According to the year, advantages and disadvantages of the models:

TABLE 14.5   The Most Important Mathematical Models for Electrospinning

Researchers	Model	Year	Ref.
Taylor, G. I. Melcher, J. R.	Leakydielectricmodel ✓ Dielectric fluid ✓ Bulk charge in the fluid jet considered to be zero ✓ Only axial motion ✓ Steady state part of jet	1969	[223]
Ramos	Slender body ✓ Incompressible and axi-symmetric and viscous jet under gravity force ✓ No electrical force ✓ Jet radius decreases near zero ✓ Velocity and pressure of jet only change during axial direction ✓ Mass and volume control equations and Taylor expansion were applied to predict jet radius	1996	[224]
Saville, D. A.	Electrohydrodynamic model ✓ The hydrodynamic equations of dielectric model was modified ✓ Using dielectric assumption ✓ This model can predict drop formation ✓ Considering jet as a cylinder (ignoring the diameter reduction) ✓ Only for steady state part of the jet	1997	[225]
Spivak, A. Dzenis, Y.	Spivak and Dzenis model ✓ The motion of a viscose fluid jet with lower conductivity were surveyed in an external electric field ✓ Single Newtonian Fluid jet ✓ The electricfieldassumed to be uniformand constant, unaffected by the charges carried by the jet ✓ Use asymptotic approximation was applied in a long distance from the nozzle ✓ Tangential electric force assumed to be zero ✓ Using nonlinear rheological constitutive equation(Ostwald de Waele law), nonlinear behavior of fluid jet were investigated	1998	[150]

Jong Wook	Droplet formation model ✓ Droplet formation of charged fluid jet was studied in this model ✓ The ratio of mass, energy and electric charge transition are the most important parameters on droplet formation ✓ Deformation and break-up of droplets were investigated too ✓ Newtonian and Non-Newtonian fluids ✓ Only for high conductivity and viscous fluids	2000	[226]
Reneker, D. H. Yarin, A. L.	Reneker model ✓ For description of instabilities in viscoelastic jets ✓ Using molecular chain theory, behavior of polymer chain of spring-bead model in electric field was studied ✓ Electric force based on electric field cause instability of fluid jet while repulsion force between surface charges make perturbation and bending instability ✓ The motion paths of these two cases were studied ✓ Governing equations: momentum balance, motion equations for each bead, Maxwell tension and columbic equations	2000	[227]
Hohman, M. Shin, M.	Stability theory ✓ This model is based on a dielectric model with some modification for Newtonian fluids. ✓ This model can describe whipping, bending and Rayleigh instabilitiesandintroducednewballooninginstability. ✓ Four motion regions were introduced: dipping mode, spindle mode, oscillating mode, precession mode. ✓ Surfacecharge density introduced as the most effective parameter on instability formation. ✓ Effect of fluid conductivity and viscosity on nanofibers diameter were discussed. ✓ Steady solutionsmay be obtained only if the surface charge density at thenozzle is set to zero or a very low value	2001	[60]

Feng, J. J	✓ ModifyingHohmanmodel ✓ ForbothNewtonianandnon-Newtonianflu-ids ✓ UnlikeHohmanmodel, theinitialsur-facechargedensitywasnotzero, sothe "ballooninginstability" didnotaccrue. ✓ Onlyforsteadystatepartofthejet ✓ Simplifyingtheelectricfieldequation, which HohmanusedinordertoeliminateBallooning instability.	2002	[153]
Wan-Guo-Pan	Wan-Guo-Pan model ✓ They introduced thermo-electro-hydro dynamics model in electrospinning process ✓ This model is a modification on Spivak model which mentioned before ✓ The governing equations in this model: Modified Maxwell equation, Navier-Stocks equation, and several rheological constitutive equation	2004	[175]
Ji-Haun	AC-electrospinning model ✓ Whipping instability in this model was dis-tinguished as the most effective parameter on uncontrollable deposition of nanofibers ✓ Applying AC current can reduce this insta-bility so make oriented nanofibers ✓ This model found a relationship between axial distance from nozzle and jet diameter ✓ This model also connected AC frequency and jet diameter	2005	[172]
Roozemond (Eindhoven University and Technology)	Combination of slender body and dielectric model ✓ In this model, a new model for viscoelastic jets in electrospinning were presented by combining these two models ✓ All variables were assumed uniform in cross section of the jet but they changed in during z direction ✓ Nanofiber diameter can be predicted	2007	[228]
Wan	Electromagnetic model ✓ Results indicated that the electromagnetic field which made because of electrical field in charged polymeric jet is the most important reason of helix motion of jet during the process	2012	[229]

Dasri	Dasri model ✓ This model was presented for description of unstable behavior of fluid jet during electrospinning ✓ This instability causes random deposition of nanofiber on surface of the collector ✓ This model described dynamic behavior of fluid by combining assumption of Reneker and Spivak models	2012	[230]

The most frequent numeric mathematical methods, which were used in different models, are listed in Table 14.6.

TABLE 14.6   Applied Numerical Methods for Electrospinning

Method	Ref.
Relaxation method	[153, 159, 231]
Boundary integral method (boundary element method)	[199, 226]
Semi-inverse method	[159, 172]
(Integral) control-volume formulation	[224]
Finite element method	[223]
Kutta-Merson method	[232]
Lattice Boltzmann method with finite difference method	[233]

14.9   ELECTROSPINNING SIMULATION

Electrospun polymer nanofibers demonstrate outstanding mechanical and thermo-dynamic properties as compared to macroscopic-scale structures. These features are attributed to nanofiber microstructure [234, 235]. Theoretical modeling predicts the nanostructure formations during electrospinning. This prediction could be verified by various experimental condition and analysis methods, which called simulation. Numerical simulations can be compared with experimental observations as the last evidence [149, 236].

Parametric analysis and accounting complex geometries in simulation of electrospinning are extremely difficult due to the nonlinearity nature in the problem. Therefore, a lot of researches have done to develop an existing electrospinning simulation of viscoelastic liquids [231].

14.10   ELECTROSPINNING SIMULATION EXAMPLE

In order to survey of electrospinning modeling application, the main equations were applied for simulating the process according to the constants, which summarized in Table 14.7.

Mass and charge conservations allow v and σ to be expressed in terms of R and E, and the momentum and E-field equations can be recast into two second-order ordinary differential equations for R and E. The slender-body theory (the straight part of the jet) was assumed to investigate the jet behavior during the spinning distance. The slope of the jet surface (R') is maximum at the origin of the nozzle. The same assumption has been used in most previous models concerning jets or drops. The initial and boundary conditions, which govern the process, are introduced as:

Initial values ($z=0$):

$$R(0) = 1$$
$$E(0) = E_0$$
$$\tau_{prr} = 2r_\eta \frac{R_0'}{R_0^3}$$
$$\tau_{pzz} = -2\tau_{prr}$$

(129)

Feng [153] indicated that E(0) effect is limited to a tiny layer below the nozzle, which its thickness is a few percent of R_0. It was assumed that the shear inside the nozzle is effective in stretching of polymer molecules as compared with the following elongation.

Boundary values ($z=\chi$):

$$R(\chi) + 4\chi R'(\chi) = 0$$
$$E(\chi) = E_\chi$$
$$\tau_{prr} = 2r_\eta \frac{R_\chi'}{R_\chi^3}$$
$$\tau_{pzz} = -2\tau_{prr}$$

(130)

The asymptotic scaling can be stated as [153]:

$$R(z) \propto z^{-1/4}$$

(131)

Just above the deposit point ($z=\chi$), asymptotic thinning conditions applied. R drops towards zero and E approaches E_∞. The electric field is not equal to E_∞, so we assumed a slightly larger value, E_χ.

TABLE 14.7 Constants Which Were Used in Electrospinning Simulation

Constant	Quantity
Re	2.5×10^{-3}
We	0.1×10^{-3}
Fr	0.1×10^{-3}
Pe	0.1×10^{-3}
De	10×10^{-3}
E	1×10^{-3}
β	40×10^{-3}
χ	20×10^{-3}
E_0	0.7×10^{-3}
E_χ	0.5×10^{-3}
r_η	0.9×10^{-3}

The momentum, electric field and stress equations could be rewritten into a set of four coupled first order ordinary differential equations (ODE's) with the above-mentioned boundary conditions. The numerical relaxation method is chosen to solve the generated boundary value problem.

The results of these systems of equations are presented in Figs.14.4 and 14.5,which matched quite well with the other studies that have been published [60, 78, 153, 154, 159].

The variation prediction of R, R', ER², ER²' and E versus axial position (z) are shown in Fig.14.4. Physically, the amount of conductible charges reduces with decreasing jet radius. Therefore, to maintain the same jet current, more surface charges should be carried by the convection. Moreover, in the considered simulation region, the density of surface charge gradually increased. As the jet gets thinner and faster, electric conduction gradually transfers to convection. The electric field is mainly induced by the axial gradient of surface charge, thus it is insensitive to the thinning of the electrospun jet:

$$\frac{d(\sigma R)}{dz} \approx -\left(2R \frac{dR}{dz} \right) / Pe \tag{132}$$

Therefore, the variation of E versus z can be written as:

$$\frac{d(E)}{dz} \approx \ln \chi \left(\frac{d^2 R^2}{dz^2} \right) / Pe \qquad (133)$$

Downstream of the origin, E shoots up to a peak and then relaxes due to the decrease of electrostatic pulling force in consequence of the reduction of surface charge density, if the current was held at a constant value. However, in reality, the increase of the strength of the electric field also increases the jet current, which is relatively linear [78, 153]. As the jet becomes thinner downstream, the increase of jet speed reduces the surface charge density and thus E, so the electric force exerted on the jet and thus R' become smaller. The rates of R and R' are maximum at $z=0$, and then relaxes smoothly downstream toward zero [153, 159]. According to the relation between R, E and z, ER^2 and $ER^{2'}$ vary in accord with parts (c) and (d) in Fig. 14.25.

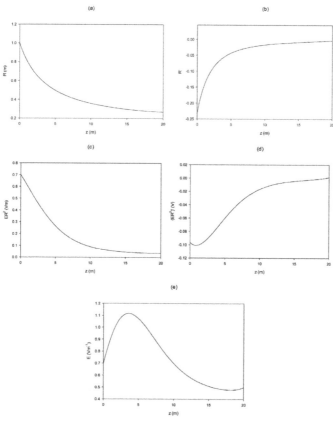

FIGURE 14.25   Solutions given by the electrospinning model for (a) R; (b) R'; (c) ER^2; (d) $ER^{2'}$ and (e) E.

Figure 14.26 shows the changes of axial, radial shear stress and the difference between them, the tensile force (T) versus z. The polymer tensile force is much larger in viscoelastic polymers because of the strain hardening. T also has an initial rise, because the effect of strain hardening is so strong that it overcomes the shrinking radius of the jet. After the maximum value of T, it reduces during the jet thinning. As expected, the axial polymer stress rises, because the fiber is stretched in axial direction, and the radial polymer stress declines. The variation of T along the jet can be nonmonotonic, however, meaning the viscous normal stress may promote or resist stretching in a different part of the jet and under different conditions [153, 159].

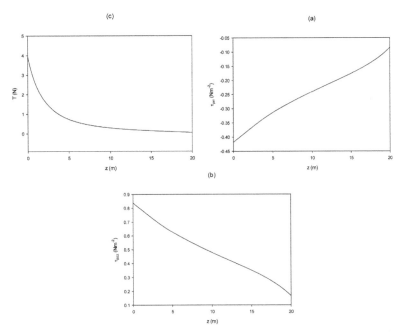

FIGURE 14.26   Solutions given by the electrospinning model for (a)τ_{prr}; (b)τ_{pzz} and (c) T.

14.11   APPLIED NUMERICAL METHODS FOR ELECTROSPINNING

Mathematics is indeed the language of science and the more proficient one is in the language the better. There are three main steps in the computational modeling of any physical process: (i) problem definition;(ii) mathematical model; and(iii) computer simulation [237–239].

(i)   The first natural step is to define an idealization of our problem of interest in terms of a set of affiliate quantities,which it would be wanted to measure. In defining this idealization we expect to obtain a well-posed problem, this is one that has a unique solution for a given set of parameters. It might not always be possible to insure the integrity of the realization since, in some instances; the physical process is not entirely conceived.

(ii) The second step of the modeling process is to represent the idealization of the physical reality by a mathematical model: the governing equations of the problem. These are available for many physical phenomena. For example, in fluid dynamics the Navier–Stokes equations are considered to be an accurate representation of the fluid motion. Analogously, the equations of elasticity in structural mechanics govern the deformation of a solid object due to applied external forces. These are complex general equations that are very difficult to solve both analytically and computationally. Therefore, it needs to introduce simplifying assumptions to reduce the complexity of the mathematical model and make it amenable to either exact or numerical solution.

(iii) After the selection of an appropriate mathematical model, together with suitable boundary and initial conditions, it could be proceed to its solution. In this part, it will be considered the numerical solution of mathematical problems,which are described by partial differential equations (PDEs).

The sequential steps of analyzing physical systems can be summarized as presents in Fig. 14.27.

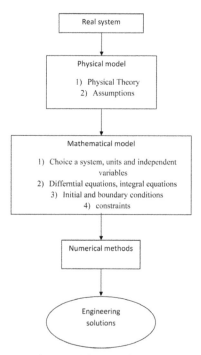

FIGURE 14.27   The diagram of sequential steps of analyzing physical systems.

Numerical analysis is the study of algorithms, which uses numerical approximations for the problems of mathematical analysis. The overall goal of the field of numerical analysis is the design and analysis of techniques to give approximate but accurate solutions to hard problems. Numerical analysis naturally finds applications in all fields of engineering and the physical science. Numerical analysis is also concerned with computing (in an approximate way) the solution of differential equations, both ordinary differential equations and partial differential equations. Ordinary differential equations appear in celestial mechanics (planets, stars and galaxies), numerical linear algebra is important for data analysis, stochastic differential equations and Markov chains are essential in simulating living cells for medicine and biology [240, 241]. PDEs are used to describe most of the existing physical phenomena. The most general and practical way to produce solutions to the PDEs is to use numerical methods and computers. It is essential that the solutions produced by the numerical scheme are accurate and reliable. PDEs are solved by first discretizing the equation, bringing it into a finite-dimensional subspace. This can be done by several methods such as a finite element method, a finite difference method, or (particularly in engineering) a finite volume method. The theoretical justification of these methods often involves theorems from functional analysis. This reduces the problem to the solution of an algebraic equation [242, 243]. In Table 14.8, the advantages of simulation illustrate by comparison between doing experiments and simulations.

TABLE 14.8   The Advantages of Simulation Versus Doing Experiments

Experiments	Simulations
Expensive	Cheap (er)
Slow	Fast (er)
Sequential	Parallel
Single-purpose	Multiple-purpose

Now, some numeric methods, which applied by researchers in the investigation on electrospinning models and simulations are over looked.

14.11.1   KUTTA-MERSON METHOD

One hundred years ago, Runge was completing his famous research paper. This work, published in 1895, extended the approximation method of Euler to a more elaborate scheme, which was able to have a greater accuracy by using Taylor series expansion. The idea of Euler was to bring up the solution of an initial value problem forward by a sequence of small time-steps. In each step, the rate of change of the solution is treated as constant and is found from the formula for the derivative evaluated at the beginning of the step [244, 245].

The system of ODEs arising from the application of a spatial discretization of a system of PDEs can be very large, especially in three-dimensional simulations. Consequently, the constraints on the methods used for integrating these systems are somewhat different from those,which have taken much of the advance of numerical methods for initial value problems. On their high accuracy and modest memory requirements, the Runge–Kutta methods have become popular for simulations of physical phenomena. The classical fourth-order Runge–Kutta method requires three memory locations per dependent variable but low-storage methods requiring only two memory locations per dependent variable can be derived. This feature is easily achieved by a third-order Runge–Kutta method but an additional stage is required for a fourth-order method. Since the primary cost of the integration is in the evaluation of the derivative function, and each stage requires a function evaluation, the additional stage represents a significant increase in expense. For the same reason, error checking is generally not performed when solving very large systems of ODEs arising from the discretization of PDEs [245–247].

In 2004 and in 2006, Reznik et al. [248, 249] experimentally and theoretically studied on the shape evolution of small compound droplets in the normal form and at the exit of a core-shell system in the presence of a sufficiently strong electric field,respectively.

In the first study they considered an axisymmetric droplet of an incompressible conducting viscous liquid on an infinite conducting plate. It was neglected in the gravity effects so the stationary shape of the droplet is spherical. The droplet shaped as a spherical segment rests on the plate with a static contact angle of the liquid/air/solid system (Fig. 14.28).

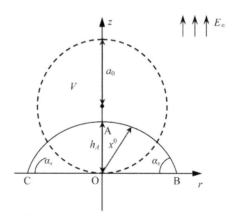

FIGURE 14.28   The initial shape of the droplet at the moment when the electric field is to be applied.

They estimated the relevant characteristic dimensionless parameters of the problem and used the Stokes equations for the liquid motion within the droplet then the equations solved numerically using the Kutta–Merson method [249].

In the second work, a core-shell nozzle (Fig. 14.29) consists of a central cylindrical pipe and a concentric annular pipe surrounding it. When the process took place without an applied electric field, the outer surface of the droplet and the interface between its components acquire near-spherical equilibrium shapes owing to the action of the surface and interfacial tension, respectively. If after the establishment of the equilibrium shape an electric field is applied to the compound nozzle and droplet attached to an electrode immersed in it, with a counter electrode, say a metal plate, located at some distance from the droplet tip, the latter undergoes stretching under the action of Maxwell stresses of electrical origin. At first, the problem was defined then the droplet evolution determined by time stepping using the Kutta-Merson method.

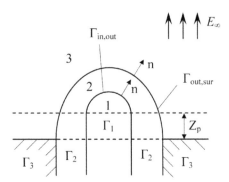

FIGURE 14.29   Core-shell droplet at the exit of a core-shell nozzle.

Both the core and the shell fluids are considered as leaky dielectrics, whose electric relativepermittivities and conductivities are denoted. The electric boundary conditions describe the jump in the normal component of the electric field and electric induction at the boundaries. The governing equationsrendered dimensionless and subject to the boundary conditionswere solved with the aid of equivalent boundary integral formulations in which a set of corresponding integral equations was solved [248].

In 2010, Holzmeister et al. [232] considered a simple two-dimensional model can be used for describing the formation of barb electrospun polymer nanowires with a perturbed swollen cross-section and the electric charges "frozen" into the jet surface. This model is integrated numerically using the Kutta-Merson method with the adoptable time step. The result of this modeling is explained theoretically as a result of relatively slow charge relaxation compared to the development of the secondary electrically driven instabilities,which deform jet surface locally. When the disparity of the slow charge relaxation compared to the rate of growth of the secondary electrically driven instabilities becomes even more pronounced, the barbs transform in full-scale long branches. The competition between charge relaxation and rate of growth of capillary and electrically driven secondary localized perturbations of the jet surface is affected not only by the electric conductivity of polymer solutions but also by their viscoelasticity. Moreover, a nonlinear theoretical model was able to resemble the main morphological trends recorded in the experiments.

14.11.2 FINITE ELEMENT METHOD

The finite element method is a numerical analysis technique for obtaining approximate solutions to a vast variety of engineering problems. All finite element methods involve dividing the physical systems, such as structures, solid or fluid continua, into smallsubregions or elements. Each element is an essentially simple unit, the behaviorofwhich can be readily analyzed. The complexities of the overall systems are accommodated by using large numbersofelements, rather than by resorting to the sophisticated mathematics requiredby many analytical solutions [250, 251].

A typical finite element analysis on a software system requires the following information [252]:
1.   nodal point spatial locations (geometry);
2.   elements connecting the nodal points;
3.   mass properties;
4.   boundary conditions or restraints;
5.   loading or forcing function details;
6.   analysis options.
And the FEM Solution Process [253]:
1.   Divide structure into pieces (elements with nodes)(discretization/meshing).
2.   Connect (assemble) the elements at the nodes to form an approximate system of equations for the whole structure (forming element matrices).
3.   Solve the system of equations involving unknown quantities at the nodes (e.g., displacements).
4.   Calculate desired quantities (e.g., strains and stresses) at selected elements.
One of the main attractions of finite element method is the ease with which they can be applied to problems involving geometrically complicated systems. The price that must be paid for flexibility and simplicity of individual elements is in the amount of numerical computation required. Very large sets of simultaneous algebraic equationshavetobesolved, andthis canonly be done economically with the aid of digital computers [254]. Therefore, the advantage of this method is that for a smooth problem where the derivatives of the solution are well behaved, the computational cost increases algebraically while the error decreases exponentially fast

and the disadvantage of it is that the method leads to nonsingular systems of equations that can easily solve by standard methods of solution. This is not the case for time-dependent problems where numerical errors may grow unbounded for some discretization [238].

The finite element method is used for spatial discretization. Special numerical methods are used and developed for viscoelastic fluid flow, including the DEVSS and log-conformation techniques. For describing the moving sharp interface between the rigid particle and the fluid both ALE (Arbitrary Lagrangian Euler) and XFEM (extended finite element) techniques are being developed and employed [255]. In the development of numerical algorithms for the stable and accurate solution of viscoelastic flow problems, like electrospinning process, applying the finite element method to solve constitutive equations of the differential type is useful [256].

In 2011, Chitral et al. analyzed the electrospinning process based on an existing electrospinning model for viscoelastic liquids using a finite element method. Four steady-state equations concluding Coulomb force, an electric force imposed by the external electric field, a viscoelastic force, a surface tension force, a gravitational force, and an air drag force were solved as a set of equations with this method [257].

14.11.3 BOUNDARY INTEGRAL METHOD (BOUNDARY ELEMENT METHOD)

Boundary integral equations are a classic tool for the analysis of boundary value problems for partial differential equations. The term "boundary element method"(BEM) denotes any method for the approximate numerical solution of these boundary integral equations (Fig. 14.30). The approximate solution of the boundary value problem obtained by BEM has the distinguishing feature that it is an exact solution of the differential equation in the domain and is parameterized by a finite set of parameters living on the boundary [258].

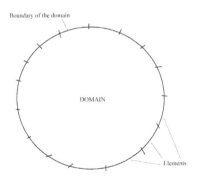

FIGURE 14.30   The idea of boundary element method.

The BEM has some advantages over other numerical methods like finite element methods(FEM) or finite differences [258–260]:

1. Only the boundary of the domain needs to be discretized. Especially in two dimensions where the boundary is just a curve this allows very simple data input and storage methods.
2. Exterior problems with unbounded domains but bounded boundaries are handled as easily as interior problems.
3. In some applications, the physically relevant data are given not by the solution in the interior of the domain but rather by the boundary values of the solution or its derivatives. These data can be obtained directly from the solution of boundary integral equations, whereas boundary values obtained from FEM solutions are in general not very accurate.
4. The solution in the interior of the domain is approximated with a rather high convergence rate and moreover, the same rate of convergence hold for all derivatives of any order of the solution in the domain. There are difficulties, however, if the solution has to be evaluated close to, but not on the boundary.

Some main difficulties with BEM are the following [258, 260–261]:

1. Boundary integral equations require the explicit knowledge of a fundamental solution of the differential equation. This is available only for linear partial differential equations with constant or some specifically variable coefficients. Problems with inhomogeneities or nonlinear differential equations are in general not accessible by pure BEM. Sometimes, however, a coupling of FEM and BEM proves to be useful.
2. For a given boundary value problem there exist different boundary integral equations and for each of them several numerical approximation methods. Thus, every BEM application requires that several choices be made. To evaluate the different possibilities, one needs a lot of mathematical analysis. Although the analysis of BEM has been a field of active research in the past decade, it is by no means complete. Thus there exist no error estimates for several methods that are widely used. From a mathematical point of view, these methods, which include very popular ones for which computer codes are available, are in an experimental state, and there might exist problems of reliability.
3. The reason for the difficulty of the mathematical analysis is that boundary integral equations frequently are not ordinary Fredholm integral equations of the second kind. The classical theory of integral equations and their numerical solution concentrates on the second kind integral equations with regular kernel, however. Boundary integral equations may be of the first kind, and the kernels are in general singular. If the singularities are not integrable, one has to regularize the integrals,which are then defined in a distributional sense. The theoretical framework for such integral equations is the

theory of pseudodifferential operators. This theory was developed 20 years ago and is now a classic part of Mathematical Analysis, but it is still not very popular within Applied Mathematics.

4. If the boundary is not smooth but has corners and edges, then the solution of the boundary value problem has singularities at the boundary. This happens also if the boundary conditions are discontinuous, e.g. in mixed boundary value problems. BEM clearly has to treat these singularities more directly than FEM. Because the precise shape of the singularities frequently contains important information, for example, stress intensity factors in fracture mechanics, this is a positive aspect of BEM. But besides practical problems with the numerical treatment of these singularities, nonsmooth domains also present theoretical difficulties. These have so far been satisfactorily resolved only for two-dimensional problems. The analysis of BEM for three-dimensional domains with corners and edges is still in a rather incomplete stage.

The research paper of Kowalewski,whichis published in 2009, can be effectively addressed by use of a boundary element method (BEM). In essence, the boundary element method is a statement of the electrostatic problem (Poisson equation) in terms of boundary integrals, as such, it involves only the discretization of boundary surfaces, which in our case would be the electrode surfaces and the outer shell of the fiber. The model was, essentially, a time-dependent three-dimensional generalization of known slender models with the following differences:

(i)   the electric field induced by the generator and by the charges on the fiber is explicitly resolved, instead of being approximated from local parameters;

(ii) electrical conductivity is neglected.

Indeed, the convection of surface charges is believed to strongly overcomes bulk conduction at locations distant from the Taylor cone by a few fiber radii, since we are mostly interested in the description of the bending instability, this assumption appears reasonable [262].

14.11.4 (INTEGRAL) CONTROL-VOLUME FORMULATION

All the laws of mechanics are written for a system, which is defined as an arbitrary quantity of mass of fixed identity. Everything external to this system is denoted by the term surroundings, and the system is separated from its surroundings by its boundaries. The laws of mechanics then state what happens when there is an interaction between the system and its surroundings [263].

Typically, to understand how a given physical law applies to the system under consideration, one first begins by considering how it applies to a small, control volume, or "representative volume." There is nothing special about a particular control volume, it simply represents a small part of the system to which physical laws can be easily applied. This gives rise to what is termed a volumetric, or volume-wise formulation of the mathematical model [264–265].

In fluid mechanics and thermodynamics, a control volume is a mathematical abstraction employed in the process of creating mathematical models of physical processes. In an inertial frame of reference, it is a volume fixed in space or moving with constant velocity through which the fluid (gas or liquid) flows. The surface enclosing the control volume is referred to as the control surface [266, 267].

Control-volume analysis is "more equal," being the single most valuable tool to the engineer for flow analysis. It gives "engineering" answers, sometimes gross and crude but always useful [263].

The advantage of this method, over that of the finite-element method, is the following. In the finite-element method, it is necessary to construct a grid over the flaw as well as the entire region surrounding the flaw, and to solve for the fields at all points on the grid. In contrast, in the volume-integral method, it is only necessary to construct a grid over the flaw and solve for the currents in the flaw; the Green's function takes care of all regions outside the flaw. This removes the complicated gridding requirements of the finite-element method, and reduces the size of the problem tremendously. That is the reason which this method can obtain much more accurate probe responses, and in much less time than a finite-element code, while running on a small personal computer or workstation. And without the complicated gridding, problems can be set up much more quickly and easily [268].

The famous researcher in electrospinning process, Feng[269] considered the steady stretching process is important in that it not only contributes to the thinning directly for Newtonian flows. The jet is governed by four steady-state equations representing the conservation of mass and electric charges, the linear momentum balance, and Coulomb's law for the electric field,which used control-volume balance for analyzing them.

14.11.5   RELAXATION METHOD

Relaxation method, an alternative to the Newton iteration method, is a method of solving simultaneous equations by guessing a solution and then reducing the errors that result by successive approximations until all the errors are less than some specified amount. Relaxation methods were developed for solving nonlinear systems and large sparse linear systems, which arose as finite-difference discretizations of differential equations [270, 271]. These iterative methods of relaxation should not be confused with "relaxations" in mathematical optimization, which approximate a difficult problem by a simpler problem, whose "relaxed" solution provides information about the solution of the original problem [272]. In solving PDEs problem with this method, it's necessary to turn them to the ODEs equations. Then ODEs have to be replaced by approximate finite difference equations. The relaxation method determines the solution by starting with a guess and improving it, iteratively. During

the iterations the solution is improved and the result relaxes towards the true solution. Notice that the number of mesh points may be important for the properties of
the numerical procedure. In relax-setting, increase the number of mesh points if the
convergence towards steady state seems awkward [273, 274].

The advantage of this method is a relative freedom in its implementation. It
can be used for smooth problem. This is useful in particular when an analytical
solution to the model is not available and can handle models,whichexhibit saddle-
point stability. Therefore, a "large stepping" in the direction of the defect is possible, while the termination point is defined by the condition that the vector field
becomes orthogonal. The relaxation algorithm can easily cope with a large number
of problems,which arise frequently in the context of multidimensional, infinite-time
horizon optimal control problems [275, 276].

A researcher in a thesis named "A Model for Electrospinning Viscoelastic Fluids" studied the electrospinning process of rewriting the momentum, electric field
and stress equations as a set of six coupled first order ordinary differential equations.
It made a boundary value problem, which can best be solved by a numerical relaxation method. Within relaxation methods the differential equations are replaced by
finite difference equations on a certain mesh of points covering the range of integration. During iteration (relaxation) all the values on the mesh are adjusted to bring
them into closer agreement with the finite difference equations and the boundary
conditions [277].

14.11.6   LATTICE BOLTZMANN METHOD WITH FINITE
DIFFERENCE METHOD

Lattice Boltzmann methods (LBM)(or Thermal Lattice Boltzmann methods
(TLBM)) are a class of computational fluid dynamics (CFD) methods for fluid
simulation. Instead of solving the Navier–Stokes equations, the discrete Boltzmann
equation is solved to simulate the flow of a Newtonian fluid with collision models such as Bhatnagar-Gross-Krook (BGK). LBM is a relatively new simulation
technique for complex fluid systems and has attracted interest from researchers
in computational physics. Unlike the traditional CFD methods, which solve the
conservation equations of macroscopic properties (i.e., mass, momentum, and energy) numerically, LBM models the fluid consisting of fictive particles, and such
particles perform consecutive propagation and collision processes over a discrete
lattice mesh. Due to its particulate nature and local dynamics, LBM has several
advantages over other conventional CFD methods, especially in dealing with complex boundaries, incorporating of microscopic interactions, and parallelization of
the algorithm. A different interpretation of the lattice Boltzmann equation is that
of a discrete-velocity Boltzmann equation. By simulating streaming and collision
processes across a limited number of particles, the intrinsic particle interactions
evince a microcosm of viscous flow behavior applicable across the greater mass.

The LBM is especially useful for modeling complicated boundary conditions and multi phase interfaces [278–280]. The idea that LBE is a discrete scheme of the continuous Boltzmann equation also provides a way to improve the computational efficiency and accuracy of LBM. From this idea, the discretization of the phase space and the configuration space can be done independently. Once the phase space is discretized, any standard numerical technique can serve the purpose of solving the discrete velocity Boltzmann equation. It is not surprising that the finite difference, finite volume, andfinite element methods have been introduced into LBM in order to increase computational efficiency and accuracy by using nonuniform grids [281]. The lattice Boltzmann method has a more detailed microscopic description thana classical finite difference scheme because the LBM approach includes a minimal set of molecular velocities of the particles. In addition, important physical quantities, such as the stress tensor, or particle current, is directly obtained from the local information. However, the LBM scheme may require more memory than the corresponding finite difference scheme. Another motivation is that the boundary conditions are more or less naturally imposed for a given numerical scheme[282]. In the ordinary LBM, this discrete BGK equation is discretized into a special finite difference form in which the convection term does not include the numerical error. But considering the collision term, this scheme is of second-order accuracy. On the other hand, the discrete BGK equation can be discretized in some other finite-difference schemes or in the finite-volume method and other computational techniques for the partial differential equations, and these techniques are considered as a natural extension of numerical calculations [283–285].

There is a method which couples lattice Boltzmann method for the fluid with a molecular dynamics model for polymer chain. The lattice Boltzmann equation using single relaxation time approximations (Fig. 14.31).

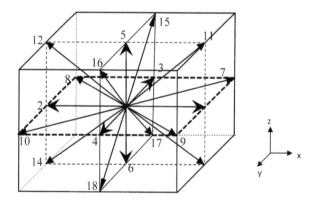

FIGURE 14.31   Lattice scheme.

In a thesis named "Modeling Electrospinning Process and a Numerical Scheme Using Lattice Boltzmann Method to Simulate Viscoelastic Fluid Flows," the researcher wrote the conservation of mass and momentum based on the lattice scheme whichare presented above. It was assumed that the fluid is incompressible and the continuum/macroscopic governing equations were written then forward step finite difference method in time applied for solving obtained equations [286].

14.11.7   SEMI-INVERSE METHOD

From the mathematical point of view, for the problems which the analytical solution may be very hard to attain, even in the simplest boundary value problem, because the set of equation forms a nontrivial system of nonlinear, partial differential equations generating often non unique solutions. To solve the resulting boundary-value problems, inverse techniques can be used to provide simple solutions and to suggest experimental programs for the determination of response functions. Two powerful methods for inverse investigations are the so-called inverse method and the semi-inverse method. They have been used in elasticity theory as well as in all fields of the mechanics of continua [287–289].

In the inverse method, a solution is found in priori such that it satisfies the governing equation and boundary condition. We can obtain the solution through this way for a luck case, but it is not a logical way. For the semi-inverse method, certain assumptions about the components of displacement strain are made at the beginning. Then, the solution is confined by satisfying the equations of equilibrium and the boundary conditions [289].

In the semi-inverse method is systematically studied and many examples are given to show how to establish a variational formulation for a nonlinear equation. From the given examples, we found that it is difficult to find a variational principle for nonlinear evolution equations with nonlinear terms of any orders [290].

In the framework of the theory of Continuum Mechanics, exact solutions play a fundamental role for several reasons. They allow investigating in a direct way the physics of various constitutive models to understand in depth the qualitative characteristics of the differential equations under investigation and they provide benchmark solutions of complex problems.

The Mathematical method used to determine these solutions is usually called the semi-inverse method. This is essentially a heuristic method that consists in formulating a priori a special ansatz on the geometric and/or kinematical fields of interest, and then introducing this ansatz into the field equations. Luck permitting, these field equations reduce to a simple set of equations and then some special boundary value problems may be solved. Another important aspect in the use of the semi-inverse method is associated with fluid dynamics with the emergence of secondary flows and in solid mechanics with latent deformations. It is clear that "Navier-Stokes fluid" and an "isotropic incompressible hyperelastic material" are intellectual constructions [287, 291].

The most interesting features of this method are its extreme simplicity and concise forms of variational functionals for a wide range of nonlinear problems [292, 293]. An advantage of the semi-inverse method is that it can provide a powerful mathematical tool in the search for variational formulations for a rather wide class of physic problems without using the well-known Lagrange multipliers, which can result in variational crisis (the constrains can not eliminated after the identification of the multiplier or the multipliers become zero), furthermore, to use the Lagrange multipliers, we must have a known variational principle at hand, a situation which not always occurs in continuum physics [294, 295]. Although, using this method has not any special difficulties, there are primary difficulties with ill-posed problems [296, 297].

In 2004, Wan et al.studied in a model of steady state jet,which introduced by Spivak but they considered the couple effects of thermal, electricity, and hydrodynamics. Therefore, the model consists of modified Maxwell's equations governing electrical field in a moving fluid, the modified Navier-Stokes equations governing heat and fluid flow under the influence of electric field, and constitutive equations describing behavior of the fluid. The set of conservation laws can constitute a closed system when it is supplemented by appropriate constitutive equations for the field variables such as polarization. The most general theory of constitutive equations determining the polarization, electric conduction current, heat flux, and Cauchy stress tensor has been developed by Eringen and Maugin. By the semi-inverse method, it could be obtained various variational principles for electrospinning. After obtaining a set of equations, they were solved by the semi-inverse method [298].

14.12   CONCLUDING REMARKS OF ELECTROSPINNING MODELING

Comprehensive investigation on the nanoscience and the related technologies is an important topic these days. Due to the rising interest in nanoscale material properties, studying on the effective methods of producing nanomaterial such as the electrospinning process plays an important role in the new technologies progress. Since the electrospinning process is dependent on a lot of different parameters, changing them will lead to significant variations in the process. In this paper, we have attempted to analyze the each part of process in detail and investigate some of the most important relationships from ongoing research into the fundamental physics that govern the electrospinning process. The idea has been to provide a few guiding principles for those who would use electrospinning to fabricate materials with the initial scrutiny. We have outlined the flow procedure involved in the electric field and those associated with the jetting process. The relevant processes are the steady thinning jet, whose behavior can be understood quantitatively using continuum equations of electrohydrodynamics, and the ensuing fluid dynamical instabilities that give rise to whipping of the jet. The current carried by the jet is of critical importance to the process and some scaling aspects concerning the total measured current have been discussed. The basics of electrospinning modeling involve mass conservation, electric charge conservation, momentum balance, coulomb's law and constitutive equations (Ostwald-de Waele power law, Giesekus equation and Maxwell equation),which are discussed in detail. These relations play an important role in setting final features. The dominant balance of forces controlling the dynamics of the electrospinning process depends on the relative magnitudes of each underlying physical effect entering the set of governing equations. Due to the application of the obtained nano web and material for different usages, surveying on microscopic and macroscopic properties of the product is necessary. Therefore, for achieving to this purpose, the relationships are pursued. Also for providing better understanding of electrospinning simulation, a model has been implemented for a straight jet in a relaxation method and simulated for the special viscoelastic polymers. The model capability to predict the behavior of the process parameters was demonstrated using simulation in the last part. The plots obviously showed the changes of each parameter versus axial position. During the jet thinning, the electric field shoots up to a peak and then relaxes. The tensile force also has an initial rise and then reduces. All the plots show similar behavior with the results of other researchers. At the end of this study, various numerical methods which applied in the different electrospinning simulations were revised.

14.13 NUMERICAL STUDY OF COAXIAL ELECTROSPINNING

Electrospinning providesa simple and highly versatilemethod for the large-scale fabrication ofnanofibers. The technique has attracted significant attention in the last decade, since it allows for a straightforward, relatively easy and cheap method of manufacturing polymer nanofibers.Recently, advances in thetechnique of electrospinning have allowedthis method to be used to directly fabricatecore shell, coresheath and hollow nanofibers ofcomposites, ceramics, and polymers withdiameters ranging from 20 nm–1 mm. By replacing the single capillary with acoaxial spinneret in electrospinning set up, it is possible to generate core-sheathand hollow nanofibers made of polymers, ceramicsor composites [299–302]. These nanofibers, as a kind of onedimensionalnanostructure, have attracted specialattention in several research groups, because thesestructures could further enhance material propertyprofiles and these unique core–sheath structuresoffer potential in a number of applications includingnanoelectronics, microfluidics, photonics,and energy storage [300, 303, 304]. This process leads to encapsulation of core material (not necessarily polymeric),which is beneficial for storage and drug delivery of bioactive agents. The shell material which is polymeric substance can be carbonized or calcinated.

Coaxial electrospinning is of particular interest for materials that will not easily form fibers via electrospinning on their own [305].

Similarly to electrospinning, co-electrospinning employs electric forces acting on polymer solutions in dc electric fields and resulting in significant stretching of polymer jets due to a direct pulling and growth of the electrically driven bending perturbations [306]. In this method, a plastic syringe with two compartments containing different polymer solutions or a polymer solution (shell) and a nonpolymeric Newtonian liquid or even a powder (core) is used to initiate a core-shell jet. At the exit of the core-shell needle attached to the syringe appears a core-shell droplet, which acquires a shape similar to the Taylor cone due to the pulling action of the electric Maxwell stresses acting on liquid. Liquid in the cone, being subjected to sufficiently strong (supercritical) electric field, issues a compound jet, which undergoes the electrically driven bending instability characteristic of the ordinary electrospinning process. Strong jet stretching resulting from the bending instability is accompanied by enormous jet thinning and fast solvent evaporation. As a result, the core-shell jet solidifies and core-shell fibers are depositing on a counter electrode(Fig. 14.32)[299, 306–308].

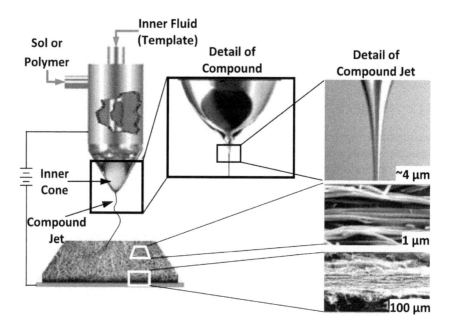

FIGURE 14.32   Experimental setup for co-electrospinning to produce core-shell nanofibers [309].

14.13.1   LITERATURE REVIEW OF COAXIAL ELECTROSPINNING

Coaxial electrospinning or co-electrospinning of core-shell micro and nanofibers was born about 10 years ago as a branch of nanotechnology,which bifurcated from a previously known electrospinning. Through electrospinning, co-electrospinning inherited roots in polymer science and electrohydrodynamics, while some additional genes from textile science and optical fiber technology were spliced in addition. It also engulfed emulsion electrospinning. Co-electrospinning rapidly became widely popular and its applications proliferated into such fields as biotechnology, drug delivery and nanofluidics. It also triggered significant theoretical and experimental efforts directed at a better understanding and control of the process. The literature dealing with co-electrospinning already consists of hundreds of papers, while its simulation and modeling have not been investigated sufficiently.

Greiner's group [301]hasdemonstratedthat core–sheath nanofibers could be fabricatedby co-electrospinning two different polymersolutions through a spinneret comprising two coaxialcapillaries. They manufactured nano and micro fibers in two stages process,which started with ordinary electrospinning of core polymer (stage one), and followed by the coating deposition of the shell polymer (stage two)[310]. Li and Xia preparedTiO$_2$/polymer composites and ceramic hollownanofibers by electrospinning two immiscible liquidsthrough a coaxial, two-capillary spinneret, followed byselective removal of the cores [302].Using a similar setup,Loscertales and co-workers have demonstrated thatpolymer-free, inorganic nanotubes could also be fabricatedby co-electrospinning an aged inorganic sol andan immiscible (or poorly miscible) liquid such as oliveoil or glycerin, followed by selective removal of theinner liquid [309].

Recently, both coaxial electrospray and coaxial electrospinning of two immiscible liquids became interesting research fields besides single-liquid electrospray and electrospinning. An electrified coaxial liquid jet can break into compound droplets in which one substance is coated with the other, or can make core-shell fibers on the submicrometer scale under different conditions [311–314]. In the case of electrospraying, the jets should be rapidly atomized into tiny core-shell droplets, with no viscoelasticity or jet bending involved, whereas in the case of co-electrospinning, polymer jet should stay intact to make nanofibers, and viscoelasticity and jet bending are the dominant phenomena. Loscertales et al. [309]initially proposed that a coaxial spinneret could be used to produce a core sheath jet. However, in their experimental studies, instability led to the breakup of their jet and the formation of core-sheath nanoparticles. Mccann et al. [299] solved the instability problem by co-spinning two immiscible solutions, followed by gelation (or cross-linking) and stabilization of the sheath.

14.13.2   COAXIAL ELECTROSPINNING MODELING

As discussedabove, when two immiscible liquids are injected at appropriate flow rates through two electrified capillary needles, one of them inside the other, the menisci of both liquids adopt conical shapes with an outer meniscus surrounding the inner one. A liquid thread is issued from each one of the vertex of the two menisci in such a way that a compound, coaxial jet of two co-flowing liquids is formed downstream [315, 316].

Several theoretical works dealt with different aspects of co-electrospinning with the goal of better understanding its underlying physics and enhancing its efficiency. Using core-shell nozzles does not necessarily imply formation of core-shell jets. The core material may be not entrained into the shell jet, which results in monolithic instead of core-shell jets. The detailed numerical simulations of flows developing in a core-shell droplet at the exit of a core-shell nozzle under the action of the pulling electric Maxwell stresses showed that the electric charges very rapidly escape to the outer surface of the forming jet [317]. As a result, core entrainment is possible only due to viscous tractions. The core entrainment was predicted to be facilitated by a core nozzle protruding from a coaxial shell nozzle, which was demonstrated experimentally [317, 318].

A complete knowledge of the governed equations of the electrified coaxial jets generated via EHD as afunction of the flow rates and the physical properties of the two liquids would be highly desirable; mainly, the current transported by them, the diameters of the inner and outer jets, and the size of the droplets resulting from their break up. However, the task of finding the above dependences is undoubtedly extremely complex, much more complex for coaxial electrospinning than for ordinary one since the number of parameters and unknowns is larger in the former problem than in the later one [315].

Because of its complexity, even the knowledge of simple cone jet electrospinning process is still incomplete, in spite of the fact that the equations and boundary conditions that govern its electro-hydrodynamic behavior are well known. The existence of very disparate scales in the cone–jet problem, a free interface whose position must be consistently calculated from the solution of the problem, and the time-dependent break up of the jet are different aspects of the problem which greatly contribute to its numerical complexity.

Competing models on the emitted current and jet diameter in single electrospinning, based on different simplifying hypothesis, are found in the literature. Researchers predict different expressions for the emitted current except for a nondimensional constant,which either depends on the dielectric constant of the liquid [148, 319] or it is independent of it[313]. Even more unsatisfactory is the prediction of the jet diameter for which the existing models give different dependences on the flow rate and on the physical properties of the liquid. Unfortunately, the existing experimental measurements do not suffice to either completely support or reject some of these models since, for example, the jet diameter predicted by the different models are comparable to the accuracy with which it can be experimentally sized [320]. Therefore, more precise knowledge of the electrospinning must come from either numerical simulations or more refined experimental measurements. A numerical simulation of the transition zone between the cone and the jet,which can throw light on this complex problem has been carried out by [321].

Lopez-Herrera et al. [315] applied an experimental investigation on the electrified coaxial jets of two immiscible liquids issuing from a structured Taylor cone. The effect of the flow rates of both liquids on the current transported by these coaxial jets and on the size of the compound droplets and final coaxial jet diameter was investigated. Their suggested scaling laws fitted well the scaling law reported by Refs. [319] and [148], but also exhibit a good agreement with the other competing scaling law jet size derived analytically by [311].A draw back of this model is that the small differences between the jet diameters predicted by the two models are comparable to the accuracy with which can be experimentally sized. Therefore, their measurements did not provide a definitive proof for ruling out the wrong model [315].

Artana [322] carried out a temporal linear instability analysis of a cylindrical electrified jet flowing inside a cylindrical coaxial electrode. Hohman [60, 61] performed a complete instability analysis of a charged jet using an asymptotic expansion of radial direction to obtain a mechanism of electrospinning. Feng [153] introduced non-Newtonian rheology into a theoretical model for electrospinning. Loscertales [316] first presented an electrospray method to generate the compound droplets by using two immiscible coaxial liquid jets. The electrified coaxial jet can be divided into outer-driving and inner-driving states. The outer-driving state means that outer fluid has higher conductivity and free charges are located on the outer interface, and the innerdriving state means inner fluid has higher conductivity and free charges are located on the inner interface.

López-Herrera [315] carried out an experiment on coaxial jets generated from an electrified Taylor cone and studied the scaling law between electric current and drop size under the influence of flow rate and other liquid properties. Chen [312] found experimentally a series of different modes in the coaxial jet electrospray with outer driving liquid. Sun et al.[301] Li and Xia, [302] and Yu et al.[323] investigated the mechanism of coaxial electrospinning and successfully produced composite or hollow nanofibers of different materials. However, little has been done so far in the theoretical analysis of the instability of the electrified coaxial jet. Li et al.[324] performed temporal linear instability analysis of an electrified coaxial jet inside a coaxial electrodeand the analytic dispersion equation is derived in their model.

Boubaker presenteda model to core-shell structured polymer nanofibers deposited via coaxial electrospinning. His investigations were based on a modified Jacobi-Gauss collocation spectral method, proposed along with the Boubaker Polynomials Expansion Scheme (BPES), for providing solution to a nonlinear Lane-Emden-type equation [325].

14.13.3   ANALYTICAL AND NUMERICAL MODELS

Coaxial electrospray and electrospinning are important branches of electrohydro-dynamicsthat concern the fluid dynamics with the electric force effects [223, 315, 316, 326, 327]. Tocreate and maintain a steady cone–jet mode in coaxial processis a complex process that requires appropriate equilibrium of differentforces. Since co-axial electrospinning has many potential advantagesover othermicroencapsulation/nanoencapsulation processes, it is ofgreat interest to study the multiphysical mecha-nism and derive theanalytical and numerical models for this process.some of the most importan t equations govern the process are mentioned in following sections.

14.13.3.1   FORMULATION

To better understand the governing mechanism of coaxial electrospray, it is impor-tant to establish a theoretical framework in advance. The complete set of govern-ing equations can be given by the fluid dynamic equations (i.e., the Navier-Stokes equations) and the electrical equations (i.e., the Maxwell's equations). For Newto-nian fluids of uniform constitution, the governing equations for each phase can be expressed as:

$$\frac{1}{\rho}\frac{d\rho}{dt} + \nabla u = 0 \qquad (134)$$

$$\rho\frac{du}{dt} = -\nabla p + \mu\nabla^2 u + \rho g + f \qquad (135)$$

$$\nabla D = q, \nabla B = 0, \nabla \times E = -\frac{\partial B}{\partial t}, \nabla \times H = J + \frac{\partial D}{\partial t} \qquad (136)$$

where the notation d/dt = 9/9t + u∇ is the material derivative and the quantities ρ, u, p, μ, g, f$_c$, D, q, B, E, H and J stand for the density, velocity vector, pressure, dynamic viscosity, gravitational acceleration, electric force, electric displacement vector, free charge density, magnetic induction, electric intensity field, magnetic intensity field and conduction current density, respectively. The Eq. (134)represents the conserva-tion of mass; Eq. (135) expresses the momentum equation and Eq.(136) show the well- known Maxwell's equations.

The above equations can be simplified with reasonable assumptions. In most studies, the liquids and the gas can be considered as incompressible fluids. Thus Eq.(134) is simplified as ∇u=0. For simplicity, the linearly constitutive relations D=μE, B=ÇH and J=KE are introduced, with μ, Ç and K standing for permittiv-ity, permeability and electrical conductivity, respectively. Since the magnetic field is very weak, the Maxwell's equations can be simplified as the following: $\nabla D = q$, $\nabla E = 0$, $dp/dt + \nabla j = 0$. The detailed derivation of the governing equations can be found in Refs. [328, 329].

Artana and his co-workers considered a liquid jet flowing vertically downwards out from an injector and into a gas at room pressure. Depending on the velocity of the liquid, one obtains different kinds of jets. The one of interest here corresponds to the second wind regime or to the regime of atomization that is to say for very high velocity (about 100 m/s). For these regimes the aspect of the jet looks like a pulverization shaped as a cone composed of sparse droplets in most of its volume, except in the region of its revolution axis where the density of the droplets is very high. This jet flows through one coaxial cylindrical electrode brought to a certain potential that we will consider positive. The potential of the electrode V_0 is maintained constant and the injector is earthed (see Fig. 14.33). The analysis of stability is done with an infinite jet inside an infinite electrode.

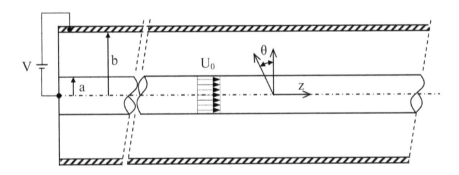

FIGURE 14.33   Schema of analyzed problem.

They assumed the two fluids to be incompressible and the motion to be irrotational. They neglect the effects of gravity, magnetic fields, viscosity and mass transfer at the interface. Liquid and gas are considered as isothermal and incompressible and their electrical properties are those of an ohmic conductor for the liquid phase and of a perfect dielectric for the gaseous phase, both having uniform conductivity and dielectric constant. The electric charge on the jet was at the jet surface and there was no free charge source in the bulk of the liquid or of the gaseous phase. The velocity profile in a typical stream wise station was invariant with the axial coordinate. To solve the problem they considered three regions: the regions of the liquid, gas and the interface. They used Gibbs's model [330] for the interface and we formulate the conservation equations as Slattery[331].

14.13.3.2 FLUID MECHANICS EQUATIONS

The mass conservation leads to each phase[322]:

$$\nabla \cdot U_i = 0 \tag{137}$$

With Ui the velocity of the phase i=1 for the liquid and i=2 for the gas. At the interface:

$$\nabla_\zeta \cdot U_\zeta = 0, \tag{138}$$

where U_ζ is the velocity vector of the interface and ∇_ζ the surface divergence.

As the motion being irrotational, Artana et al.[322] considered a potential function for the velocity φ_i and they obtained the Laplace equation $\Delta\varphi_i$=0.

As there is *no mass transfer* between phases

$$\left(U_i - U_\zeta\right) \cdot n = 0, \tag{139}$$

or

$$\left(\nabla_1 - U_2\right)n = 0, \tag{140}$$

n being normal to the interface. As for *momentum conservation*, it leads to in each phase:

$$\rho_i \frac{dU_i}{dt} - \nabla^\circ T_i = 0 \tag{141}$$

ρi and **T**i are the mass density and Maxwell's constraint tensor of the phase i and Eq. (13.9) is the material derivative.

$$\frac{d}{dt} = \frac{\partial}{\partial_t} + U_i \cdot \nabla \tag{142}$$

At the interface:

$$-\nabla_\zeta{}^\circ\left(T_\zeta\right) - \left(T_1 - T_2\right)^\circ n = 0, \tag{143}$$

where$\mathbf{T}_f$ is the constraint tensor in this region. The expression of the Maxwell constraint tensor is:

$$T_i = -p_i I + T_i^{el} \tag{144}$$

where*pi* is the static pressure and **I** the identity tensor. The components of $\mathbf{T}^{el}{}_i$, using Einstein's notation, are:

$$\left(T_i^{el}\right)_{jk} = \varepsilon_i \left(E_i\right)_k \left(E_i\right)_j - \left(\frac{\varepsilon_i}{2}\right)\delta_{kj}\left(E_i\right)_m\left(E_i\right)_m, \qquad (145)$$

with εi and $(Ei)j$ the permittivity and the electric field in the j-th direction, of the phase i. The expression of the surface constraint tensor is given by:

$$T_\zeta = \gamma_{i-j}\left(I - n\otimes n\right) \qquad (146)$$

According to [322]:

$$\nabla_\zeta{}^{\circ}T_\xi = \gamma_{i-j}\langle v_n\rangle n + grad\zeta\lambda_{i-j}, \qquad (147)$$

$<v_n>$ being the mean curvature and γ_{i-j} the surface tension. As in our problem γ_{i-j} is constant, grad ζ the surface gradient of this magnitude vanishes.

14.13.3.3   ELECTRICAL EQUATIONS

The Maxwell's equations can be simplified with some assumptions. In the bulk, it can be written that [322]:

$$\nabla \cdot D_i = 0 \qquad (148)$$

$$\nabla \times E_i = 0, \qquad (149)$$

where Ei and Di are the electric field strength and the dielectric displacement. By using an electrical potential function V, it can be expressed:

$$E_i = -\nabla V_i \qquad (150)$$

That leads to Laplace equation $\Delta V=0$.

At the interface, the continuity of the tangential electric field, Gauss law at the surface and conservation of charge considering the rate of change of a surface element can be expressed as [332]:

$$n\times\|E_i\| = 0, \qquad (151)$$

$$\|\varepsilon_i E_i\| \cdot n = q_\zeta, \qquad (152)$$

$$\frac{dq_\zeta}{dt} + q_\zeta\left(U_\zeta \cdot n\right)\nabla_n + \nabla_\zeta J_\zeta + n\|J_i\| - \left(U_\zeta \cdot n\right)\|q_i\| = 0 \tag{153}$$

with q_ζ the surface charge, J_ζ the surface current density, J_i the volume current density, q_i the volume charge density and we notice a jump in a magnitude from a phase to another with the symbol $\|\ \|$.

14.13.3.4   THE ELECTRIC FIELD OF THE NONPERTURBED JET

In the nonperturbed case negative electrical charges are uniformly placed on the surface of the jet, which is consequently isopotential. In the liquid media this leads immediately to

$$V_1\left(r,\theta,z\right) = 0 \ and \ E_1\left(r,\theta,z\right) = -\nabla V_1 = 0 \tag{154}$$

In the gas, the resolution of Poisson equation

$$\Delta V_2 = 0 \ with \ V_2\left(a,\theta,z\right) = 0 \ and \ V_2\left(b,\theta,z\right) = V_0, \tag{155}$$

which gives,

$$V_2\left(r,\theta,z\right) = V_0 \frac{\ln\left(r/a\right)}{\ln\left(b/a\right)} \tag{156}$$

The electric field is then

$$E_2\left(r,\theta,z\right) = -\frac{V_0}{\ln\left(b/a\right)}\frac{i_r}{r} \tag{157}$$

Artana et al.[322] studied the linear instability analysis of an electrified coaxial jet. In fact, the mentioned nonperturbed state is totally ideal as it never occurs because of the high instability character of the flow and the many possible sources of perturbation like, for instance, the roughness of the orifice of the injector which induces modifications on the ideal velocity and potential.

They simulated the real phenomenon, by artificially perturbing the considered solution. Hence, they perturbed the interface between the liquid and the gas and determined the modifications induced on the velocity, the electrical potential and the pressure. The perturbation, which was arbitrary, was decomposed into a Fourier series and was considered as a set of small elementary waves propagating on the surface of the jet. As they considered a linear analysis, they accepted that there is no interaction between the different modes (or waves) and so the analysis can be undertaken for each individual mode. The amplitude of some of them diminished and for others it grew until it became large enough to give rise to droplets. Therefore, the study of the creation of the droplets came down to the one of the stability of the jet submitted to a perturbation, whatever its origin Refs. [322, 333]. Artana et al.[322] supposed that the flow was subjected to a modification at its interface. The coordinates of each point at the interface:

$$OM_\zeta = (a, \theta, z, t) \tag{158}$$

Here, Eq. (158) became Eq. (159) after modification.

$$OM_\zeta = (r_s(\theta, z, t), \theta, z) \tag{159}$$

where:

$$r_s = a + \eta, and\, \eta \tag{160}$$

where r_s is the perturbation that depends on the space variables and on time t. Because they analyzed the initial stages of growth or decay, they studied directly a single mode and η was written as:

$$\eta = \eta_0 \exp\left[\left(\omega t + i(kz + n\theta)\right)\right] \tag{161}$$

At the interface, the normal n to the interface at anypoint is

$$n = \left(1, -\left(\frac{in}{a}\right)\eta, -ik\eta\right) \tag{162}$$

and the mean curvature $\langle v_n \rangle$ is:

$$\langle v_n \rangle = \nabla \cdot n = \frac{1}{r_s} - \frac{1}{r_s^2}\frac{\partial^2 \eta}{\partial \theta^2} - \frac{\partial^2 \eta}{\partial z^2} \tag{163}$$

The perturbation of the interface led to solutions for the velocity and for the electric field, which were different from the ones obtained in the nonperturbed case. The governing equations of the fluid mechanics problem still were $\Delta \varphi_i = 0$ and $\Delta v_i = 0$. After that, they solved them with a moving interface (boundaries changing with time). Since they confined their investigations to the linear stability analysis, they neglected all the terms proportional to η^2 or to higher exponents.

14.13.3.5   THE PERTURBED STATE

14.13.3.6   VELOCITY FIELD

Using Slattery's formulation [331]and considering that $U_2(b, \theta, z)=0$ and that the velocity is finite $(A)M$, they arrived to:

$$U_1 = C_1 \nabla \left(I_n (kr) \eta \right) \quad \text{and} \quad U_2 = U_0 + C_2 \nabla \left(\left(\frac{I_n (kr)}{C_3} + \frac{K_n (kr)}{C_4} \right) \eta \right) \tag{164}$$

with

$$C_1 = \frac{\omega}{k I_n' (ka)}, \quad C_2 = \frac{\omega - i U_0 k}{k}, \quad C_1 = I_n' (ka)(1 - \lambda) \quad \text{and}$$

$$C_4 = K_n' (ka)\left(1 - \frac{1}{\lambda}\right) \tag{165}$$

with

$$\lambda = \frac{I_n' (kb) K_n' (ka)}{K_n' (kb) I_n' (ka)} \cdot E_1 = 0 \tag{166}$$

where I_n and K_n are the modified Bessels functions of first and second kind, I_n' and K_n' being their derivatives with respect to the variable r.

14.13.3.7   ELECTRIC FIELD

The perturbed solution depends on the electrical state of the jet, generally. Artana and his co-workers perturbed a cylindrical column of liquid flowing through a cylindrical coaxial electrode and that, in the nonperturbed state, the electrical charges were in electrical equilibrium on the surface of the liquid. The electrical state of the jet after perturbation depends on the way the electrical charges can move when the surface is perturbed. They considered two cases [322]:

In the first one the electrical relaxation time of charges was small enough compared to a certain characteristic time of the deformation (in their case, the period of the perturbation). "Relaxation is quicker than deformation." The jet remains isopotential at any time.

In the second case, which was the opposite extreme of the first one, the charges were in a way "linked" to the fluid particles and follow the dilatation and stretching of the surface of the jet which consequently does not remain isopotential. This was the case of jet with charges on its surface supporting a perturbation with a large growth rate (very high-velocity jet). They considered these two cases separately, it being understood that the reality of the electrical state of the surface was somewhere between them.

14.13.3.7.1 ISOPOTENTIAL CASE

If the surface of the liquid remains in electrical equilibrium despite the motion induced by the perturbation, the boundary conditions needed for the determination of the electric field in the gas are $V_2(b)=V_0$ at the electrode and $V_2(a+\eta)=0$ at the interface.

Gauss law and the continuity of the tangential component of the electric field at the interface immediately gave the solution in the liquid. They obtained

$$E_1 = 0 \text{ and } E_2 = C_5 \nabla \left[\ln\left({r}/{a} \right) - \left(\frac{I_n(kr)}{C_6} + \frac{K_n(kr)}{C_7} \right) \eta \right] \qquad (167)$$

With

$$C_5 = \frac{V_0}{\ln\left({b}/{a} \right)}, C_6 = aI_n(ka)(1-\beta) \text{ and } C_7 = aK_n(ka)\left(1 - {1}/{\beta} \right), \qquad (168)$$

Being

$$\beta = \frac{K_n(ka)I_n(kb)}{I_n(ka)K_n(kb)} \qquad (169)$$

14.13.3.7.2 NONISOPOTENTIAL CASE

As it was mentioned above, in this case it is supposed that the electrical charges, uniformly distributed on the surface of the jet in the nonperturbed state, move by following the fluid particles of the surface. Now the distribution of the electrical charges on the surface of the jet is not uniform any longer. From a physical point of view, the extremely rapid motion of the jet surface inhibits the rearrangement of the charges at the surface motivated by the electric forces. This assumption is equivalent to considering a problem in which the mobilities of the charges in the bulk and the interface are null.

So in the equation of charge conservation, any contribution of the current density in the rate of change of the surface charge density can be disregarded. The contribution of the surface current density is reduced to the convection term $\nabla_\varsigma(q_\varsigma U_\varsigma)$. However, as the linear terms are kept only, this convection term has no contribution either. The linearized equation can finally be written as

$$\frac{\partial q_\zeta}{\partial t} + q_\zeta \left(U_\zeta \cdot n\right) \nabla_n = 0 \tag{170}$$

The integration in time of this equation then gives the following solution

$$q_\zeta(\theta, z, t) = q_0 \frac{a}{r_s}, \tag{171}$$

where q_0 is the surface density of the electrical charges of the nonperturbed state ($t=0$). The boundary conditions for the solution of Laplace equation that determines the value of the electric field are deduced from the following considerations:

- The potential at any point of the liquid (especially for $r=0$) is finite.
- In the gas we have $V_2(b) = V_0$.
- At the interface we use the traditional results deduced from Maxwell's equations:continuity of tangential component of the electric field and discontinuity of the normal component of the displacement vector.

Considering the electrical potential functions

$$V_i = A_i \ln\left(\frac{r}{a}\right) + R_i(r) \exp\left(i(kz + n\theta) + \omega t\right) \text{ with } R_i(r) = B_{i1} I_n(kr) + B_{i2} K_n(kr) \tag{172}$$

and with the cited boundary conditions, we have

$$E_1 = C_s \nabla\left(\frac{I_n(kr)}{c_8} \eta\right) \text{ and } E_2 = -C_s \nabla\left(\ln\left(\frac{r}{a}\right) - \frac{K_n(kr)}{c_9} \eta\right) \tag{173}$$

with

$$C_8 = a I_n(ka)\left(1 - \frac{\varepsilon_1}{\varepsilon_2} \frac{I'_n(ka) K_n(ka)}{I_n(ka) K'_n(ka)}\right), \tag{174}$$

$$C_9 = a K_n(ka)\left(\frac{\varepsilon_2}{\varepsilon_1} \frac{K'_n(ka) I_n(ka)}{K_n(ka) I'_n(ka)} - 1\right), \tag{175}$$

where ε_1 and ε_2 being the permittivities of the two media. In the expressions of the electric field in the gaseous phase Artana et al. assume that the product kb is large enough to consider the quotient $K_n(kb)/I_n(kb) \approx 0$, condition usually verified.

14.13.3.8. PRESSURE FIELD

Knowing the velocity and electric fields with the equation of conservation of linear momentum the pressure in both media can be determined. The term dU_i/dt can be written as

$$\frac{dU_i}{dt} = \frac{\partial U_i}{\partial t} + (U_i \circ \nabla) U_i, \tag{176}$$

where

$$(U_i \circ \nabla) U_i = (\nabla \times U_i) \times U_i + \frac{1}{2} \nabla (U_i^2) \tag{178}$$

The motion supposed to be irrotational and the linearized Navier-Stokes equation then becomes

$$\rho_i \frac{dU_i}{dt} - \nabla \circ (p_i I + T_{iel}) \tag{179}$$

The solution of Navier-Stokes equation in terms of pressure is

$$p_1 = C_{10} I_n (kr) \eta \text{ and } p_2 = C_{11} + C_{12} \left(\frac{I_n (kr)}{C_{13}} + \frac{K_n (kr)}{C_{14}} \eta \right) \tag{180}$$

With

$$C_{10} = -\rho_1 \frac{\omega^2}{kI_n'(ka)}, \ C_{11} = -\rho_2 \frac{U_0^2}{2}, C_{12} = -\rho_2 \frac{(\omega - iU_0 k)^2}{k} \tag{181}$$

$$C_{13} = I_n'(ka)(1 - \lambda) \text{ and } C_{14} = K_n'(ka)\left(1 - \frac{1}{\lambda}\right) \tag{182}$$

14.13.3.9   THE DISPERSION EQUATION

The dispersion equation is obtained by substitution of all terms in the momentum conservation equation at the interface. This equation can be expressed as:

$$\frac{\gamma_{i-j}}{a^2} k (1 - n^2 - (ak)^2) + \rho_1 \frac{I_n (ka)}{I_n'(ka)} \omega^2 - \rho_2 \alpha \omega^2 (\omega - ikU_0)^2 - \frac{\varepsilon_0 V^2}{a^3 \ln^2 (b/a)} k (1 + ak\varsigma) = 0 \tag{183}$$

Considering a temporal analysis, it can be written:

$$\omega_r^2 = \frac{k}{a^2 (1 - \chi) \rho_2 \alpha} \left[-\gamma_{i,j} (1 - n^2 - (ak)^2) + \varepsilon_0 \frac{V^2}{a \ln^2 (b/a)} (1 + ak\varsigma) \right] - \left(\frac{U_0 k}{(1 - \chi)} \right)^2 \chi \tag{184}$$

$$\omega_i = \frac{U_0 k}{1 - \chi} \tag{185}$$

where

$$\chi = \frac{\rho_1 I_n(ak)}{\rho_2 I_n'(ak)\alpha} \quad \text{and} \quad \alpha = \frac{I_n(ka)}{I_n'(ka)(1-\lambda)} + \frac{K_n(ka)}{K_n'(ka)\left(1 - \frac{1}{\lambda}\right)} \tag{186}$$

and where ω_i and ω_r are, respectively, the real part and the imaginary part of the growth rate u. In the equipotential case,

$$\varsigma = \frac{I_n'(ka)}{I_n(ka)(1-\beta)} + \frac{K_n'(ka)}{K_n(ka)\left(1 - \frac{1}{\beta}\right)} \tag{187}$$

and in the nonequipotential one,

$$\varsigma = -\left(\frac{K_n(ka)}{K_n'(ka)} - \frac{\varepsilon_2 I_n(ka)}{\varepsilon_1 I_n'(ka)}\right)^{-1} \tag{188}$$

These equations, taken in the same conditions as the ones of the articles of,respectively, Levich [334], Taylor [335], Melcher [336], Bailey [337]or Lin and Kang [338], give the same equation that each of these authors has found.

Reznik et al.[339] introduced a numerical model which can predict velocity profile of flow at the central pipe which assume the axisymmetric Poiseuille velocity profile and the corresponding flow in the annular pipe, which were imposed according to the analogous boundary conditions:

$$u_z = \frac{2Q_1}{\pi r_{in}^2}\left(1 - \left(\frac{r}{r_{in}}\right)^2\right) \quad 0 < r < r_{in} \tag{189}$$

$$u_z = \frac{2Q_2}{\pi r_{out}^2 \omega}\left(1 - \left(\frac{r}{r_{out}}\right)^2 - \left(1 - \left(\frac{r_{in}}{r_{out}}\right)^2\right)\frac{\ln\left(\frac{r}{r_{out}}\right)}{\ln\left(\frac{r_{in}}{r_{out}}\right)}\right) \quad r_{in} < r < r_{out} \tag{190}$$

$$\omega = 1 - \left(\frac{r_{in}}{r_{out}}\right)^4 + \frac{\left(1 - \left(\frac{r_{in}}{r_{out}}\right)^2\right)^2}{\ln\left(\frac{r_{in}}{r_{out}}\right)} \tag{191}$$

whererin and rout are the radii of the core and shell nozzles and Q1 and Q2 are the corresponding volumetric flow rates. In their model, the fluid motion is governed by the Stokes equations, the electric tractions are related to the Maxwell stresses equations, and the continuity of the tangential component of the electric field and the balance of the free charge at the surfaces were considered as governing equations. Using dimensionless parameters and applyingthe Kutta-Merson method as a numerical method, velocity profiles were achieved[317].

In electrohydrodynamics problems, the electric force affects the movement of fluids, which changes the distribution of charges within the fluids and on the interfaces. Therefore, the electric force is coupled with surface/interface tension, viscous force, inertia force and so on. Considering that there are two interfaces in a coaxial electrospray problem, the kinematic, dynamic and electrical boundary conditions are required for each interface. The kinematic interface condition for each interface can be written as:

$$\frac{\partial F}{\partial t} + u\nabla F = 0$$

(192)

where F is the interface function. The dynamic boundary condition for each interface can be expressed as:

$$\left\|T^m + T^e\right\|n = \gamma\nabla n + (\delta - nn)\nabla\gamma$$

(193)

where || and || indicated the jump of corresponding quantity across the interface. T^m, T^e, n, γ and δ represent the hydrodynamic stress tensor, electrical Maxwell tensor, normal unit vector, surface tension and identity matrix, respectively. The electrical boundary conditions for each interface can be given by:

$$n\|D\| = q_s$$

(194)

$$n\|E\| = 0$$

(195)

The Eq. (194) expresses the Gauss law in which the surface charge density q_s, satisfies the surface charge conservation law. The Eq. (195)represents the continuity of the tangential component of the electric field. For viscous fluids, the tangential component of the velocity should be continuous on the interface:

$$n \times \|u\| = 0$$

(196)

In addition to the above boundary conditions, there are othersolution-dependent boundary conditions such as the finiteness ofvelocity and electric intensity field at the symmetric axis. Thesegoverning equations and boundary conditions are applicable inmost cases no matter whether the fluids are liquids or gasses.It must be pointed out that solving the above problem is verydifficult because of the following challenges:

The two interfacesof the problem are unknown, the length scales of the capillaryneedles and the jets are very disparate, and the breakup of jetsis time dependent. Therefore, modeling a coaxial electrosprayprocess is acomplicated procedure involving a larger number ofunknowns and parameters. Further simplifications and assumptionsare necessary in order to study the coaxial electrified jetunder either the outer-driving or the inner-driving flow conditions.

Lopez-Herrera et al.[315] introduced two additional governing equations in order to explore the influence on the spraying process of the properties of the liquids: i.e. the electrical conductivity K, dielectric constant β, interfacial tension of the liquid couple γ, viscosity μ, etc.

The first equation was the normal balance across the outer interfaces:

$$\frac{\gamma_{01}}{R+\delta} \approx P_1 - P_0 + \frac{1}{2}\varepsilon_0\left[E_{n0}^2 - \beta E_{n1}^2\right] \qquad (197)$$

And the second one was the normal balance across the inner interfaces:

$$\frac{\gamma_{12}}{R} \approx P_2 - P_1 + \frac{1}{2}\varepsilon_0\beta E_{n1}^2 \qquad (198)$$

where δ is the thickness of the layer oil, R is the radius of the inner interface, p is the pressure, E_n is the normal component of the electric field, and subscripts 0, 1, and 2 refer to air, oil and driving liquid, respectively. In Eqs.(197) and (198), it was assumed that E_{n2} is much smaller than E_{n0} and E_{n1}. By adding these two equations and neglecting δ as compared to R, one arrives at [315]:

$$\frac{\gamma_{12} + \gamma_{01}}{R} \approx P_2 - P_0 + \frac{1}{2}\varepsilon_0 E_{n0}^2 \qquad (199)$$

This is different from a single-axial electrospray processwhere analytical and numerical models can be obtained and thekey process parameters can be analyzed systemically.

14.13.4   TWO-PHASE FLOW PATTERNS MODEL IN COAXIAL ELECTROSPINNING

By comparison, electrospinning composite nanofiberswithdistinct layers give rise to extra problems of fluid dynamics.Among them, flow patterns of two-phase flow-swithin capillary tubes are often studied first of all as a basisfor advanced research. Figure 14.34 illustrates two flowpatterns—bubbly, slug and annular—generally observed inliquid–liquid (e.g., water-oil) capillary pipe flows or gas–liquidexperiments conducted in reduced gravity environments(or conditions of similarity) to understand gas–liquidbehavior in outer space. Sometimes two transitional flowpatterns are particularly labeled: bubbly slug and slug annular[340].

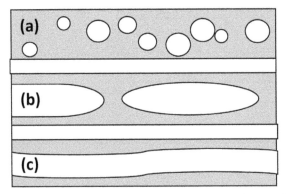

FIGURE 14.34 Two-phase flow patterns: (a) bubbly, (b) slug and(c) annular.

Flow pattern maps have been drawn from experiments,but a widely accepted theory is not accessibleyet. A flow pattern model based on the Weber number,however, is regarded as relatively effective in distinguishingannular flows from slug flows. For gas-liquid two-phaseflows, slug flows change to slug-annular transitional flows atunity of the gas Weber number and annular flows occur if thegas Weber number is greater than 20 [340, 341].

Hu et al. [340]presented a mathematic physical model to study the flow patterns of two-phase flows in coaxially electrospinning composite nanofibers, where shear-thinning non-Newtonian properties and strong electric fields were two characteristics that are of particular concern. They applied numerical simulations using commercial CFD software as a preprocessor. By means of these, they tried to study the mechanism of coaxial electrospinning and provide useful and least heuristic information for the development of advanced composite nanofibers.

According to this two phase flow pattern, Fig.14.35 illustrates a two-tube system ready to coaxially electrospin: The regions I and II are, respectively, filled with core and shell liquids. A coaxial two-phase flow may form if an electric field is activated in the x direction. Only steady flows are concerned to avoid theoretical obstacles at the start up of electrospinning, for instance, strong singularity at the outer tube exit [340].

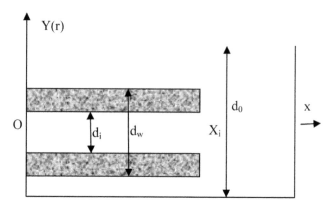

FIGURE 14.35 Shematic arrangement of the two-tube system ready to coaxially electrospin in both rectangular coordinates (x, y, z) and the cylindrical polar coordinates (r, θ, x).

14.13.4.1 TWO-PHASE PATTERN MODEL FOR LAMINAR FLOW

For incompressible liquids of constant density flowing inan electric field, the continuum and momentum equationscan read)neglecting gravity), where V, ρ, and p are velocity vector, liquid density, andpressure, in the order given; $T^{\leftrightarrow}$and $\sigma^{\leftrightarrow M}$are stress tensors dueto viscosity and the electric field, respectively [342].

$$\nabla \cdot \vec{V} = 0 \tag{200}$$

$$\rho \frac{D\vec{V}}{Dt} = -\nabla p + \nabla \cdot \ddot{\vec{T}} + \nabla \cdot \ddot{\vec{\sigma}}^{M} \tag{201}$$

Since most liquids suitable for electrospinning nanofibersare shear-thinning non-Newtonian fluids, [343] the viscousstress tensor $T_j$in Eq. (201) is modeled according to the powerlaw

$$\ddot{T} = \alpha \dot{\gamma}^{n-1} \ddot{D} \tag{202}$$

in which the deformation rate tensor $D^{\leftrightarrow}$has components:

$$\ddot{T} = \alpha \dot{\gamma}^{n-1} \ddot{D} \tag{203}$$

the shear rate

$$\dot{\gamma} = \sqrt{\vec{D} : \vec{D}} \tag{204}$$

α is the measureof the average viscosity of a liquid; n is the measure ofthe deviation of a liquid from Newtonian fluids.

The electric stress tensor σ⁺ᴹin Eq. (205(is related to thedensity of free charges ρ^e, the gradient of an isotropic electriccontribution, and dielectric permeability εin the electricfield E)ε$_0$ taken as absolute permeability(.

$$\nabla \cdot \vec{\sigma}^M = \rho^e \vec{E} + \nabla\left[\frac{1}{2}\varepsilon_0\rho\left(\frac{\partial \varepsilon}{\partial \rho}\right)_T \vec{E} \cdot \vec{E}\right] - \frac{1}{2}\varepsilon_0 \vec{E} \cdot \vec{E}\nabla \varepsilon \tag{205}$$

In Eq.(205(, the second term at the right-hand side,

$$\nabla\left[\frac{1}{2}\varepsilon_0\rho\left(\frac{\partial \varepsilon}{\partial \rho}\right)_T \vec{E} \cdot \vec{E}\right] \tag{206}$$

vanishes in incompressible flows withconstant density. To simplify the last term,

$$-\frac{1}{2}\varepsilon_0 \vec{E} \cdot \vec{E}\nabla \varepsilon \tag{207}$$

where the permeabilityis assumed to be homogenous in each liquid but itsgradient exists on the core-shell interface. By introducing aphase function c to indicate the two liquids, Δ$_\varepsilon$ is related toεΔc. Because the permeability gradient usually resists liquidsto flow,−Δ$_\varepsilon$= εΔc is reasonable for c=1 as the coreand c=0 as the shell. Hence, a more feasibleform of the last term in Eq.(205) becomes

$$-\frac{1}{2}\varepsilon_0 \vec{E} \cdot \vec{E}\nabla \varepsilon = \frac{1}{2}\varepsilon_0 \vec{E} \cdot \vec{E}\nabla c \tag{208}$$

The phase indicator c, an unknown function tracking interfacemovement, is considered to satisfy the convective equation

$$\frac{\partial c}{\partial t} + \vec{V} \cdot \nabla c = 0 \tag{209}$$

Solved by the volume of fluid method, the phase indicator $c$acts as a fractional volume function in Eq. (209) the effect of surface tension is addressed by adding asource term into the momentum.

$$\vec{F} = \frac{2\bar{\rho}}{\rho_{core} + \rho_{shell}}\sigma\kappa\nabla c \tag{210}$$

where σ is the surface tension coefficient, k is the interface curvature, and

$$\bar{\rho} = c\rho_{core} + (1-c)\rho_{shell} = c\rho_{core} + (1-c)\rho_{shell} \qquad (211)$$

Can represent the liquid density of the core—ρ_{core}, the shell—ρ_{shell}, or the mixture of the two, depending on the c value. With these treatments above, the final momentum equation for the laminar two-phase flow in the process of coaxially electrospinning composite nanofibers yields [340]:

$$\bar{\rho}\frac{D\vec{V}}{Dt} = -\nabla p + \nabla \cdot \left(\alpha\dot{\gamma}^{n-1}\vec{D}\right) + \rho^e\vec{E} + \frac{1}{2}\varepsilon\varepsilon_0\vec{E}\cdot\vec{E}\nabla c + \frac{2\bar{\rho}}{\rho_{core}+\rho_{shell}}\sigma\kappa\nabla c \qquad (212)$$

In Eq. (212) $\rho^e E$ acts as a body force to drive liquids to move, whereas all other forces, whereas all other forces, $\nabla\cdot\left(\alpha\dot{\gamma}^{n-1}\vec{D}\right)$ (caused by viscosity), $\frac{1}{2}\varepsilon\varepsilon_0\vec{E}\cdot\vec{E}\nabla c$ (induced by electric charges on interfaces), and $\frac{2\bar{\rho}}{(\rho_{core}+\rho_{shell})}\sigma\kappa\nabla c$ (due to surface tension), resist liquids to flow.

For study limited to steady flows, time derivatives in Eqs.(211) and (212) are necessary for numerical computational techniques instead of physical description. Zero velocity and hydrostatic pressure are assigned as initial conditions. Boundary conditions are set at tube walls, inlet at $x=0$ and outlet at $x=x_0$.

Walls: $\vec{V} = 0$,

$$x = 0 : \partial\vec{V}/\partial x = 0,\ \partial p/\partial x = 0,\ c = 1 \text{ (core) and 0 (shell)} \qquad (213)$$

$$x = x_0 : \partial\vec{V}/\partial x = 0,\ \partial p/\partial x = 0,\ \partial c/\partial x = 0$$

Thus a two-phase laminar flow problem is posed by the governing Eqs. (206) and (212), the interface convective Eq. (211), and boundary conditions (213).

14.13.5 SCALING LAW

Since the cone and the jet are in disparate scales, they are usuallystudied separately. As for the cone, Marin et al.[344]studied thecoaxial electrospray within a bath containing a dielectric liquid andobserved that a sharp tip in the inner dielectric meniscus would beformed without mass emission. They presented an analyticalmodel of the flow in the inner and the outer menisci based ondifferentsimplifying hypotheses. The fluid dynamic equations weresimplified in a low Reynolds number limit and under the assumptionthat the electrical effects inside the liquid bulk were negligible. The electrical equations were also reduced into the Laplace equationof the electric field. After the boundary conditions on thetwo interfaces were applied and the assumptions of self-similarityand very thin conductive layer were made, the velocity, the pressure,fields and the electrical shear stress at the outer surface werefinally calculated. As for the jet and the resulting droplets, Lopez-Herrera et al.derived the scaling laws of the diameter of the coaxialelectrified jet and the current transported throughout the jet byexperimenting with different liquids, such as water, sunflower oil,ethylene-glycol and Somos [315]. The dimensionless parameters weredefined based on the reference characteristic values of the flow rateQ_0, the current I_0, and the diameter d_0, as given by:

$$Q_0 = \frac{r_{eff} s_0}{\rho \sigma}, I_0 = \left(\frac{r_{eff}^2 s_0}{\rho}\right)^2, d_0 = \left(\frac{Q_0 s_0}{\sigma}\right)^{1/3} \tag{214}$$

where y_{eff} denotes the effective value of the surface tension.The results indicated that the current I/I_0 on the driving flowrate Q/Q_0 closely followed a power law of $(Q/Q_0)^{1/2}$, similar tothat in single-axial electrospinning. It was also found that the mean diameter of the droplets resulted from the breakup of the coaxialjets scaled linearly with both inner and outer flow rates in the case of outer driving; whereas that diameter was closely dependent ofthe ratio of inner and outer flow rates in the case of inner driving.Marín et al. obtained the diameter of the coaxial jets d as afunction of the flow rate Q [344]. They found that the experimentalresults fitted in the $Q^{1/2}$ law as below:

$$\frac{d}{d_0} = 1.25(Q/Q_0)^{1/2} \tag{215}$$

Mei et al.found the particle encapsulation conditions relevant with the flow rates and the material properties in the case of the inner driving flow [327]. Let r* be the charge relaxation length and R* be the inertial length:

$$r^* = (Q\varepsilon_0/\sigma)^{1/3}, R^* = \left(\frac{\rho Q^2}{\gamma}\right)^{1/3} \tag{216}$$

The particle encapsulation conditions were therefore expressed as:

$$\frac{r^*_O}{r^*_I} < 500, \frac{R^*_O}{R^*_I} < 0.015 \tag{217}$$

where the subscripts O and I indicate the outer liquid and the inner liquid, respectively. Furthermore, the flow rates of the inner needle and outer needle may affect the range of the stable cone–jet, and thus affect the jet size and the particle size. Chen et al. found that the working range for the stable cone–jet could be expanded by increasing the inner liquid flow rate and by decreasing the outer liquid flow rate in the case of outer driving [345]. It has been shown previously that the particle size decreases as the applied voltage increases in a stable cone–jet mode. Similar reduction of the particle size can also be achieved by reducing the flow rate, which can be explained by easier overtaking of the electrical force over the hydrodynamic forces in reduction of flowing materials [316]. In practice, stable cone–jet mode should be adjusted at the higher applied voltage and lower flow rates in order to get the smaller particle size.

14.13.6   INSTABILITY ANALYSIS

Coaxial electrospraying is a new effective technique to form micro/nano capsul-esthat are monodisperse and controllable. It has many applications in the drug industry,food additives, paper manufacture, painting and coating processes. In experiments,when two immiscible liquids are emitted from two homocentric capil-lary tubes,respectively, under appropriate flow and electric field conditions, a sta-tionary Taylorcone is formed. At the tip of the Taylor cone, a steady axisymmetric coaxial jet withnearly uniform diameter arises. The coaxial jet is intrinsically unsta-ble, and breaksup into micro compound droplets at some distance downstream. This is called thecone-jet coaxial electrospraying mode [316].

Recently, many experiments have been carried out to investigate the mecha-nismof coaxial electrospraying and the scaling laws between important quantities, e.g.Loscertales et al. [316], Lopez-Herrera et al. [315], Chen et al. [312] and Mari-net al. [344]. Theoretical and numerical work by Li and Yin & Yin [346, 347] has analyzed the linear instability of an inviscid coaxial jet having a conductingannular liquid under a radial electric field, whereboth theequipotential andnonequipotential cases were studied; Higuera [348] performed a brief but valuablenumerical simula-tion of a stationary electrified coaxial jet in the framework of theleaky dielectric model and quasi-uni-directional approximation.

The breakup process of a liquid jet under an electric field is closely related to thegrowth and propagation of unstable disturbance waves at the interface between fluids.Therefore, instability analysis is useful in predicting the breakup modes of liquid jets,and also in predicting the intact jet length and droplet size [349–351]. In the instability analysis of electrically charged singleliquid jets, two kinds of electric field, that is, radial and axial, are usually encountered. Also,the electrical properties of liquids may be of a perfect conductor, perfect dielectric orleaky dielectric (dielec-tric with finite conductivity). Thus, manycases can arise due tothe variation of the electric field imposed on the jet and electrical properties of theliquid. For instance, Turnbull [352], analyzed the temporal linear stability ofconducting and insulating liquid jets in the presence of both radial and axial electricfields; Saville [225] and Mestel [353] studied the stability of a charged leakydielectric liquid jet under a tangential electric field, paying particular attention tothe effect of liquid viscosity; Lopez-Herrera, Riesco Chueca and Ganan-Calvo [148, 311, 313] researched the stability of a viscous leaky dielectric liquid jet under a radial electricfield, taking into account the effect of ambient air flow. Garcıa et al.,andGonzalez, Garciaand Castellanos [354]investigated the effect of AC radial electricfields on the instability of liquid jets. Huebner and Chu [355] and Son and Ohba [356],respectively, explored the jet instability under axisymmetric and nonaxisymmetricdisturbances. More recently, nonlinear effects have been specially studied in order toexplain the experimental phenomena, such as the formation of satellite droplets (e.g.,Refs. [313, 315, 357–361]).

In most of the published reports, the liquid jet is usually assumed to beeither perfectly conducting or perfectly dielectric. In practice, most liquids usedin experiments are leaky dielectric. Unlike prefect conductors or dielectrics, for leakydielectrics free charge may occur in the fluid bulk and therefore electromechanicalcoupling occurs not only at interface but probably also in the bulk. Furthermore,electric stresses on an interface are no longer perpendicular to it, because free chargeaccumulated on the interface may modify the electric field. From this point of view,electric stresses tangential to the interface are inevitable, and must be balanced byviscous stresses. For perfect conductors or perfect dielectrics, the tangential componentof electric stress vanishes, because free charge is reset instantaneously on the interfaceto keep the interface equipotential for the former, and it is absent for the latter [225].

In Li et al.studies [346, 347], the assumption that the innerand outer liquids are inviscid was made in the instability analysis of the electrifiedcoaxial jet. However, with such an assumption the tangential component of electricstress on the interface cannot be balanced due to the absence of liquidviscosity.Moreover, liquid viscosity mayplay an important role in the jet instability, becausethe diameter of the coaxial jet generated in electrospraying experiments is very small,usually of the order of several tens of micrometers. Therefore, from this physicalpoint of view, it is incorrect to neglect the viscosity of liquid.

To gain further insight into the outer-driving coaxial electrospraying and to study theaxisymmetric instability behavior of the charged coaxial jet under a radial electricfield,Li et al. [346, 347] proposed a viscous leaky dielectric model, based on the theory of the Taylor–Melcher leaky dielectric model [225, 336].

Accordingly the outer liquid was assumed to be a leaky dielectric, acting as the driving liquid. Furthermore, the conductivity of the outer liquid was assumed to be large enough so that free charge is relaxed to the interface instantaneously. The inner liquid in the outer-driving coaxial electrospraying, for a generic case, should be considered as a leaky dielectric too. Though the conductivity of the inner fluid is much smaller than that of the outer liquid, it is sufficient to have charge relaxed and transferred to the interface, where the distribution of charge density is determined by the surface charge conservation equation. In such a leaky dielectric case, both the conductivity ratio of inner to outer liquid and the electrical relaxation time of the inner liquid are taken into account. The initial steady state may be solved following the method provided by Higuera [348]. In the outer-driving electrospraying experiments the electrical conductivity of the inner liquid is at least two or three orders of magnitude smaller than that of the outer liquid [315]. Also, in the numerical simulation of Higuera [348], it was found that the charge density at the inner interface is about two orders smaller than that at the outer interface. Therefore, the conductivity of the inner liquid is negligible, and free charge can be supposed to lie approximately only on the outer air–liquid interface in theoretical studies. In Li et al.[346, 347] model, the inner liquid was approximated as a perfect dielectric and there was no free charge at the inner liquid–liquid interface. Such an approximation was deduced from the generic leaky dielectric model, assuming that the outer-to-inner conductivity ratio approaches infinite.

It is well known that the breakup of liquid jets is closely associated with the jet instability [362–364]. Therefore, the hydrodynamic instability theory can be used for coaxial electrospray analysis and has successfully predicted the experimental observations [365, 366]. The instability theory deals with the mathematical analysis of the response of disturbances with small amplitudes superposed on a laminar basic flow. If the flow returns to its original laminar state, it is recognized as stable. However, if the disturbance grows and the flow changes into a different state, it is recognized as unstable. When analyzing the instability problems, the governing equations and the boundary conditions described above are used and the classical method of expansion of normal modes is usually implemented. This method analyzes the development of perturbations in space only, in time only, or in both space and time. The analysis results may provide theoretical insight and practical guidance for coaxial electrospray process control.

For instability analysis in a cylindrical coordinate (z, r, θ), the arbitrary and independent perturbations are typically decomposed into Fourier series like exp ωt+i ($kz+n\theta$), where θ, k andn stand for the frequency, the axial wave number and the azimuthal wave number, respectively. In the case of coaxial electrospray, a local temporal method is used for instability analysis.

This method assumes a real axial wave number and pursues a complex frequency since its linear dispersion relation is relatively easy to solve. To the best of the authors' knowledge, this is the most commonly used method so far for instability analysis of coaxial electrospray. Other methods are waiting for development in the future. Li et al. [366] studied the instability of an inviscid coaxial jet under an axial electric field. They also studied the axisymmetric and nonaxisymmetric instabilities of a viscous coaxial jet under a radial electric field [324]. These studies solved the governing equations and boundary conditions based on a number of assumptions and simplifications, such as: the inner and the outer liquids were assumed to be perfect conductors, perfect dielectrics or leaky dielectrics; the free charges were relaxed to the interface instantaneously, and the effects of gravitational acceleration and temperature were ignored. Figure 14.36 sketches a simplified physical model for coaxial electrospray. It consists of acylindrical inner liquid 1 of radius R_1, an annular outer liquid 2 of outer radius R_2, and an ambient gas 3 that is stationary air in an unperturbed state. The basic flows should be assumed uniform or with specific shapes. Si et al. used the uniform velocities U_i with i = 1, 2, 3 for the inner liquid, the outer liquid and the ambient gas (in this case U_3 = 0) and the uniform axial electric field E_0[367]. The corresponding dispersion relations were derived and written in an explicit analytical form, and the eigenvalues were computed by numerical methods. In general, the dispersion relation could be expressed in the form of:

$$f(\omega, k, n, U_i, E_0, K, ...) = 0 \qquad (218)$$

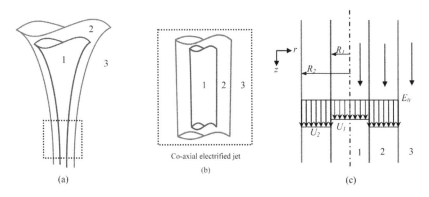

FIGURE 14.36 Schematic descriptions of (a) the coaxial cone–jet configuration; (b) thecoaxial electrified jet; (c) the simplified theoretical model.1: the inner liquid; 2: the outer liquid; 3: the ambient gas.

The involved dimensionless parameters include: dimensionless wave number $\alpha = kR_2$, dimensionless frequency $\beta = \omega R_2/U_2$, Weber number $We = \rho_2 U_2^2 R_2/\gamma_2$, dimensionless electrostatic force $E = \varepsilon E_0^2/\rho_2 U_2$, density ratios $S = \rho_1/\rho_2$ and $Q = \rho_3/\rho_2$, velocity ratio $U = U_1/U_2$, diameter ratio $R = R_1/R_2$, electric permittivity ratios $\varepsilon_{1p} = \varepsilon_1/\varepsilon_3$ and $\varepsilon_{2p} = \varepsilon_2/\varepsilon_3$, conductivity ratio $K = K_1/K_2$, and interfacial tension coefficient ratio $\gamma = \gamma_1/\gamma_2$.

The instability analysis yielded the following three unstable modes: paravaricose mode, parasinuous mode and transitional mode. The paravaricose mode occurs when the phase difference of initial perturbations at the inner and the outer interfaces was about 180°. The parasinuous mode occurs at a phase difference of approximately 0°. The transitional mode occurs when the initial perturbation is changed from in phase to out of phase. In particular, the maximal growth rate of dimensionless frequency β_{rmax} dominates the jet breakup because the perturbation for β_{rmax} grows most quickly (i.e., the perturbation grows exponentially in the dimensional form of exp $\{\omega_r t\}$). The corresponding axial wave number plays an important role in fabricating MPs and NPs because the wave number is closely associated with the wavelength of perturbations (i.e., $\alpha = 2\pi R_2/\lambda$, where λ stands for the wavelength). The larger max is, the smaller the size of resulting MPs and NPs becomes. As a result, Eq.(218) allows us to study the effects of the electric field, the electrical conductivity, the electrical permittivity and the other important hydrodynamic parameters on the instability of the coaxial jet. It also allows us to predict the different flow modes and the corresponding transitions.

14.13.7   INSTABILITY MODEL FOR A VISCOUS COFLOWING JET

Consider an infinitely long coflowing jet with two immiscible liquids surrounded by the ambient air, as sketched in Fig. 14.37. The inner liquid cylinder has a radius R_1, and the outer liquid annulus has an exterior radius R_2. An earthed annular electrode of radius R_3 is positioned surrounding the jet, and a voltage V_0 is imposed on the jet surface. The electrical property of the outer liquid is assumed to be leaky dielectric; the inner liquid and air are perfect dielectrics. A basic radial electric field of magnitude $-V_0/[r \ln (R_2/R_3)]$ is thus formed in the air. Free charge is assumed tobe relaxed on the interface between the air and outer liquid instantaneously, owing to the conductivity of the outer liquid [346, 347]. The density of surface charge on the unperturbed air–liquid interface is $-\varepsilon_3 V_0/[R_2 \ln (R_2/R_3)]$.

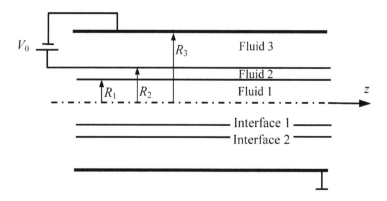

FIGURE 14.37   Schematic of description of the theoretical model.

The liquids and air are considered to be incompressible and Newtonian and the flow assumes to be axisymmetric. Effects of gravitational acceleration, magnetic field and temperature are ignored [346, 347]. The governing equations of the flow are:

$$\nabla \cdot u_i = 0, \quad i = 1,2,3,$$

(219)

$$\rho_i\left(\frac{\partial u_i}{\partial t}+u_i \cdot \nabla u_i\right)=-\nabla p_i + \mu_i\nabla^2 u_i, \quad i=1,2,3, \tag{220}$$

where u, ρ, p and μ are the velocity, density, pressure and dynamic viscosity, respectively. The subscripts 1, 2 and 3 denote the inner liquid, the outer liquid and the ambient air, respectively.

When the jet is perturbed by an arbitrary disturbance, both the inner liquid–liquid interface and the outer air–liquid interface depart from their original equilibrium positions. For infinitesimal disturbances, their new positions can be expressed as r =R$_j$+ η_j, j = 1, 2, whereηj is the displacement of the interface from R$_j$, and the subscripts 1 and 2 denote the inner and outer interfaces, respectively [346, 347]. Considering only the axisymmetric instability,we have:

$$\eta_j = \eta_j(z,t) \tag{221}$$

The boundary conditions include the no-slip condition at the electrode, that is,

$$u_3 = 0 \text{ at } r = R_3; \tag{222}$$

the continuity of the velocity at the inner and outer interfaces, that is,

$$u_2 = u_3 \text{ at } r = R_2 + \eta_2; \tag{223}$$

$$u_1 = u_2 \text{ at } r = R_1 + \eta_1; \tag{224}$$

the finiteness of the velocity at the symmetric axis, that is,

$$u_1 < \infty \text{ at } r = 0; \tag{225}$$

the kinematic boundary conditions at the interfaces, that is,

$$u_{1,2} = \left(\frac{\partial}{\partial t}+u_{1,2}\cdot\nabla\right)\eta_1 \text{ at } r = R_1 + \eta_1, \tag{226}$$

$$u_{2,3} = \left(\frac{\partial}{\partial t}+u_{2,3}\cdot\nabla\right)\eta_2 \text{ at } r = R_2 + \eta_2, \tag{227}$$

where u is the radial velocity component; and the dynamic boundary conditions at the interfaces, that is,

$$(T_2 - T_1)\cdot n_1 - \gamma_1(\nabla \cdot n_1)n_1 = 0 \text{ at } r = R_1 + \eta_1, \tag{228}$$

$$(T_3 - T_2) \cdot n_2 - \gamma_2 (\nabla \cdot n_2) n_2 = 0 \text{ at } r = R_2 + \eta_2, \qquad (229)$$

where T is the stress tensor, γ is the surface tension, n is the normal unit vector and $\nabla \cdot n$ is the curvature. For the axisymmetric case,

$$n_j = \frac{(1, -\eta_{jz})}{\sqrt{1 + \eta_{jz}^2}} \text{ and } \nabla \cdot n_j = \frac{1}{\sqrt{1 + \eta_{jz}^2}} \left(\frac{1}{R_j + \eta_j} - \frac{\eta_{jzz}}{1 + \eta_{jz}^2} \right), \ j = 1,2, \qquad (230)$$

where η_z and η_{zz} are the first- and second-order partial derivatives of η with respect to z, respectively. In the presence of an electric field, the stress tensor T includes not only the hydrodynamic stress tensor but also the electrical Maxwell tensor, that is,

$$T = T^h + T^e, \text{ with } T^h = -p\delta + \mu\left[\nabla u + (\nabla u)^T\right] \text{ and } T^e = \varepsilon E - \frac{1}{2}\varepsilon E \cdot E\delta, \qquad (231)$$

where δ is the identity matrix and E is the electric field intensity.
Identity matrix and E is the electric field intensity.
The governing equations and boundary conditions related to the electric field are needed to close the problem. As free charge is absent in the bulk, the Maxwell equations in the liquids and air reduce to [346, 347]:

$$\nabla \cdot E_i = 0 \text{ and } \nabla \times E_i = 0, \ i = 1,2,3, \qquad (232)$$

Introduce an electrical potential function ψi, satisfying the Laplace equation:

$$\nabla^2 \psi_i = 0, \ i = 1,2,3, \qquad (233)$$

and the electric field intensity $Ei = -\nabla \psi i$. The electrical boundary conditions are:
(a) zero electrical potential at the annular electrode, that is,

$$\psi_3 = 0 \text{ at } r = R_3; \qquad (234)$$

(b) The finiteness of the electric field at the symmetric axis, that is,

$$E_1 = 0 \text{ at } r = 0; \qquad (235)$$

(c) Continuity of the tangential component of the electric field at the inner and outer interfaces, that is,

$$n_j \times [E] = 0 \text{ at } r = R_j + \eta_j, \ j = 1,2, \qquad (236)$$

where the symbol [·] indicates the jump of the corresponding quantity across the interface; (d) continuity of the normal component of the electric displacement at the inner interface, that is,

$$n_1 \cdot \left(\varepsilon_2 E_2 - \varepsilon_1 E_1 \right) = 0 \text{ at } r = R_1 + \eta_1 ; \tag{237}$$

where the surface charge density qs satisfies the surface charge conservation equation

$$\frac{\partial q_s}{\partial t} + u \cdot \nabla q_s - q_s n \cdot (n \cdot \nabla) u + [\sigma E] \cdot n = 0 \tag{238}$$

The four terms on the left-hand side of Eq. (238) represent the contributions of charge accumulation, surface convection, surface dilation and bulk conduction, respectively [225].

Before the instability analysis, the basic velocity profile of the jet in the unperturbed state should be obtained. As the jet is perfectly cylindrical and the flow is axisymmetric, the basic velocity field is unidirectional, that is, $u=W(r) \, ez$, where W is the axial velocity component and e_z is the unit vector in the axial direction. Therefore, the momentum equation (220) reduces to

$$\left(\frac{d^2 W_i}{dr^2} + \frac{1}{r} \frac{dW_i}{dr} \right) = -\frac{G_i}{\mu_i}, \ i = 1,2,3, \tag{239}$$

where $G_i = -\partial pi/\partial z$ is the negative of the stream wise pressure gradient [351]. According to the balance of forces on the interfaces, the pressure gradients in the liquids and air should be equal, that is, $G_i = G$. Integrating (239) and using the continuity conditions of the velocity and shear force on the interfaces, the solutions of W are obtained. Choosing μ_2, R_2, and U_2(the velocity of the jet at the outer interface, $=G(R^2_3 - R^2_2)/(4\mu_3)$) as the scales of dynamic viscosity, length and velocity, respectively, the solutions W_i in the following dimensionless form become:

$$W_2 = 1 + \mu_{r3} \frac{1-r^2}{b^2 - 1}, \tag{240}$$

$$W_2 = 1 + \mu_{r3} \frac{1-r^2}{b^2 - 1}, \tag{241}$$

$$W_1 = 1 + \mu_{r3} \frac{1-a^2}{b^2 - 1} + \frac{\mu_{r3}}{\mu_{r1}} \frac{a^2 - r^2}{b^2 - 1}, \tag{242}$$

where two radius ratios are $a = R_1/R_2$, $b = R_3/R_2$, and two viscosity ratios are μ_{r1} $= \mu_1/\mu_2$, $\mu_{r3} = \mu_3/\mu_2$. Note that the logarithm function $\ln r$ included in general solution of Eq.(239) by Chen [351], vanishes owing to the absence of the gravitational acceleration. It can be seen from Eqs.(240)–(242) that the basic velocity profile is closely associated with the relative viscosity of the inner liquid and outer liquid and that of air and the outer liquid. First, suppose the ambient air is almost inviscid, that is, $\mu_{r3} \leq 1$. According to Eq.(241), the basic axial velocity of the outer liquid is nearly uniform, that is, $W_2 \approx 1$. For the inner liquid, there are two cases: if its viscosity is of the order of the air viscosity (i.e., $\mu_{r1} \sim O(\mu_{r3})$), such as a dense gas (without lose of generality, the inner liquid can be gas.), it can be seen from Eq.(2.19c) that the corresponding basic velocity profile is still parabolic; conversely, if its viscosity is much larger than that of the air (i.e., $\mu_{r1} \geq \mu_{r3}$), such as various polymers and oils, the basic velocity profile is also approximately uniform with the same magnitude as the outer liquid. Therefore, in general, according to the magnitude of the inner liquid viscosity, two cases are involved [346].

Case I: $\mu_{r1} \sim O(\mu_{r3})$, the basic axial velocity profile is

$$W_1 = 1 + \mu_{r3}\frac{1-a^2}{b^2-1} + \frac{\mu_{r3}}{\mu_{r1}}\frac{a^2-r^2}{b^2-1}, \tag{242}$$

where the relative velocity ratio $\Lambda = U_0/U_2 - 1$ (U_0 is the velocity of the jet at the symmetric axis $r = 0$, $\approx 1 + (\mu_{r3}/\mu_{r1})(a^2/(b^2- 1))$). Case II: $\mu_{r1} \geq \mu_{r3}$, the basic velocity profile is

$$W_3 = 1 + \frac{1-r^2}{b^2-1}, \; W_2 \cong 1, \; W_1 \cong 1. \tag{243}$$

Figure 14.38 shows two typical basic velocity profiles; the axial velocity profile of the inner liquid is apparently parabolic for $\mu_{r1} = 0.018$, corresponding to case I, andappears to be uniform for $\mu_{r1} = 43$, corresponding to case II. In the following instability analysis we assume that the outer liquid is viscous and the air is inviscid in both cases and that the inner liquid is inviscid in case I and is viscous in case II. The advantage of this is that an analytical dispersion relation can be obtained. Moreover, the continuity of tangential force seems to be satisfied inherently for the simplified velocity profiles in both cases. Numerical studies of the instability of a viscous coaxial jet with leaky dielectric liquids in a radial electric field have to our knowledge not be reported before [346].

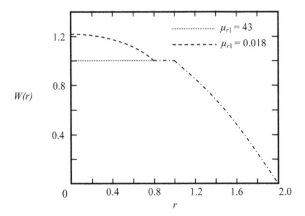

FIGURE 14.38 The influence of the relative viscosity of the inner liquid on the basic velocity profile.For µrs=0.018, a=0.8 and b=2.

If an axisymmetric jet of a viscous liquid is subjected to an axial electric field, its basic axial velocity profile is essentially parabolic, owing to the action of electrical shear stress [353]. However, as is well known, if both the viscosity and the nonuniform velocity profile are considered, the dispersion relation in an analytical form is beyond reach. In such a case either the velocity profile is assumed to be uniform [225, 315], or the viscosity of the liquid is assumed to be low or high [353], or the wavelength of the disturbance is assumed to be long [352]. But in this paper a radial electric field not an axial electric field is studied. In the presence of a radial electric field the electrical shear stress in the unperturbed state is absent as discussed above, and so the basic velocity profile for viscous liquids can reasonably be considered to be uniform.

The Froude number Fr $=U^2/gL$ (U is the characteristic velocity; g is the gravitational acceleration; L the characteristic length) measures the relative effect of gravity. If the characteristic length of the coaxial jet is not so small, Fr is finite. In such a case, gravity may be as important as the axial electric field. Both gravity and the axial electric field induce nonuniformity of the basic axial velocity, but their effects may be opposite: if the axial electric field is in the same direction as gravity (which is in accordance with most experimental situations), the electrical shear force makes the velocity at the interface larger than in the liquid bulk, and with the spatial evolution of the jet its effect is diffused into the bulk owing to liquid viscosity; on the other hand, gravity, acting as a pressure gradient, induces a larger velocity in the liquid center. Consequently, the action of the axial electric field and gravity may reach an equilibrium state as the jet evolves spatially and the well-developed jet has an axial uniform velocity. In such a case, the uniform velocity profile approximation is appropriate [347].

For infinitesimal axisymmetric disturbances, the perturbation of the interface (and also of the other physical quantities), is decomposed into the form of a Fourier exponential, that is, $\eta j\,(z,\,t) = \hat{\eta} j\,\exp(\omega t + ikz)$, $j=1, 2$, where $\hat{\eta} j$ is the initial amplitude of the perturbation at the interface, ω is the complex wave frequency, the real and imaginary parts of which are the temporal growth rate and frequency, respectively, k is the real axial wavenumber related to the wavelength by $k = 2\pi/\lambda$, and the imaginary unit $i = \sqrt{-1}$. Substituting the perturbation expression into the governing equations and boundary conditions the dispersion relation between ω and k is obtained. The derivation details are given in Appendix A.

Choosing ρ_2, γ_2, $\rho_2 U_2^2$, ε_3 and $-V_0/[R_2\,\ln(R_2/R_3)]$ as the characteristic scales of density, surface tension, pressure, electrical permittivity and electric field intensity, respectively, the dispersion relation is written in the following dimensionless form:

$$D(k,\omega) = \left(\frac{\hat{\eta}_2}{\hat{\eta}_1}\right)_1\left(\frac{\hat{\eta}_1}{\hat{\eta}_2}\right)_2 - 1 = 0, \tag{244}$$

For case I, where $(\hat{\eta}_2/\hat{\eta}_1)_1$ and $(\hat{\eta}_1/\hat{\eta}_2)_2$ are the amplitude ratio of the initial disturbances at the interfaces, respectively, expressed as

$$\left(\frac{\hat{\eta}_2}{\hat{\eta}_1}\right)_1 = \left[\frac{l^2+k^2}{a}\Delta_4\left(\frac{Eu\,Re\varsigma}{\omega}+1+\frac{l^2}{k^2}\right)-\frac{2k^2}{a}\Delta_3\left(\frac{Eu\,Re\varsigma}{\omega}+2\right)\right]^{-1}$$
$$\times\left[Re^2\,H_1\Delta_3\Delta_4 - 4lk^2\Delta_3\left(\Delta_6 - \frac{1}{la}\Delta_4\right)+\Delta_4\left((l^2+k^2)\Delta_1 - \frac{2k}{a}\Delta_3\right)\frac{l^2+k^2}{k}+\left(\frac{2k^2}{a}\Delta_3 - \frac{l^2+k^2}{a}\Delta_4\right)\frac{Eu\,Re\varsigma}{\omega}\right] \tag{245}$$

and

$$\left(\frac{\hat{\eta}_1}{\hat{\eta}_2}\right)_2 = \left[4k^2\Delta_3 - \frac{(l^2+k^2)^2}{k^2}\Delta_4 + k\left((l^2+k^2)\Delta_2\Delta_4 - 2lk\Delta_3\Delta_5\right)\frac{Eu\,Re\xi}{\omega} - Re^2\,\Delta_3\Delta_4 Euk\xi J\right]^{-1}$$
$$\times\left[Re^2\,H_2\Delta_3\Delta_4 - k\Delta_4\left((l^2+k^2)\Delta_2 + 2k\Delta_3\right)\left(\frac{Eu\,Re\varsigma}{\omega}+1+\frac{l^2}{k^2}\right)+2lk^2\Delta_3\left(\Delta_5 - \frac{1}{l}\Delta_4\right)\left(\frac{Eu\,Re\varsigma}{\omega}+2\right)\right] \tag{246}$$

The amplitude ratio $(\hat{\eta}_2/\hat{\eta}_1)_1$ comes mainly from the dynamic balance at the inner interface and $(\hat{\eta}_1/\hat{\eta}_2)_2$ from the dynamic balance at the outer interface. The symbols appearing in the dispersion equation are

$$H_1 = \frac{iS\omega}{k}\left(i\omega\frac{I_0(ka)}{I_1(ka)} - \frac{2\Lambda}{a}\right)+\frac{\Gamma}{Wea^2}\left(1-(ka)^2\right) \tag{247}$$

$$H_2 = \frac{iQ\omega}{k}\left(i\omega L - \frac{2}{b^2-1}\right)+Eu\left(1+k(\varsigma+1)J\right)-\frac{1}{We}\left(1-k^2\right) \tag{248}$$

$$J = \frac{I_1(k)K_0(kb) + K_1(k)I_0(kb)}{I_0(k)K_0(kb) - K_0(k)I_0(kb)} \tag{249}$$

$$\varsigma = \frac{-k\left(1 + \frac{l^2}{k^2}\right)\frac{\Delta_2}{\Delta_3} + 2l\frac{\Delta_5}{\Delta_4} + \frac{\operatorname{Re}\omega J}{k}}{\frac{\operatorname{Re}}{k}\left(\varepsilon_{r2}(\omega + \tau)\kappa - \omega J\right) + \frac{Eu\operatorname{Re}}{\omega}\left(k\frac{\Delta_2}{\Delta_3} - l\frac{\Delta_5}{\Delta_4}\right)} \tag{250}$$

$$\xi = \frac{\left(1 + \frac{l^2}{k^2}\right)\frac{1}{\Delta_3} - \frac{2}{\Delta_4}}{\frac{\operatorname{Re}}{k}\left(\varepsilon_{r2}(\omega + \tau)\kappa - \omega J\right) + \frac{Eu\operatorname{Re}}{\omega}\left(k\frac{\Delta_2}{\Delta_3} - l\frac{\Delta_5}{\Delta_4}\right)} \tag{251}$$

$$L = \frac{I_0(k)K_1(kb) + K_0(k)I_1(kb)}{I_1(k)K_1(kb) - K_1(k)I_1(kb)}, \kappa = -\frac{I_1(ka)\Delta_1 - \frac{\varepsilon_{r2}}{\varepsilon_{r1}}I_0(ka)\Delta_3}{I_1(ka)\Delta_0 - \frac{\varepsilon_{r2}}{\varepsilon_{r1}}I_0(ka)\Delta_2} \tag{252}$$

And $\Delta_0 - \Delta_6$ are

$$\Delta_0 = I_0(ka)K_0(k) - K_0(ka)I_0(k), \quad \Delta_1 = I_0(ka)K_1(k) + K_0(ka)I_1(k), \tag{253}$$

$$\Delta_2 = I_1(ka)K_0(k) + K_1(ka)I_0(k), \Delta_3 = I_1(ka)K_1(k) - K_1(ka)I_1(k), \tag{254}$$

$$\Delta_4 = I_1(la)K_1(l) - K_1(la)I_1(l), \Delta_5 = I_1(la)K_0(l) + K_1(la)I_0(l), \tag{255}$$

$$\Delta_6 = I_0(la)K_1(l) + K_0(la)I_1(l), \tag{256}$$

where $In(x)$ and $Kn(x)(n = 0, 1)$ are the nth-order modified Bessel functions of the first and second kinds. Here the wavenumber k and complex frequency ω have been normalized by $1/R_2$ and U_2/R_2, respectively. In the following numerical section, we mainly take case I as an example to illuminate the unstable modes and behaviors of a viscous coflowing jet under a radial electric field. Case II is calculated when studying the effect of the inner liquid viscosity and in part of the thin layer approximation. The dispersion relation for case II is given in Appendix B.

The dimensionless parameters involved in the dispersion relation include:
- the density ratios $S = \rho_1/\rho_2$ and $Q = \rho_3/\rho_2$,
- the interfacial tension coefficient ratio $\Gamma = \gamma_1/\gamma_2$,
- the electrical permittivity ratios $\varepsilon_{r1} = \varepsilon_1/\varepsilon_3$ and $\varepsilon_{r2} = \varepsilon_2/\varepsilon_3$,
- the relative electrical relaxation time $\tau = R_2\sigma_2/U_2\varepsilon_2$,
- the Reynolds number $\mathrm{Re} = \rho_2 U_2 R_2/\mu_2$,
- the Weber number $\mathrm{We} = \rho_2 U_2^2 R_2/\gamma_2$,
- the electrical Euler number $Eu = \varepsilon_3 V_0^2/\rho_2 U_2^2 R_2^2 \ln^2(R_3/R_3)$.

The last three dimensionless parameters represent the relative magnitudes of the viscous force, surface tension and electrical force to the inertia force, respectively. In most experiments, R_3 is much larger than R_2, that is, the radius ratio $b \geq 1$. In the dispersion relation (244), the parameter b appears only in J and L, for which we have the limits:

$$J\big|_{b\to\infty} = -\frac{K_1(k)}{K_0(k)} \text{ and } L\big|_{b\to\infty} = -\frac{K_0(k)}{K_1(k)} \tag{257}$$

For large Reynolds numbers ($\mathrm{Re} \geq 1$), the dispersion relation (244) and the amplitude ratios (245) and (246) can be simplified dramatically to:

$$D(k,\omega) = ak^2 \left(T_1\Delta_3 - \Delta_1\right)\left(T_2\Delta_3 + \Delta_2\right) + 1 = 0, \tag{258}$$

and

$$\left(\frac{\hat{\eta}_2}{\hat{\eta}_1}\right)_1 = -ak\left(T_1\Delta_3 - \Delta_1\right), \left(\frac{\hat{\eta}_1}{\hat{\eta}_2}\right)_2 = k\left(T_2\Delta_3 + \Delta_2\right) \tag{259}$$

with

$$T_1 = -\omega^{-2}\left[\frac{iS\omega}{k}\left(i\omega\frac{I_0(ka)}{I_1(ka)} - \frac{2\Lambda}{a}\right) + \frac{\Gamma}{Wea^2}\left(1 - (ka)^2\right)\right] \tag{260}$$

$$T_2 = -\omega^{-2}\left[iQ\omega\left(-i\omega L + \frac{2}{b^2 - 1}\right) - Euk(1 + kJ) + \frac{k}{We}\left(1 - k^2\right)\right] \tag{261}$$

If the relative velocity ratio $\varLambda=0$, the dispersion relation is consistent with that for the electrified coaxial jet in the equipotential case where$\varLambda$ is fixed to 1[346].

In addition, the dispersion relation (244) can be reduced to that for a single-liquid jet in a simple way. If the radius ratio a approaches zero, the jet consists only of the outer leaky dielectric liquid. The inner interface vanishes, and the numerator of Eq. (246) is zero, yielding

$$\mathrm{Re}^2 H_2 - k\left[\left(l^2+k^2\right)\frac{\Delta_2}{\Delta_3}+2k\right]\left(\frac{Eu\,\mathrm{Re}\varsigma}{\omega}+1+\frac{l^2}{k^2}\right)+2lk^2\left(\frac{\Delta_5}{\Delta_4}+\frac{1}{l}\right)\left(\frac{Eu\,\mathrm{Re}\varsigma}{\omega}+2\right)=0 \qquad (262)$$

Where $\dfrac{\Delta_5}{\Delta_4}=\dfrac{-I_0(l)}{I_1(l)}$, $\dfrac{\Delta_5}{\Delta_4}=\dfrac{-I_0(l)}{I_1(l)}$ and

$$\varsigma=\frac{k\dfrac{I_0(k)}{I_1(k)}\left(1+\dfrac{l^2}{k^2}\right)-2l\dfrac{I_0(l)}{I_1(l)}\dfrac{\Delta_5}{\Delta_4}+\dfrac{\mathrm{Re}\,\omega J}{k}}{\dfrac{\mathrm{Re}}{k}\left(\varepsilon_{r2}\left(\omega+\tau\right)\kappa-\omega J\right)+\dfrac{Eu\,\mathrm{Re}}{\omega}\left(-k\dfrac{I_0(k)}{I_1(k)}+l\dfrac{I_0(l)}{I_1(l)}\right)} \qquad (263)$$

After some algebra, the above equation is written in a clearer form:

$$\varsigma\omega^2+\frac{2\omega}{\mathrm{Re}}\left(2k^2\varsigma-1\right)+\frac{4k^2}{\mathrm{Re}^2}\left(k^2\varsigma-l^2\varsigma_v\right)+T+\left(1+\frac{Euk\varsigma}{EJ\omega^2}\left(l^2\varsigma_v-k^2\varsigma\right)\right)^{-1}$$
$$\left[\frac{2Eu\varsigma k^2}{E\,\mathrm{Re}\,\omega}\left(2+\frac{1}{kJ}\right)\left(k^2\varsigma-l^2\varsigma_v\right)+\frac{Eu\varsigma}{E}\left(2k^2\varsigma+\frac{k\varsigma l^2\varsigma_v}{J}+kJ\right)\right]=0 \qquad (264)$$

Where

$$\varsigma(k)=\frac{I_0(k)}{kI_1(k)},\ \varsigma_v(l)=\frac{I_0(l)}{lI_1(l)}$$

$$E=\frac{\varepsilon_{r2}}{kJ}\left(1+\frac{\tau}{\omega}\right)-\varsigma,\ T=\frac{iQ\omega}{k}\left(i\omega L-\frac{2}{b^2-1}\right)+Eu\left(1+kJ\right)-\frac{1}{We}\left(1-k^2\right) \qquad (265)$$

The Eq.(264) is the dispersion relation for a viscous jet with a leaky dielectric liquid, which exactly corresponds with Lopez-Herrera et al. [315, 368].

14.13.7.1   NUMERICAL RESULTS OF VISCOUS COFLOWING JET MODEL

The dimensionless dispersion relation (244) is a quartic equation for the complex frequency ω. Given an axial wavenumber k, there are generally four eigenvalues corresponding to four different modes, but only two of the modes are unstable in the Rayleigh regime [351]; these are usually called the para-sinuous mode and the para-varicose mode. Suppose the interface perturbation amplitude ratio $\hat{\eta}_1/\hat{\eta}_2 = |\hat{\eta}_1/\hat{\eta}_2|\exp(i(\theta_1 - \theta_2))$, where $|\hat{\eta}_1/\hat{\eta}_2|$ is the relative magnitude of the amplitudes, and $\theta = \theta_1 - \theta_2$ is the corresponding phase difference. The para-sinuous mode means that the inner liquid–liquid interface and the outer air–liquid interface are perturbed almost in phase, that is, θ approaches $0°$; and the para-varicose mode means that the two interfaces are perturbed nearly out of phase, that is, θ approaches $180°$. Under most experimental situations [315], the para-sinuous mode is less stable than the para-varicose one and is dominant in the jet instability, promoting the formation of compound droplets. However, as long as the unstable para-varicose mode exists, coaxial electrospraying is negatively influenced by it. On the other hand, under a sufficiently intense electric field and sufficiently large flow rate, those nonaxisymmetric modes may become comparable to the axisymmetric ones, and even become dominant [346, 347, 356].

The theoretical model for the axisymmetric instability and maintaining the axisymmetric modes dominant through controlling the values of the dimensionless parameters was studied by Li et al. [347]. The non-axisymmetric instability deserves special study. They solved Eq. (244) numerically in order to study the behavior of the coaxial jet under the radial electric field, mainly taking case I as an example. For convenience of calculation and comparison, a set of dimensionless parameters was chosen as a reference set. In case I, they took water and air as the outer and inner liquids, respectively, because water is the most common leaky dielectric and air is a common perfect dielectric, corresponding to the case of the outer-driving coaxial electrospraying [347]. Their physical properties can be found in Ref. [315]. The reference dimensionless parameters are $Q=0.001$, $S=0.001$, $a=0.8$, $b=10$, $\Lambda=0.2$, $Re =10$, $We=10$, $\Gamma=1$, $Eu = 0.15$, $\varepsilon r_1 =1$, $\varepsilon r_2 =80$ and $\tau = 1$. In the calculation, the dimensionless parameters were fixed to the reference values unless stated otherwise.

14.13.7.2   EFFECT OF LIQUID VISCOSITY AND COMPARISON WITH THE INVISCID MODEL

In order to study the effect of the viscosity of the outer liquid, Li et al. [347] first established an inviscid model. In the inviscid model, both the inner liquid and the outer liquid are assumed to be inviscid, and the velocities of the inner and outer liquids in the basic state are assumed to be uniform with a discontinuity at the inner and outer interfaces. Denoting the base axial velocities of the inner and outer liquids by

U_1 and U_2, respectively, a new dimensionless parameter $\Lambda\dagger = U_1/U_2 - 1$ is obtained. For leaky dielectrics, the dimensionless dispersion relation of this inviscid model is also expressed in the form of Eq. (244), but with different perturbation amplitude ratios:

$$\left(\frac{\hat{\eta}_2}{\hat{\eta}_1}\right)_1 = \frac{k^2 a H_1 \Delta_3 + k a \omega^2 \Delta_1}{\omega^2} \tag{266}$$

$$\left(\frac{\hat{\eta}_1}{\hat{\eta}_2}\right)_2 = -\frac{k^2 H_2 \Delta_3 - k \omega^2 \Delta_2}{\omega^2 + E u k^3 J \xi \Delta_3} \tag{267}$$

where

$$H_1 = -\frac{S\left(\omega + i\Delta^1 k\right)^2 I_0(ka)}{k I_1(ka)} + \frac{T}{Wea^2}\left(1 - (ka)^2\right) \tag{268}$$

$$H_2 = -\frac{Q\omega^2}{k} L + Eu\left(1 + k(\xi+1)J\right) - \frac{1}{We}\left(1 - k^2\right) \tag{269}$$

$$\zeta = \frac{k\omega(J - \Delta_2/\Delta_3)}{k\varepsilon_{r2}(\omega + r)k - k\omega J} \tag{270}$$

$$\xi = \frac{\omega/\Delta_3}{k\varepsilon_{r2}(\omega + \tau)k - k\omega J} \tag{271}$$

The other symbols are the same as in the viscous model. The coordinate system in this inviscid model is still moving with velocity U_2.

Note that the dispersion relation for the inviscid model is a little different from for the case of large Reynolds numbers, because the tangential dynamic continuity condition at the outer interface is treated differently in these two cases. In the large-Re case the tangential component of the electrical stress at the outer interface is very small, or even vanishes. In such a case the jet surface can be regarded to be approximately equipotential. However, in the inviscid leaky dielectric case, the tangential dynamic condition is not satisfied because the viscous shear is not taken into account, and the jet may be nonequipotential. The dispersion relation given by Eqs. (266) and (267) reduces to that for the equipotential case only if the relative electrical relaxation time τ approaches infinity [346]. In addition, if τ approaches zero it is reduced to the dispersion relation for the nonequipotential case [347].

The influence of the viscosity of the outer liquid on the unstable modes is shown by Li and his co-workers studies [342, 347] in Fig. 14.39, where the relative velocity ratio $\Lambda = 0$. For comparison, the curves for the large Reynolds number limit (marked with '∞') and the inviscid leaky dielectric model with the relative velocity ratio $\Lambda \dagger = 0$ (dashed) are also plotted. It is clear thatthe growth rates of both the para-varicose mode in Fig. 14.39(a) and the para-sinuousmode in Fig. 14.39(b) are greatly enhanced as the Reynolds number increases, indicatingthat the viscosity of the outer liquid has a remarkable stabilizing effect on the jetinstability.

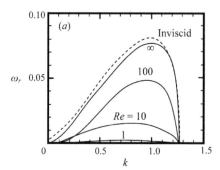
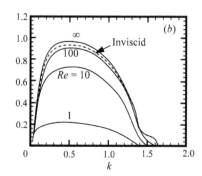

FIGURE 14.39 The influence of the Reynolds number on the growth rates of (a) the para varicose mode and (b) the para sinuous mode, for $\Lambda=0$. The dashed curves are for the inviscid leaky dielectric model.

For the para-sinuous mode, the large-Re limit, that is, the equipotential case isthe most unstable. However, for the para-varicose mode, the inviscid leaky dielectricmodel possesses the maximum growth rate. In general these two special cases are veryclose. The reason may be that the reference value of the relative electrical permittivityof the outer liquid is so high ($\varepsilon r_2 =80$) that the denominators of Eqs. (270) and (271) are large enough to make the effects of ζ and ξ negligible. On the other hand, it is shown in Fig. 14.39 that the viscosity of the outer liquid has no apparent effect on the range of the unstable axial wavenumbers. The cut-off wavenumber kc is approximately 1.25 for the para-varicose mode and 1.6 for the para-sinuous mode.

In case I, the basic velocity profile of the inner liquid is parabolic, where a dimensionless parameter Λ is involved. Apparently, this parameter represents the effect of the shear in the axial velocity on the jet instability. As the density ratio S (=0.001) in the reference state is so small that the effect of Λ cannot be shown clearly, in the calculation the value of S is chosen to be 0.1. Although the density ratio S is increased, the para-varicose mode is little influenced, the para-sinuous mode is stabilized slightly as Λ increases. If the density of the inner liquid is comparable to that of the outer liquid, the effect of Λ becomes more significant. The calculation shows that the growth rate of the para-varicose mode is also reduced as Λ increases. It is well known that the para-sinuous and para-varicose modes are associated primarily with the inner and outer interfaces [346, 347, 351], respectively. Apparently, the parameter Λ influences the mode closely connected with the inner interface more strongly. In general, Li and his co-workers result are in accordance with the fact that the shear in the axial velocity has a stabilizing effect on the jet instability [353]. Similar to the liquid viscosity, the shear in the axial velocity has a negligible effect on the unstable wavenumber range. The critical wavenumber is approximately the same as shown in Fig. 14.39.

In coaxial electrospraying experiments with two liquids, the inner one is usually highly viscous [312, 368]. However, in case I the viscosity of the inner liquid is neglected. Therefore, it is necessary to study the jet instability in case II, in which the inner liquid viscosity is allowed to be comparable to or larger than that of the outer liquid. We take sunflower oil as the inner dielectric liquid. Its physical properties can be found in Lopez-Herrera et al. [368].

14.13.7.3 DISCUSSION ON THE ELECTRICAL RELAXATION TIME AND TWO LIMITING CASES

For EHD leaky dielectrics, there are two important characteristic times, that is, the electrical relaxation time and the hydrodynamic time. The electrical relaxation time $\tau_e \sim \varepsilon/\sigma$ measures the speed of charge relaxation, that is, the contribution of conduction to charge transportation. The hydrodynamic time can have several definitions, such as the capillary time $\tau_C \sim (\rho L^3/\gamma)^{1/2}$, the viscous diffusion time $\tau_V \sim \rho L^2/\mu$, and the convective flow time $\tau_F \sim L/U$. In the most models, the convective flow time $\tau_F \sim R_2/U_2$ is chosen as the characteristic hydrodynamic time, which measures the contribution of convection to charge transportation. As a result, a dimensionless parameter $\tau = (R_2\sigma_2)/(U_2\varepsilon_2)$ measuring the relative magnitude of the hydrodynamic time and the electrical relaxation time, that is, the relative importance of conduction and convection, is involved in the dispersion relation (244). In this section, two limiting cases are derived according to the relative magnitude of the electrical relaxation time and the hydrodynamic time in a more generic sense. Selecting an appropriate hydrodynamic time τ_h, the surface charge conservation equation (238) is nondimensionalized as follows:

$$\frac{\partial q_s}{\partial t} + u.\nabla q_s - q_s n.(n.\nabla)u + \frac{\tau_h}{\tau_e}[\sigma E].n = 0. \tag{272}$$

For well-conducting liquids, the electrical relaxation time τ_e is usually several orders of magnitude smaller than the hydrodynamic time τ_h, that is, τ_h/τ_e, indicating that charge is transported mainly by conduction and the effect of convection is negligible. This case is called the small electrical relaxation time limit (SERT). In this limit the surface charge conservation equation (272) reduces to

$$[\sigma E].n = 0. \tag{273}$$

Note that the surface charge density is absent in Eq.(273). It can be obtained through the boundary conditions. In discussed model, the condition of Eq.(273), which is obeyed at the outer air–liquid interface, implies that the outer liquid is equipotential (i.e., the equipotential case). Conversely, for relatively imperfectly conducting liquids with a relatively high velocity, the electric relaxation time τ_e may be much larger than the hydrodynamic time τ_h, that is, τ_h/τ_e. In such a case charge convection becomes significant and the effect of conduction is negligible. It is called the large electrical relaxation time limit (LERT). In this limit,Eq.(272) reduces to

$$\frac{\partial q_s}{\partial t} + u_s . \nabla q_s - q_s n_s . (n_s . \nabla) u_s = 0. \tag{274}$$

where only the bulk conduction disappears, corresponding to the nonequipotential case. In case LERT, charge at the interface cannot be reset instantaneously to maintain the interface equipotential. The Eq. (274)may serve as the surface charge conservation equation for this limit. In the study of the interfacial instability of a conducting liquid jet under a radial electric field [322, 347], both the equipotential (SERT) and nonequipotential (LERT) cases are considered. According to these two limits, Eqs.(245) and (246) are reduced, with the dispersion relation (244) unchanged in form. For SERT, the amplitude ratio of the initial disturbances is

$$\left(\frac{\hat{n}_2}{\hat{n}_1}\right)_1 = \frac{akRe^2 H_1 \Delta_8 \Delta_4 + a(l^2+k^2) \ ^2\Delta_1\Delta_4 - 4alk^3\Delta_8\left(\Delta_6 - \frac{2}{ia}\Delta_4\right) - 2k(l^2+k^2)\Delta_8\Delta_4}{(l^2+k^2)\Delta_4/k - 4k^3\Delta_8} \tag{275}$$

$$\left(\frac{\hat{n}_1}{\hat{n}_2}\right)_2 = \frac{kRe^2 H_2 \Delta_8\Delta_4 - (l^2+k^2) \ ^2\Delta_2\Delta_4 + 4lk^3\Delta_8\left(\Delta_8 + \frac{2}{i}\Delta_4\right) - 2k(l^2+k^2)\Delta_8\Delta_4}{4k^3\Delta_8 - (l^2+k^2) \ ^2\Delta_4/k} \tag{276}$$

whereH_1 is the same as Eq. (2.24a) and

$$H_2 = \frac{iQ\omega}{k}\left(i\omega L - \frac{2}{h^2-1}\right) + Eu(1 + kJ) - \frac{1}{W_a}(1 - k^2) \tag{277}$$

For LERT, the amplitude ratio of the initial disturbances has the same form as Eqs.(247)
and (248), but with

$$\zeta = \frac{-k\left(1+\frac{l^2}{k^2}\right)\frac{\Delta_2}{\Delta_3} + 2l\frac{\Delta_5}{\Delta_4} + \frac{Re\ \omega J}{k}}{\frac{Re\ \omega}{k}(\varepsilon_{r2}k - J) + \frac{Eu\ Re}{\omega}\left(k\frac{n_2}{n_3} - l\frac{n_5}{n_4}\right)}. \tag{278}$$

$$\xi = \frac{\left(1+\frac{l^2}{k^2}\right)\frac{1}{\Delta_3} - \frac{2}{\Delta_4}}{\frac{Re\ \omega}{k}(\varepsilon_{r2}k - J) + \frac{Eu\ Re}{\omega}\left(k\frac{n_2}{n_3} - l\frac{n_5}{n_4}\right)} \tag{279}$$

Figure 14.40 illustrates the growth rates of the unstable modes for these two limitcases, solid curves for SERT (i.e., $\tau \to 0$) and dashed curves for LERT (i.e., $\tau \to \infty$) according to Li investigation,where several values of the electrical Euler number are considered. It can be seenfrom Fig.14.40(a) that for the para-varicose mode only solid curves for SERT appearwhen the electric field exists. The difference between the two limit cases is remarkable,because for LERT the growth rate of the para-varicose mode decreases to zerowhen the electrical Euler number exceeds 0.01. However, for SERT the growth rate is enhanced by the electric field and the unstable region moves towards relativelyshort waves. On the other hand, for the para-sinuous mode as shown in Fig. 14.40(b), the difference between two limit cases is discernible but not significant for therange of Eu studied, indicating that the influence of the relative electrical relaxationtime τ on the jet instability is small. This may be attributed to the large relativeelectrical permittivity of the outer liquid (ε_{r2} =80), which makes $k\varepsilon_{r2}(\omega + \tau)\kappa$ in thedenominators of Eqs. (249) and (250) a large term, and consequently the influence ofτ is weakened.

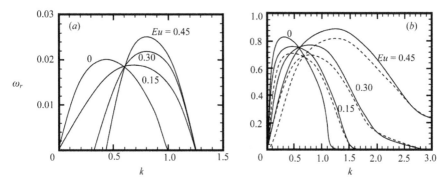

FIGURE 14.40 The growth rates of (a) the para varicose mode and (b) the para sinuous mode, for the limit case SERT (solid curve) and LERT (dashed curve) for different electrical Euler numbers.

14.13.7.4   EFFECT OF THE ELECTRIC FIELD TOGETHER WITH THE OTHER PARAMETERS ON THE JET INSTABILITY

For the inviscid coaxial jet under a radial electric field [347], in agreement with Li et al. [342, 346] studies, the radial electric field has a dual effect on both the para-varicose and para-sinuous modes, destabilizing them greatly when the axial wavenumber k exceeds a critical value, and stabilizing them if k is smaller than that value. Using viscous model, the dual effect of the electric field is persistent for both the para-varicose mode and para-sinuous mode, as shown in Fig. 14.40.

It is well known that when arbitrary disturbances are applied to the jet, the perturbation wave with maximum growth rate and those close to it grow faster than the others, which become dominant in the jet breakup process. Although two unstable modes occur in the jet instability process, the para-sinuous mode is much less stable than the para-varicose one in most situations [311–312, 347]. Consequently, the most unstable wavenumber k_{max} comes from the para-sinuous mode. In Fig. 14.40(b), the value of k_{max} is amplified as the electrical Euler number increases, predicting that the most likely wavelength $\lambda = 2\pi R_2/k_{max}$ is diminished by the radial electric field.

It is necessary to study the effect of several important dimensionless parameters, such as the Weber number and Reynolds number, on the dominant wavenumber k_{max} and corresponding maximum growth rate ω_{max}. Figure 14.41(a, b) illustrates the effect of the Weber number and electrical Euler number on k_{max} and ω_{max}, respectively. The selected values of the Weber number are relatively small, since the coaxial jet is very thin and the surface tension is generally large in the experiments. It is found that the electric field influences k_{max} and ω_{max} slightly when the Weber number is relatively small (We<5). However, at relatively large Weber numbers (We>10), the electric field enhances k_{max} and ω_{max} distinctly. The behavior of k_{max} and ω_{max} indicates that at small Weber numbers the jet instability is dominated primarily by the capillary force, while with the increase of We, the jet instability is dominated primarily by the electrical force [312, 347]. On the other hand, both the electrical Euler number and Weber number change the cut-off wavenumber k_c significantly, as shown in Fig. 14.41(c). Obviously, k_c is enlarged as We increases, especially at large Euler numbers. So the instability region may be extended into the first wind-induced regime, which reduces the formation of monodisperse droplets in coaxial electrospraying.

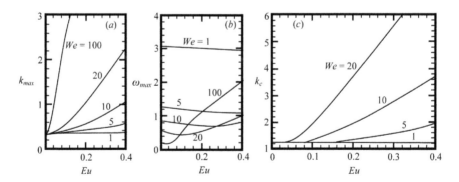

FIGURE 14.41 (a) The dominant axial wavenumber k_{max}, (b) the maximum growth rate ω_{max}, and (c) the cut off axial wave number k_c versus the electrical Euler number for different Weber numbers.

14.13.7.5   DISCUSSION ON THE THIN LAYER APPROXIMATION A →1

For coflowing jets, one extreme case is that the thickness of the outer annular liquid layer is thin. In this section, simple dispersion relations were seeked relatively applicable to the thin layer coating case $a \to 1$ for case I and case II. Some details are given in Appendix C; where case I is taken as an example. The thin layer approximation ultimately results in the dispersion relation [347]:

$$\omega^2 \left(QL - S\frac{I_0(k)}{I_1(k)} \right) + 2i\omega\left(\frac{Q}{b^2-1} - sA \right) + \frac{1+T}{We}k(1-k^2)$$

$$-Eu(1+kJ)\left[1 - \frac{\omega(1+kJ)}{\varepsilon_{r1}(\omega+\tau)\dfrac{I_1(k)}{I_0(k)} - \omega J} \right] = 0 \qquad (280)$$

for case I, and

$$\omega^2 \frac{S}{k}\frac{I_0(k)}{I_1(k)} + \frac{2\mu_r\omega}{Re}\left(2k\frac{I_0(k)}{I_1(k)} - 1 \right) + \frac{4u_r^2 k^2}{SRe^2}\left(k\frac{I_0(k)}{I_1(k)} - l_1\frac{I_0(l_1)}{I_1(l_1)} \right)$$

$$+ |\frac{iQ\omega}{k}\left(i\omega L - \frac{2}{b^2-1} \right) + Eu(1+kJ) - \frac{1+T}{We}(1-k^2)$$

$$- \frac{SEu\left[k\omega\left(J\dfrac{I_0(k)}{I_1(k)} \right) + \dfrac{2\mu_r k^2}{S\,Re}\left(k\dfrac{I_0(k)}{I_1(k)} - l_1\dfrac{I_0(l_1)}{I_1(l_1)} \right) \right]^2}{Sk\omega\left[\omega J - \varepsilon_{r1}(\omega+\tau)\dfrac{I_1(k)}{I_0(k)} \right] + k^2 Eu\left(k\dfrac{I_0(k)}{I_1(k)} - l_1\dfrac{I_0(l_1)}{I_1(l_1)} \right)} = 0 \qquad (281)$$

for case II. The above expressions appear much simpler than for the coaxial jet. However, their validity needs to be evaluated. Note that under the thin layer approximation the viscosity and dynamic force of the outer liquid, as well as the electrical permittivity of the outer liquid, have no influence. Only the electrical permittivity of the inner liquid and the electrical relaxation time of the outer liquid layer play a role [346, 347]. Especially in the limit case $\tau \to \infty$ Eqs.(280) and (281), respectively, reduce to

$$\omega^2 \left(QL - S\frac{I_0(k)}{I_1(k)} \right) + 2i\omega\left(\frac{Q}{b^2-1} - SA \right) + \frac{1+T}{We}k(1-k^2) - Eu(1+kJ) = 0 \qquad (282)$$

and

$$\omega^2 \frac{S}{k} \frac{I_0(k)}{I_0(k)} + \frac{2\mu_r \omega}{Re} \left(2k \frac{I_0(k)}{I_1(k)} - 1 \right) + \frac{4\mu_1^2 k^2}{S Re^2} \left(k \frac{I_0(k)}{I_1(k)} - l_1 \frac{I_0(l_1)}{I_1(l_1)} \right)$$

$$+ \frac{iQ\omega}{k} \left(i\omega L - \frac{2}{b^2 - 1} \right) + Eu(1 + kJ) - \frac{1+T}{We}(1 - k^2) = 0 \qquad (283)$$

The dispersion relations (282) and (283) accord with those for the single-liquid jet in the equipotential case but with double surface tension, the former for the inviscid liquid and the latter for the viscous liquid.

Under the thin layer approximation, only one unstable mode, that is, the para-sinuous mode, exists. Compared with the exact solution of the leaky dielectric model, if the viscosity of the inner liquid

is neglected the thin layer approximation is more accurate when the density of the inner liquid is comparable to that of the outer liquid, but if the inner liquid viscosity is taken into account the thin layer approximation is accurate only for a relatively small electric field [346, 347].

14.14 APPLICATION OF COAXIAL ELECTROSPUN NANOFIBERS

Due to their excellent physicochemical properties and flexible characteristics, coaxial electrospun core–sheath or hollow fibers can be used in various fields. For example, core–sheath fibers can exhibit an enhanced mechanical property compared to simple fibers, such as silk fibroin-silk sericin core–sheath fibers with a breaking strength of 1.93 MPa and a breaking energy of 7.21 J kg^{-1}, which are 82% and 92.8% higher than those of single silk fibroin fibers. To illustrate more specifically, in this section, the applications of these fibers in the fields including lithium ion batteries, solar cells, luminescence, super capacitors, photocatalytic environmental remediation and filtration are critically reviewed with examples and design principles.

14.14.1 LITHIUM ION BATTERIES

Right now, the whole society faces a serious energy challenge with traditional fossil energy being used up and other new energies not ready for large-scale deployment. The widely used internal combustion engine using fossil energy has faced a bottleneck in the energy field development due to its low energy conversion resulting from its low energy efficiency, normally less than 30%. And the combustion of fossil fuels can cause severe environmental pollution. Urged by all these problems, scientists turn to a high-grade sustainable clean energy, electricity. And among the energy storage devices, batteries have been in the spotlight attracting considerable attention. Among all the batteries, the lithium ion battery has become the primary candidate in many applications, such as communication, transportation and regenerated energy sectors due to its higher voltage (about 3.6 V, two times higher than that of aqueous batteries), gravimetric specific energy (about 240 W h kg^{-1}, six times higher than that of lead acid batteries), long duration (500–1000 cycles), wide temperature range (20 to 60°C) and minimum memory effect [117, 372]. Figure 14.42 shows the schematic diagram of a typical commercial lithium ion battery with lithium alloy compound and graphite as the cathode and the anode, respectively. After the circuit is connected, electrons will flow from the anode to the cathode through an external circuit forming current driven by the chemical potential difference between the electrode materials, at the same time, lithium ions are transported in the same direction through the electrolyte inside the battery. At the cathode, lithium ions react with cathode material and electrons, and deposit there. During the charging process, both electrons and lithium ions go back though the previous pathway driven by the applied potential difference. Through the charge–discharge process, the stored chemical energy is finally converted into electricity. The chemical reactions during the charging and discharging processes are listed [116].

$$Cathod : \frac{1}{2}Li^+ + \frac{1}{2}e^{-1} + Li_{0.5}CoO_2 \Leftrightarrow LiCoO_2$$

$$Anode : LiC_6 \Leftrightarrow Li^+ + e^- + C_6$$

During the charging process, the lithium ions will insert into the graphite layer and combine with carbon atoms.

(a)   (b)

FIGURE 14.42 Schematic illustration of a typical lithium-ion battery with graphite andLiCoO2 as anode and cathode materials, respectively.

This requires that the anode material should provide an extremely large surface area for a convenient combination. On the other hand, the insertion of lithium ions could cause electrode volume expansion, which affects the electric contact and battery capacity. And this expansion is a very serious problem in all lithium batteries. Therefore, different electrode materials have been studied aiming to eliminate the side-effect of the volume expansion and to support enough active sites. Hollow nanofibers are excellent candidates as anode material for lithium ion batteries. The hollow structure not only possesses very high specific surface area, but also buffers the volume expansion during the lithiation process. And the hollow structure can integrate various components together to promote the performance of the lithium ion battery [116, 117, 372].

Graphite carbon is a very common commercial anode material for lithium ion batteries, but due to its low capacity and safety issues, other materials including transition metals, semiconductors and lithium alloys have been studied. Generally, carbon material is an excellent choice for an electrode in the battery system and it can be classified into graphitic (soft) and nongraphitic (hard) carbon. Specifically, soft carbon possesses a well-ordered lamellar structure, while hard carbon displays a relatively turbostatic arrangement. And significant property differences as the lithium ion battery anode result due to the structural diversity. Hard carbon possesses a high capacity (400–500 mA h g^{-1}), but poor capacity retention performance, which means that the high capacity will get attenuated very soon. Compared with hard carbon, soft carbon has a lower, but reversible capacity (200–300 mA h g^{-1}), however, it shows a very serious voltage hysteresis during the dilithiation process, in which lithium ions are desorbed from the anode. A combination of the advantages of both materials can enhance the performance of lithium ion batteries and the coaxial electrospinning technique has been reported to achieve this goal. Other scientists considered adding other components to enhance the performance. With a high theoretical capacity of about 4000 mA h g^{-1}, nearly ten times that of the commercial graphite anode, silicon has been integrated into electrodes to improve the lithium ion battery performance. TiO_2 is another good choice due to its low cost, high working voltage, and structural stability during lithium insertion and extraction processes, although it has some drawbacks, it still attracts the attention. SnO_2 is also an excellent anode material due to its higher capacity (about 800 mA h g^{-1}) than that of graphite, high charge and discharge capacity, and fast electron transportation. All these materials could be fabricated into hollow composite nanofibers as anodes for lithium ion batteries to enhance their performance [116, 117, 373].

Lee et al.[374] have used coaxial electrospinning to fabricate hollow carbon nanofibers as anode materials and studied the carbonization temperature effect on the electrochemical performance. Both styrene-co-acrylonitrile(SAN) and PAN are good choices as carbon precursors and their DMF solutions serve as the core and sheath solution, respectively. The as-spun fibers experienced a one-hour stabilization process at 270–300°C in an air atmosphere and a following one-hour carbonization at 800, 1000, 1200, 1600°C in a nitrogen atmosphere. During stabilization, the linear PAN molecules were converted to the ladder structure and got carbonized in the following process, at the same time, the core phase burned out resulting in the tubular structure.

Liu et al.[375] reported a core–sheath soft–hard carbon nanofiber web, which-displayed an improved electrochemical performance as an anode in the lithium ion battery. A special terpolymer fibril (93 wt% acrylonitrile, 5.3 wt% methylacrylate and 1.7 wt% itaconic acid) in DMF served as the sheath solution and mineral oil as the core solution. The as-spun fibers were stabilized for 6.5 h at 270°C in an air atmosphere and then carbonized for 1 h at 850°C under nitrogen protection. Finally, the soft–hard core–sheath carbon nanofibers were obtained with sheath PAN converted to hard carbon and core mineral oil decomposed to amorphous soft carbon. In this anode configuration, the hard sheath could prevent the deformation of soft core, which dominated the stable reverse capacity after a long service time. And an enhanced reversible capacity was obtained, 520, 450 and 390 mA h g⁻¹ after 20 cycles at 25, 50 and 100 mA g⁻¹, respectively.

Lee et al.[374] have studied the electrochemical performance of the Si–C core–sheath fibers as the lithium ion battery anode. The fibers were prepared by following similar procedures63 and are briefly stated as follows. The Si nanoparticles with a diameter smaller than 100 nm were added to the core solution. The as-spun nanofibers were stabilized at 270–300°C for one-hour in an air atmosphere and carbonized at 1000°C for one-hour in a nitrogen atmosphere with a heating speed of 10°C min⁻¹. PAN converted to a carbon sheath and Si nanoparticles attached onto the inner wall of hollow fibers. Si–C hollow fibers are formed. Figure 14.43 illustrates the volume expansion mechanisms during the lithiation process. Due to the large d-space between turbostratic carbon layers, the electrode volume expansion could be caused by the combination of Si and lithium ions.

Hwang et al. [376] have fabricated core–sheath fibers containing different loadings of Si in the carbon core matrix as the anode for lithium ion batteries. Specifically, PAN was the sheath material, and poly(methyl methacrylate)(PMMA) and silicon nanoparticles were core materials. DMF served as a solvent for both phases. Proper quantity of acetone was also added to the core solution to prevent the mixing of core and sheath materials. PMMA worked as a stabilizer to encapsulate silicon nanoparticles and left enough space after burning out to buffer the volume expansion during the charging process. The as-spun fibers were stabilized at 280°C for 1 h in an air atmosphere and further carbonized at 1000°C for 5 h in argon. The electrode was prepared as follows: about 70 wt% electrospunfibers, 15 wt% super P and 15 wt% poly (acrylic acid)(PAA) were mixed and added to 1-methyl-2-pyrrolidinone (NMP) to form slurry, which was then pasted onto a copper current collector. The prepared electrode was dried in a vacuum oven at 70°C for 6 h and then punched into circular discs. The more weight percent of silicon was added, the higher discharge capacity the samples had demonstrated. This is due to the superior capacity of silicon itself. For comparison, two more control materials have been prepared with the same method as electrodes. The first one was bare silicon nanoparticles with super P and the second one was carbon fibers decorated with Si nanoparticles. Compared with these two control electrodes, the desired electrode exhibited a high discharge capacity around 1250 mA h g⁻¹ with a nearly 100% retention in the first 100 cycles at 0.242 A g⁻¹ current rate. With further increase in the current rate up to 2.748 and 6.89 A g⁻¹, the electrode could still display a rather stable capacity in a larger cycle period with 99% retention after 300 cycles and 80.9% retention after

1500 cycles. This improved electrode performance is due to the extremely high theoretical specific capacity of silicon, nearly 4000 mA h g^{-1}, which is about ten times higher than those of commercial graphite anodes. And the performance is also determined by the stability of the solid electrolyte interphase (SEI) and the contact between Si nanoparticles and carbon matrix. Because the Si volume expansion at high current rate was not that considerable, the SEI layers were more stable and thus more reversible charge–discharge reaction occurred.

Han et al.[377] have investigated TiO_2 as the anode for lithium ion batteries. An improved electrochemical performance of the TiO_2 hollow fibers and nitridated TiO_2 hollow fibers was reported than that of solid ones. Ti (OiPr)$_4$ was used as a titanium precursor and PVP as a stabilizer in the sheath phase, and mineral oil was used as the core phase. After natural drying, extraction of mineral oil and calcination, hollow fibers were obtained. The nitridation step was processed through another annealing treatment under ammonia and the related schematic morphology changes are shown in Fig. 14.8a–c. The solid, hollow and nitridated hollow TiO_2 nanofibers exhibited initial coulombic efficiencies of 75.8, 77.1 and 86.8%, and capacity retentions of 96.8, 98.8 and 100% after 100 cycles. And at different current rates, hollow fibers exhibited a higher capacity than solid fibers. This improved performance is due to the larger surface area (around 25%) for lithium ions to combine with active materials, and the decreased diffusion length of the lithium ions down to nearly 50%. The diffusion length is the maximum distance, through which lithium ions must diffuse to combine with anodic materials. Therefore, the longer the diffusion distance is, the more adverse it is for the performance of the lithium ion battery. And the high conductivity of the nitridated hollow nanofibers will also benefit its performance. The enhanced anode stability arises from the hollow structure, which provides space for volume expansion during lithiation.

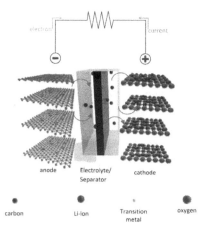

FIGURE 14.43   The schematic diagram of (a) volume expansion behavior during contact-lithiation and (b) ionization of Li atoms in the Si matrix.

14.14.2 SOLAR CELLS AND LUMINESCENCE

Other than serving as electrodes for lithium ion batteries, coaxial electrospun fibers have also been applied in other electrochemical fields, such as electrodes of solar cells and luminescence materials. Due to the nonrenewability of traditional fossil energy and its environmental impact, scientists have been searching and developing clean future energy resources, among which solar, wind, nuclear and even biomass energies have been used commercially. Particularly, solar energy has attracted extensive attention due to its environmentally friendly nature, stabilization and regeneration. Furthermore, the energy that the earth receives in just one hour can support the whole energy consumption of our world in one year. Thus solar energy has great potential to be developed for future deployments through conversion to electricity in solar cells. However, the charge recombination especially at the anode is a very serious problem that limits the efficiency[116].

Different strategies have been explored to suppress the charge recombination, for example, reduced recombination sites [378] and construction of an energy barrier. The core–sheath nanofibers are excellent candidates that can satisfy these two requirements. The tiny dimension could reduce the possibility that electrons recombine with the electrolyte and the sheath phase can naturally prevent serious combination as barriers. On the other hand, light is a very common, but very potential future energy. There are two ways to generate light, incandescence and luminescence. Incandescence describes that items such as tungsten filament can emit light when heated to a very high temperature. Compared with incandescence, luminescence refers to "cool light," such as the screens of electronic devices. Luminescence is caused by the jump of excited electrons back to the less excited or ground state. The energy difference is released in the form of "cool light"[379]. Luminescence can be classified into many different types according to input energy sources, such as chemiluminescence, electroluminescence, triboluminescence and bioluminescence. If the input energy is another kind of light, for example, ultraviolet radiation or X-ray, this luminescence is called fluorescence. Many materials have been found to be luminescent, including transition metal ions, rare earth metal complexes, heavy metals, quartz, feldspar and aluminum oxide with electron–hole pairs. Some organic materials are also luminescent. Luminescence makes materials detectable and scientists are trying to develop new signaling systems and devices with luminescent materials aiming to fabricate sensors with high efficiency and accuracy in detection, biological and medicine engineering. New luminescent materials have been prepared for LEDs, which have longer service time, lower energy consumption and stronger brightness. The utilization of core–sheath nano-fibers could increase the contact area, which is beneficial for the excitation process. On the other hand, the fiber form can integrate optical and optoelectronic devices into textile [116, 117, 380].

14.14.3   FILTRATION

Nanofibers are a perfect choice for different types of separation, such as air purification, solution filtration and osmosis separation, due to their extremely high surface area to volume ratio, highly porous structure and chemical properties in some special cases, as well as the flexibility and low cost [52, 117]. Generally, the electrospun nanofiber mats can be used for filtration directly and some researchers have used fibers for wastewater purification [381]. The unique structure of hollow nanofibers becomes the intrinsic advantage as the osmosis membrane material and the appropriate selection of sheath material could promise better performance in separation. Since the convenient fabrication of hollow fibers through the coaxial electrospinning process, it has opened a new door to this field [303]. Anka et al.[382] have reported the wonderful separation performance of the PAN hollow fiber membrane upon treatment with Indigo carmine dye and sodium chloride solutions. They used PAN–DMF solution as the sheath and mineral oil as the core. The as-spun fibers were sunk into hexane to remove the core phase. The fabrication and filtration processes are illustrated in Fig. 14.44(a) and (b). The results showed that the dye and sodium salt were rejected 100 and 97.7%, respectively.

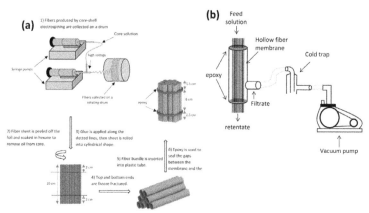

FIGURE 14.44   The schematic diagram of the hollow fiber membrane filtration setup. Reprinted with permission from American Chemical Society.

14.14.4   ELECTROSPUN CORE-SHELL NANOFIBERS AS DRUG-DELIVERY SYSTEMS

If drugs or bioactive agents are encapsulated by a shellpolymer, core-shell electrospun nanofibers can be usedfor functional drug delivery. In this regard, coaxialelectrospinning might be particularly suitable for makingbiomimetic scaffolds with drug delivery capability.The advantage is that it does not require the drug to beelectrospinnable or for it to have good physicochemicalinteraction with the carrier polymer [383–386]. In contrast, forthe cases of drugs loaded by blend electrospinning, poorinteraction between the drug and polymer [122, 123, 129] and drug nonelectrospinnability [387]both tremendously affect the drug distributionin the polymer matrix and consequently the releasebehavior.

The benefits of using core-shell nanofibers forsuch a purpose are quite obvious. Firstly, it will be able topreserve those labile biological agents such as DNA andgrowth factors from being deactivated or denatured evenwhen the applying environment is aggressive. In fact, suchprotection begins as early as during the fabrication stagebecause, unlike blend electrospinning, the aqueous solution Secondly,core-shell nanofibers belong to reservoir type drug releasedevice; therefore it will be possible to address the burstrelease problem noted in those electrospun fibers wheredrugs were usuallyincorporated through electrospinninga blend of the drug and polymer carrier [120, 123, 124, 129]. Furthermore, by manipulatingthe core-shell nano-/microstructure, desired and controlledreleasing kinetics could be achieved.

14.14.4.1   DRUG RELEASE MODELING

Modeling loading and releasing drugs will provide a basis forfurther design and optimization of processing conditions tocontrol the core-sheath nanostructure so as to achieve highlysustainable, controllable, and effective drug factorreleases. In the context of tissue engineering applications,as delivery of growth factors is indispensable in the courseof tissue regeneration, it is believed that coaxial electrospinningand the produced core-shell nanofibers will have greatpotential to locally regulate cellular process for a prolongedtime through controlled release of these appropriate growthfactors directly into the cell living microenvironment [305, 386].

related to core-shell fibers, two controlling phenomena can be considered: diffusion through the polymer shell (barrier diffusion) or partition of the drug from the core to the shell. The diffusion through the shell polymer should not be too slow; otherwise this diffusion will be rate-limiting step. In this instance, the system behaves as monolith fibers and not core-shell fibers (reservoir system). Shell porosity must also be carefully controlled since the drug from the core will be released through water-filled channels rather than through the barrier/shell polymer. Composite fibers that contain drug vehicles such as microspheres and nanoparticles are also a type of reservoir system (double barrier system) in which the drug molecules have to diffuse through longer pathways: the polymer comprising the vehicle and the "shell" polymer.

The power law equation, which was developed considering that the main mechanism for drug release is drug diffusion through the polymer or solvent diffusion inside the polymer that produces polymer relaxation/chain rearrangement is the most widely used equation in works concerning drug release:

$$\frac{me}{m_{cot}} = a_0 + kt^n \qquad (284)$$

where m_t/m_{tot} is the fractional release of the drug at time t; a_0 is a constant, representing the percentage of burst release, k is the kinetic constant and n is the release exponent, indicating the mechanism of drug release (which can either be Fickian drug diffusion or polymer relaxation and an intermediate case combining the two).

Other models consider different phenomena that control the release such as desorption due to the fact that under the assumption of diffusion control, 100% release of the drug is expected, but this was not verified experimentally. In the desorption model, the release is not controlled by diffusion, but by desorption of the drug from fiber pores or from the fiber surface. Thus, only the drug on the fiber and pore surfaces can be released, whereas the drug from the bulk can only be released when the polymer starts to degrade. These assumptions are similar to the theory of mobile agent that can be released by diffusion and the immobilized agent, which can be released through degradation. The Eq. (14) is based on a pore model, in which the effective drug diffusion coefficient, D_{eff} is considered and not the actual diffusion coefficient in water, D (with $D_{eff}/D \leq 1$) because desorption from the pore is the rate limiting step and not drug diffusion in water, which is relatively fast.

$$\frac{m_e}{m_{cot}} = a\left[1 - exp\left(-\frac{\pi^2}{8}\frac{t}{\tau_r}\right)\right] \qquad (285)$$

where the porosity factor a $\alpha= m_{s0}/ (m_{s0} + m_{b0})< 1$; with m_{s0} and m_{b0} being the initial amount of drug at the fiber surface and the initial amount of drug in the fiber bulk, respectively; m_t is the drug amount released at time t, while the total initial amount of drug in the fiber is $m_{tot} = m_{s0} + m_{b0}$ and τ_r is the characteristic time of the release process [388].

Various release kinetics exist for drug release and the most desirable one is the zero-order kinetics in which the drug is released at constant rate, independent of. Usually, zero-order kinetics is achieved for reservoir systems such as core-shell fibers or composite fibers in which the drug is properly encapsulated in the core of the fiber or in other vehicles (micro/nanoparticles).

Burst stage in this kind of system is diminished (or nonexistent) because there is no drug deposited on the surface of the fibers. As the controlling release phenomena is drug partition from one phase to another and not diffusion, there is no decrease in release rate over time as expected in a diffusion-controlled system (the release rate depends on the concentration gradient and on the length of diffusion path; as release proceeds, the concentration gradient decreases and the diffusion length increases and both contribute to slowing down the release rate).

The change of drug concentration on the fiber surface can be represented by adsorption and desorption terms with constant rates k_{ads}, k_{des}, respectively:

$$\frac{\partial c_B}{\partial_t} = k_{ada} \times C_A \times \left(C_B^{max} - C_B\right) - K_{des} \times C_A \qquad (286)$$

where C_B is drug concentration expressed in relation to the weight of the polymer material (kg/kg), C_A is drug concentration in the fluid (kg/m³).

Spatiotemporal distribution of the drug in the fluid surrounding fibers is described by diffusion equation and additional mass source term describing drug released from the fibers:

$$\frac{\partial C_A}{\partial t} = D_A \nabla^2 C_A + p_p \times \frac{\partial C_B}{\partial t} \qquad (287)$$

where ρ_p is polymer density (kg/m³).

14.15 ADVANTAGES AND DISADVANTAGES OF CO-ELECTROSPINNING PROCESS

Co-electrospinning is of particular interest for those core materials that will not form fibers via electrospinning by themselves. Here the shell polymer can serve as a template for the core material leading to cable-type structures. Cores shell fibers of this type will certainly foster applications, for example, in the field of microelectronics, optics and medicine, which justifies intensive research in this direction. Co-electrospinning can also facilitate manufacturing of polymer nanotubes and it allows one to eliminate the vapor deposition stage characteristic of three-stages TUFT-process and to develop a novel two-stage process [301, 389].

This method cannot only be used to spin the unspinnable polymers (polyaramid, nylon, and polyaniline) into ultrafine fibers, but also ensures keeping functionalizing agents like antibacterial and biomolecules agents inside nanofibers.Inspite of all advantages of coaxial electrospinning process, controlling of coelectrospinning is more difficult than simple electrospinning because supervising the flow rates and the physical properties of the two liquids, the current transported by them, the diameters of the inner and outer jets, and the size of the droplets resulting from their break up are major concerns[325].

14.16   GENERAL ASSUMPTIONS IN CO-ELECTROSPINNING PROCESS

In all offered models for coaxial electrospinning following assumptions are considered:

1. The gas and the liquids are incompressible Newtonian fluids;
2. The effects of gravity, magnetic field, and mass transfer at the interfaces are neglected.
3. The fluids are assumed inviscid, and the flow is irrotational;
4. In most cases, liquid 2 (the driving liquid) is a perfect conductor having an infinite conductivity, while liquid 1 is a perfect dielectric, which does not influence the flow because charges stay on the outer interface (jet boundary) according to electrostatics.
5. Charges are introduced from the anode and transported by means of convection and conduction of liquid 2.
6. All fluids are assumed having constant conductivity and dielectric properties.
7. Models have neglected the elastic effect due to the evaporation of the solvent.

14.17   CONCLUSIONS AND FUTURE PERSPECTIVE

Despite the intensive research in the field of nanofibers a number of unanswered questions still remain to act as a driving force for further studies. The largest challenge is a complete understanding of the electrospinning mechanism. In order to control the properties, orientation and mass production of the nanofibers, it is necessary to understand quantitatively how electrospinning transforms the fluid solution through a millimeter-sized needle into solid fibers having diameters that are four-to-five orders smaller. The next bottleneck in the electrospinning is the process efficiency and repeatability. Furthermore, the construction of a proper, 3-D scaffold remains a technological challenge, while from the point of view of drug delivery the drug loading has to be increased and the initial burst release has to be reduced in many cases.

Coaxial electrospinning is an emerging technique with only 10 years of history. Although the technique has multiple advantages over traditional microencapsulation/nanoencapsulation processes, further advancement is challenged by the complex physics of the process and a large number of design, material and process parameters contributing to the process outcome.

As an innovation of conventional electrospinning, coaxial electrospinning has attracted much attention and has been studied deeply. The parameters, including both the internal physical properties and the external operation factors have been analyzed theoretically and experimentally. In particular, the interaction between different phases due to the contact and the heterogeneousness, which is also the characteristic of coaxial electrospinning, is discussed in detail. The evaporation rate, polymer concentration, conductivity and flow rate have been systematically reviewed in achieving smooth and uniform morphology as well asfine dimension. Generally, a stronger and longer electrospinning process can result in more smooth fibers. All strategies to enhance the electrical force and elongate the electrospinning whipping stage will favor the final fibers. However, there will always be a suitable operation range for all parameters, only within whichnonbeaded fibers can be obtained. The lower limitation is the critical value, beyond which the electrospinning process can take place. And the higher limitation can prevent the strong effect on the formation of core–sheath fibers. The interaction due to the property difference should promise the formation of the core–sheath Taylor cone and the smooth electrospinning process. Too huge a difference may lead to a significantly different behavior during electrospinning. Through additional treatment, hollow nanofibers can be fabricated. The following extraction process or thermal treatment can remove the core phase, maintaining the complete sheath structure. In addition, the thermal treatment is much quicker and can stabilize and carbonize the polymer matrix at the same time, which can increase its efficiency.

Nearly all polymer solutions and their polymer composite matrices could be fabricated into core–sheath and hollow nanofibers through the coaxial electrospinning technique. Through appropriate post-treatments of the electrospun fibers, various inorganic nanofibers or nanotubes have been developed with much wider applications including photocatalysis, lithium ion batteries supercapacitors, solar cells, etc. The obtained nanofibers possess extremely high specific surface area and combined properties, which have shown superior performance over the traditional electrospun fibers. However, there are also some challenges, such as the dimension and order of the obtained nanofibers, the improved performance in their applications and the expanded applications of the core–sheath and hollow nanofibers. So far, the fine dimension of the core–sheath nanofibers is about 60 nm, which is much larger than that of the fibers through conventional electrospinning, about a few nanometers. The obtained fibers are in random order, not aligned. To obtain ultra-fine core– sheath nanofibers, the matrices should be selected properly for phases, as well as the operation parameters, collection method and post-treatment methods, such as thermal treatment. To develop the properties and performance of core–sheath and hollow nanofibers, new materials should be tested to integrate into both phases as desired, for example, biomaterials can be used as the matrix to increase the biocompatibility in applications of tissue engineering. Polymers can be integrated into a sheath phase to fabricate high strength materials with very low density. Different nanoparticles can be loaded to achieve suitable interactions during charging and discharging processes and catalysis. Core–sheath and hollow nanofibers are very promising potential materials in various applications due to their flexibility and excellent physicochemical properties. Therefore, more amazing and meaningful applications of the materials are envisioned to be achieved in the future when deployed in other fields, such as the magnetic nanocomposites, semiconductors, carbon materials with high strength and conductivity, anticorrosion and fire-proof properties.

Current research efforts on this technique focus on concept approval and feasible study, with about 100 journal papers based on individually customized experimental setups, a limited number of material combinations and empirical process parameters. Numerical study of coaxial electrospinning is rare due to the complex physical nature of the phenomenon. The researchers working in this area can be categorized into two groups.

The first group is biomedical engineers and clinical researchers who are interested in modeling encapsulating biological agents in core-shell nanofibers and simulating drug loading and releasing. The second group includes experimentalists and theorists in the field of fluid mechanics who are interested in experimental and modeling works associated with coaxial electrospinning.

Discussed models for coaxial electrospinning have some assumptions in order to simplify the process. These assumptions would restrict the accuracy of final results. So it is appropriate to reduce these approximations and make the virtual experimental environment more close to real one which can be considers in future approaches.

The coaxial electrospinning model can be extended for situations where the inner liquid has high electric conductivity and the outer liquid is dielectric. Also, the effects of gravity, magnetic field, air drag, solvent evaporation and mass transfer at the interfaces can be considered. In all offered models diameter of both layers of core shell nanojets considered to be constant, while they reduce gradually in real. This fact can be taken into account too. Future investigation could also include the effect of the viscosity of the fluids on the instability of the coaxial jet.

APPENDIXES

APPENDIX A. DERIVATION OF THE DISPERSION RELATION FOR CASE I

In case I, the inner liquid and air are inviscid, with parabolic basic velocity profiles. In such a case, the radial component of the velocity satisfies the following Bessel equation [369]:

$$\frac{d^2 \hat{u}_{1.3}}{dr^2} + \frac{1}{r}\frac{d\hat{u}_{1.3}}{dr} - \left(k^2 + \frac{1}{r^2}\right)\hat{u}_{1.3} = 0. \tag{A1}$$

where the 'hats' stand for the initial perturbation amplitudes. The solutions of Eq. (A1) are the linear combination of the modified Bessel functions. Their coefficients are determined by using the boundary conditions (221) and (235)–(237).

Suppose that the cylindrical coordinate system (r, θ, z) is moving with velocity $U2$, then the solutions are

$$\hat{u}_1 = \omega \hat{n}_1 \frac{I_1(kr)}{I_1(ka)} \quad \hat{u}_3 = \omega \hat{n}_2 \frac{I_1(kr)k_1(kb)-k_1(kr)I_1(kb)}{I_1(k)K_1(kb)-K_1(k)I_1(kb)} \qquad (A2)$$

The axial velocity and pressure can be obtained using Eqs.(219) and (220). As the outer liquid is viscous, we decompose the velocity perturbation into two terms [315, 354], that is, $u_2 = u_p + u_v$, where u_p and u_v satisfy the following linearized equations:

$$\nabla \cdot u_p = 0. \qquad \frac{\partial u_p}{\partial t} = -\nabla p_2. \quad \nabla \cdot u_v = 0. \qquad \frac{\partial u_v}{\partial t} = \frac{1}{Re} \nabla^2 u_v. \qquad (A3)$$

For u_p, a potential function $\varphi_2 (u_p = \nabla \varphi_2)$ is introduced, which satisfies the Laplace equation $\nabla^2 \varphi_2 = 0$. Therefore, the amplitude $\hat{\varphi}2$ satisfies the modified Bessel equation

$$\frac{d^2 \phi_2}{dr^2} + \frac{1}{r}\frac{d\phi_2}{dr} - k^2 \hat{\phi}_2 = 0. \qquad (A4)$$

The solution is $\hat{\varphi}_2 = A_1 I_0(kr) + A_2 K_0(kr)$, where A_1 and A_2 are coefficients to be determined by boundary conditions. The amplitudes of the radial and axial velocity components and the pressure can also be obtained [346, 347].

For the viscous part, the radial momentum equation yields

$$\frac{d^2 \hat{u}_v}{dr^2} + \frac{1}{r}\frac{d\hat{u}_v}{dr} - \left(k^2 + Re\omega + \frac{1}{r^2}\right)\hat{u}_v = 0. \qquad (A5)$$

The solution is $\hat{u}_v = A_3 I_1(lr) + A_4 K_1(lr)$, where $A3$ and $A4$ are coefficients to be determined, and $l = \sqrt{k2 + Re\omega}$. Then the continuity equation gives the solution of the axial velocity component, $\hat{w}_v = il(A_3 I_0(lr) - A_4 K_0(lr))/k$.

For the electric field, the perturbations of the electrical potentials, ψ_i, $i = 1, 2, 3$, also satisfy the Laplace equation. Further, their perturbation amplitudes $\hat{\psi}_i$, also satisfy the modified Bessel equation (A4). Consequently their solutions are the linear combinations of two modified Bessel functions $I_0(kr)$ and $K_0(kr)$, which gives five coefficients $A_5 - A_9$ to be determined.

For the present problem, there are in all 12 unknown quantities: $A_1 - A_9$, $\hat{n}_1$, $\hat{n}_2$ and $\hat{q}_s$(the perturbation amplitude of the surface charge density). On the other hand, the boundary conditions (226)–(227) and (231)–(236) provide 12 equations to solve the unknown. These equations set up a homogeneous linear system. The system has nontrivial solutions only if the determinant of its coefficient matrix is null, which provides the dispersion relation we need. However, considering the size of the coefficient matrix, it is hard to obtain an explicit expression for the dispersion relation in such a way [346, 347].

Therefore, the equations were solved step by step as outlined in the following, aiming to obtain the dispersion relation in a more compact form.

For the outer liquid, according to the continuity equation,

$$\hat{\omega}_2 = \frac{i}{k}\left(\frac{d\hat{u}_2}{dr} + \frac{\hat{u}_2}{r}\right)$$

(A6)

Differentiating the above equation with respect to r and using the momentum equation in the radial direction, we have

$$\frac{d\hat{\omega}_2}{dr} = \frac{i}{k}\left(l^2\hat{u}_2 + Re\frac{d\hat{p}_2}{dr}\right).$$

(A7)

Then using the linearized kinematic boundary conditions at the interfaces, it can be expressed as:

$$\frac{d\hat{\omega}_2}{dr}\bigg|_{r=a+n_1} = \frac{i\omega}{k}[l^2\hat{n}_1 - Rek(A_1 I_1(ka) - A_2 K_1(ka)]].$$

(A8a)

$$\frac{d\hat{\omega}_2}{dr}\bigg|_{r=1+n_2} = \frac{i\omega}{k}[l^2\hat{n}_2 - Rek(A_1 I_1(k) - A_2 k_1(k))].$$

(A8b)

Substituting the corresponding solutions into the linearized kinematic boundary conditions and the tangential dynamic boundary conditions, we obtain the expressions for A_1-A_4. The process and the expressions are omitted for brevity. On the other hand, substituting the corresponding solutions into the electric field boundary conditions, A_5-A_9 can be obtained. Finally, substituting the expressions for the corresponding quantities into the linearized normal dynamic boundary conditions, (245) and (246) are obtained, together with the dispersion relation (245).

APPENDIX B. DERIVATION OF THE DISPERSION RELATION FOR CASE II

In case II, the dispersion relation is also written in the form of equation (245), but with the amplitude ratio of the interface perturbation [346, 347]:

$$\left(\frac{\hat{n}_2}{\hat{n}_1}\right) = \left[\frac{\Delta_4}{a}\left(\frac{\pi}{H}+\omega\right)\left(\frac{EuRe\zeta}{\omega}+1+\frac{l_2^2}{k^2}\right)-\frac{\Delta_3}{a}\frac{\pi}{H}\left(\frac{EuRe\zeta}{\omega}+2\right)\right]^{-1}$$

$$\times \left[k\Delta_1\Delta_4 \left(\frac{\pi}{H} + \omega \right) \left(1 + \frac{l_2^2}{k^2} - \mu_r \left(1 + \frac{l_1^2}{k^2} \right) \right) \right.$$

$$-l_2\Delta_3\Delta_6 \frac{\pi}{H} \left(2 - \mu_r \left(1 + \frac{l_1^2}{k^2} \right) \right)$$

$$+\omega\Delta_3\Delta_4 (2(1 - \mu_r)\left(\theta - \frac{1}{a} \right) - \frac{Re\theta\pi}{k^2} \frac{\pi}{H} \quad)$$

$$+\frac{\tau}{Wea^2} (1 - (ka)^2) Re\Delta_3\Delta_4 + \frac{EuRe\xi}{\omega} \frac{1}{a} \left(\frac{\pi}{H}\Delta_3 - (\ \frac{\pi}{H} + \omega \ )\Delta_4 \right) \quad]. \quad \text{(B1a)}$$

$$\binom{\hat{n}_1}{\hat{n}_2}_2 = \left[\Delta_4 \left(1 + \frac{l_2^2}{k^2} - \mu_r \left(1 + \frac{l_1^2}{k^2} \right) \right) \left(-(l_2^2 + k^2) + \frac{Sk^2E\Delta_1}{H\Delta_3\Delta_4} \right) \right.$$

$$+\Delta_3 \left(2 - \mu_r \left(1 + \frac{l_1^2}{k^2} \right) \right) \left(2k^2 - \frac{Skl_2E\Delta_6}{H\Delta_3\Delta_4} \right)$$

$$-\frac{SRe\omega\theta E}{kH} - Re^2\Delta_3\Delta_4 Euk\xi J + \frac{EuRe\xi}{\omega} \left(\frac{SkE}{aH} \left(\frac{1}{\Delta_4} - \frac{1}{\Delta_3} \right) \right.$$

$$+k^3\Delta_2\Delta_4 \left(1 + \frac{l_2^2}{k^2} \right) - 2l_2k^2\Delta_3\Delta_5 \ ) \ ] \ -1 \left[-\frac{SkE^2}{aH\Delta_3\Delta_4} + Re^2H_2\Delta_3\Delta_4 \right.$$

$$-k\Delta_4 \left((l_2^2 + k^2)\Delta_2 + 2k\Delta_3 \right) \left(\frac{EuRe\zeta}{\omega} + 1 + \frac{l_2^2}{k^2} \right) + 2l_2k^2\Delta_3$$

$$\times \left(\Delta_5 + \frac{1}{l_2}\Delta_4 \right) - \frac{EuRe\zeta}{\omega} \frac{SkE}{aH} \left(\frac{1}{\Delta_4} - \frac{1}{\Delta_3} \right) \quad]. \quad \text{(B1b)}$$

where

$$\zeta = \left[\varepsilon_{r2}(\omega + \tau)k - \omega J + \frac{Eu}{\omega} \left(\frac{S}{aH} \left(\frac{1}{\Delta_3} - \frac{1}{\Delta_4} \right) \ 2 + k \left(k\frac{\Delta_2}{\Delta_3} - l_2\frac{\Delta_5}{\Delta_4} \right) \right) \right] \ -1$$

$$\times \left[\omega J - \frac{k^2\Delta_2}{Re\Delta_3} \left(1 + \frac{l_2^2}{k^2} \right) + \frac{2l_2k\Delta_5}{Re\Delta_4} - \frac{S}{Re \ aH} \left(\frac{1}{\Delta_5} - \frac{1}{\Delta_4} \right) \left(\frac{1}{\Delta_5} \left(1 + \frac{l_2^2}{k^2} \right) - \frac{2}{\Delta_4} \right) \right]. \text{(B2)}$$

$$\xi = \left[\varepsilon_{r2}(\omega + \tau)k - \omega J + \frac{|Eu}{\omega} \left(\frac{S}{aH} \left(\frac{1}{\Delta_3} - \frac{1}{\Delta_4} \right) \ 2 + k \left(k\frac{\Delta_2}{\Delta_3} - l_2\frac{\Delta_5}{\Delta_4} \right) \right) \right] \ -1$$

$$\times \left[\frac{k}{Re\Delta_3} \left(1 + \frac{l_2^2}{k^2} - \mu_r \left(1 + \frac{l_1^2}{k^2} \right) \right) \left(1 + \frac{S\Delta_1}{H} \left(\frac{1}{\Delta_3} - \frac{1}{\Delta_4} \right) \right) \right]$$

$$-\frac{k}{Re\Delta_4} \left(2 - \mu_r \left(1 + \frac{l_1^2}{k^2} \right) \right) \left(1 + \frac{l_2 S\Delta_6}{kH} \left(\frac{1}{\Delta_3} - \frac{1}{\Delta_4} \right) \right) - \frac{S\omega\theta}{k^2 H} \left(\frac{1}{\Delta_3} - \frac{1}{\Delta_4} \right) \tag{B3}$$

And the relative viscosity of the inner liquid $\mu r = \mu_1/\mu_2$. Note that κ, H_2 and the other symbols are the same as in case I. It is shown that the dispersion relation for case II reduces to that for case I as long as the viscosity of the inner liquid is neglected (i.e., $\mu_r=0$). The inner liquid viscosity makes the problem much more complicated [346, 347].

APPENDIX C. DERIVATION OF THE THIN LAYER APPROXIMATION A →1

In the instability analysis of an annular viscous liquid jet, the thin sheet approximation is usually derived [350, 370, 371]. If the inner and outer radii of the annular jet are supposed to approach infinity with the thickness of the liquid layer constant, a plane liquid sheet is obtained, as in Refs. [350, 370] Usually the approach of Ref. [371], is applied for expanding the dispersion relation under the thin layer limit. The derivation process is outlined below, taking case I as an example.

Define a quantity $\delta = 1-a$, the dispersion relation (2.22) is expanded under the thin layer approximation $\delta \leq 1$. First, the expansions of $\Delta_0 - \Delta_6$ are

$$\Delta_0 = -\delta - \frac{\delta^2}{2} + o(\delta^3), \quad \Delta_1 = \frac{1}{k} + \frac{k\delta^2}{2} + o(\delta^3). \tag{C1a}$$

$$\Delta_2 = \frac{1}{k} + \frac{\delta}{k} + \frac{k\delta^2}{2} \left(1 + \frac{2}{k^2} \right) + o(\delta^3), \quad \Delta_3 = -\delta - \frac{\delta^2}{2} - \frac{k^2\delta^3}{6} \left(1 + \frac{3}{k^2} \right) + o(\delta^4), \tag{C1b}$$

$$\Delta_4 = -\delta - \frac{\delta^2}{2} - \frac{l^2\delta^3}{6} \left(1 + \frac{3}{l^2} \right) + o(\delta^4), \quad \Delta_5 = \frac{1}{l} + \frac{\delta}{l} + \frac{l\delta^2}{2} \left(1 + \frac{2}{l^2} \right) + o(\delta^3), \tag{C1c}$$

$$\Delta_6 = \frac{1}{l} + \frac{l\delta^2}{2} + o(\delta^3). \tag{C1d}$$

Then, the expansions of κ, ζ and ξ are

$$k = \frac{\varepsilon_{r1} I_1(k)}{\varepsilon_{r2} I_0(k)} + k \left(1 - \frac{\varepsilon_{r1}}{\varepsilon_{r2}} \right) \left[1 + \frac{\varepsilon_{r1}}{\varepsilon_{r2}} \left(\frac{I_1(k)}{I_0(k)} \right)^2 \right] \delta + o(\delta^2). \tag{C2a}$$

$$\xi = \frac{\dfrac{k}{Re}\left(\dfrac{1}{\delta}+\dfrac{1}{2}\right)\left(\dfrac{l^2}{k^2}-1\right)+\omega J+\dfrac{k}{Re}\left(\dfrac{l^2}{k^2}-1\right)\left(\dfrac{1}{4}-\dfrac{k^2}{3}\right)\delta+o(\delta^2)}{\varepsilon_{r1}\left(\omega+\tau\right)\dfrac{I_1(k)}{I_0(k)}-\omega J+\left(\varepsilon_{r2}\left(\omega+\tau\right)ko_{(\delta)}-\dfrac{EuRek}{6}\right)\delta+o(\delta^2)} \tag{C2b}$$

$$\xi = \frac{\dfrac{k}{Re}\left(-\dfrac{1}{\delta}+\dfrac{1}{2}\right)\left(\dfrac{l^2}{k^2}-1\right)+\dfrac{k}{Re}\left(\dfrac{l^2}{k^2}-1\right)\left(\dfrac{1}{4}-\dfrac{k^2}{6}\right)\delta+o(\delta^2)}{\varepsilon_{r1}\left(\omega+\tau\right)\dfrac{I_1(K)}{I_0(k)}-\omega J+\left(\varepsilon_{r2}\left(\omega+\tau\right)ko_{(\delta)}-\dfrac{EuRek}{6}\right)\delta+o(\delta^2)}. \tag{C2c}$$

where $\kappa_{O(\delta)}$ represents a coefficient of $O(\delta)$ in the expansion of κ. Now we write the coefficients of the interface perturbation amplitudes $\hat{\eta}_1$ and $\hat{\eta}_2$ individually for each order, that is, for $O(1)$,

$$(\hat{n}_1)_1\colon (l^2-k^2)\frac{EuRe\zeta o_{\left(\frac{1}{\delta}\right)}}{\omega}, \qquad\qquad (\hat{n}_2)_1\colon (k^2-l^2)\frac{EuRe\xi o_{\left(\frac{1}{\delta}\right)}}{\omega}, \tag{C3}$$

$$(\hat{n}_1)_2\colon (k^2-l^2)\frac{EuRe\xi o_{(1/\delta)}}{\omega}, \qquad\qquad (\hat{n}_2)_2\colon (l^2-k^2)\frac{EuRe\xi o_{(1/\delta)}}{\omega}, \tag{C4}$$

for $O(\delta)$,

$$(\hat{n}_1)_1\colon Re^2 EukJ\zeta o_{(1/\delta)}+(l^2-k^2)\frac{EuRe\zeta o_{(1)}}{\omega}+\frac{(l^2+k^2)^2}{k^2}$$

$$-4k^2+\frac{3}{2}(l^2-k^2)\frac{EuRe\zeta o_{(1/\delta)}}{\omega} \tag{C5}$$

$$(\hat{n}_2)_1\colon -Re^2 EukJ\xi o_{(1/\delta)}+(k^2-l^2)\frac{EuRe\xi o_{(1)}}{\omega}+\frac{(l^2+k^2)^2}{k^2}$$

$$-4k^2+\frac{3}{2}(k^2-l^2)\frac{EuRe\xi o_{(1/\delta)}}{\omega} \tag{C6}$$

$$(\hat{n}_1)_2\colon (k^2-l^2)\frac{EuRe\zeta o_{(1)}}{\omega}-\frac{(l^2+k^2)^2}{6}+4k^2+\frac{1}{2}(k^2+l^2)\frac{EuRe\zeta o_{(1/\delta)}}{\omega}, \tag{C7}$$

$$(\hat{n}_2)_2\colon (l^2-k^2)\frac{EuRe\xi o_{(1)}}{\omega}-\frac{(l^2+k^2)^2}{k^2}+4k^2+\frac{1}{2}(l^2-k^2)\frac{EuRe\xi o_{(1/\delta)}}{\omega}, \tag{C8}$$

for $O(\delta^2)$,

$(\hat{n}_1)$ $_1$: $Re^2(H_2)$ $_{o(1)}$ $+ \left(\dfrac{(l^2 - k^2)}{6}^{\ 2} + 2(l^2 - k^2) \right) \dfrac{EuRe\zeta o_{(1/\delta)}}{\omega} + \dfrac{3}{2}(l^2 - k^2) \dfrac{EuRe\zeta o_{(1)}}{\omega}$

$$+ (l^2 - k^2) \dfrac{EuRe\zeta o_{(\delta)}}{\omega} + Re^2 EukJ\zeta o_{(1/\delta)} + \dfrac{3}{2}\dfrac{l^4 - k^4}{k^2} + l^2 - k^2, \qquad (C9)$$

$(\hat{n}_2)$ $_1$: $- Re^2 EukJ \left(\xi o_{\left(\frac{1}{\delta}\right)} + \zeta o_{(1)} \right) - \left(\dfrac{(l^2 - k^2)}{6}^{\ 2} + 2(l^2 - k^2) \right) \dfrac{EuRe\xi o_{(1/\delta)}}{\omega}$

$$+ \dfrac{3}{2}(k^2 - l^2) \dfrac{EuRe\xi o_{(1)}}{\omega} + (k^2 - l^2) \dfrac{EuRe\xi o_{(\delta)}}{\omega} + \dfrac{1}{2} \left(\dfrac{(l^2 + k^2)}{k^2}^{\ 2} - 4k^2 \right). \qquad (C10)$$

$(\hat{n}_1)$ $_2$: $\left(-(l^2 + k^2) \left(\dfrac{l^2}{6} + \dfrac{1}{2} \right) + 2k^2 \left(\dfrac{k^2}{6} + \dfrac{1}{2} \right) \right) \dfrac{EuRe\xi o_{(1/\delta)}}{\omega}$

$$+ \dfrac{k^2 - l^2}{2} \dfrac{EuRe\zeta O_{(1)}}{\omega} + (k^2 - l^2) \dfrac{EuRe\zeta O_{(\delta)}}{\omega} - \dfrac{(l^2 + k^2)}{2K^2}^{\ 2} + 2k^2, \qquad (C11)$$

$(\hat{n}_2)$ $_2$: $Re^2(H_1)$ $_o$ $_{(1)}$ $+ \left((l^2 + k^2) \left(\dfrac{l^2}{6} + \dfrac{1}{2} \right) - 2k^2 \left(\dfrac{k^2}{6} + \dfrac{1}{2} \right) \right) \dfrac{EuRe\xi o_{(1/\delta)}}{\omega}$

$$+ \dfrac{l^2 - k^2}{2} \dfrac{EuRe\xi o_{(1)}}{\omega} + (l^2 - k^2) \dfrac{EuRe\xi O_{(\delta)}}{\omega} - \dfrac{(l^2 + k^2)}{2k^2}^{\ 2} + 2k^2, \qquad (C12)$$

Substituting the above expansions into the dispersion relation, it is found that for $O(1)$ and $O(\delta)$ the dispersion relation is inherently satisfied. Therefore, the dispersion relation for the thin layer approximation is given by $O(\delta^2)$. The process of simplification is straightforward and is omitted for brevity. Ultimately the dispersion relation in the form of Eq.(279) is obtained.

KEYWORDS

- **axial/coaxial electrospinning**
- **coaxial jet**
- **core-shell**
- **core–sheath**
- **Coulomb's law**
- **electrospinning mechanism**
- **microencapsulation/nanoencapsulation processes**
- **nanocoatings**
- **Ostwald–de Waele power law**
- **polymer matrix**
- **pyrolysis**
- **scaling**
- **thermolysis**
- **thinning jet**
- **whipping jet**

REFERENCES

1. Poole, C. P., & Owens, F. J. (2003). Introduction to Nanotechnology, New Jersey: Hoboken, Wiley, 400.
2. Nalwa, H. S. (2001). Nanostructured Materials and Nanotechnology: Concise Edition: Gulf Professional Publishing. 324.
3. Gleiter, H. (1995). Nanostructured Materials: State of the Art and Perspectives. Nanostructured Materials, 6(1), 3–14.
4. Wong, Y. et al. (2006). Selected Applications of Nanotechnology in Textiles, AUTEX Research Journal 6(1), 1–8.
5. Yu, B., & Meyyappan, M. (2006). Nanotechnology: Role in Emerging Nanoelectronics. Solid-State Electronics, 50(4), 536–544.
6. Farokhzad, O. C., & Langer, R. (2009). Impact of Nanotechnology on Drug Delivery ACS Nano, 3(1), 16–20.
7. Serrano, E., Rus, G., & Garcia-Martinez, J. (2009). Nanotechnology for Sustainable Energy, Renewable and Sustainable Energy, Reviews 13(9), 2373–2384.
8. Dreher, K. L. (2004). Health and Environmental Impact of Nanotechnology: Toxicological Assessment of Manufactured Nanoparticles. Toxicological Sciences, 77(1), 3–5.
9. Bhushan, B. (2010). Introduction to Nanotechnology, in Springer Handbook of Nanotechnology, Springer 1–13.
10. Ratner, D., & Ratner, M. A. (2004). Nanotechnology and Homeland Security: New Weapons for New Wars, Prentice Hall Professional 145.

11. Aricò, A. S. et al. (2005). Nanostructured Materials for Advanced Energy Conversion and Storage Devices, Nature Materials, *4(5)*, 366–377.

12. Wang, Z. L. (2000). Nanomaterials for Nanoscience and Nanotechnology, Characterization of Nanophase Materials, 1–12.

13. Gleiter, H. (2000). Nanostructured Materials: Basic Concepts and Microstructure. Acta Materialia, *48(1)*, 1–29.

14. Wang, X. et al. (2005). A General Strategy for Nanocrystal Synthesis Nature, *437(7055)*, 121–124.

15. Kelsall, R. W. et al. (2005). Nanoscale Science and Technology, New York: Wiley Online Library, 455.

16. Engel, E. et al. (2008). Nanotechnology in Regenerative Medicine: The Materials Side. Trends in Biotechnology, *26(1)*, 39–47.

17. Beachley, V., & Wen, X. (2010). Polymer Nano fibrous Structures: Fabrication, Bio-functionalization, and Cell interactions. Progress in Polymer Science, *35(7)*, 868–892.

18. Gogotsi, Y. (2006). Nanomaterials Handbook, New York: CRC press 779.

19. Li, C., & Chou, T. (2003). A Structural Mechanics Approach for the Analysis of Carbon Nanotubes, International Journal of Solids and Structures, *40(10)*, 2487–2499.

20. Delerue, C., & Lannoo, M. (2004). Nanostructures: Theory and Modeling, Springer 304.

21. Pokropivny, V., & Skorokhod, V. (2007). Classification of Nanostructures by Dimensionality and Concept of Surface Forms Engineering in Nano-material Science, Materials Science and Engineering C, *27(5)*, 990–993.

22. Balbuena, P., & Seminario, J. M. (2006). Nano-materials: Design and Simulation: Design and Simulation, *18*, Elsevier. 523.

23. Kawaguchi, T., & Matsukawa, H. (2002). Numerical Study of Nanoscale Lubrication and Friction at Solid Interfaces, Molecular Physics, *100(19)*, 3161–3166.

24. Ponomarev, S. Y., Thayer, K. M., & Beveridge, D. L. (2004). Ion Motions in Molecular Dynamics Simulations on DNA, Proceedings of the National Academy of Sciences of the United States of America, *101(41)*, 14771–14775.

25. Loss, D., & DiVincenzo, D. P. (1998). Quantum Computation with Quantum Dots, Physical Review A, *57(1)*, 120–125.

26. Theodosiou, T. C., & Saravanos, D. A. (2007). Molecular Mechanics Based Finite Element for Carbon Nanotube Modeling Computer Modeling in Engineering and Sciences, *19(2)*, 19–24.

27. Pokropivny, V., & Skorokhod, V. (2008). New Dimensionality Classifications of Nanostructures Physical E: Low-dimensional Systems and Nanostructures, *40(7)*, 2521–2525.

28. Lieber, C. M. (1998). One-dimensional Nanostructures: Chemistry, Physics & Applications. Solid State Communications, *107(11)*, 607–616.

29. Emary, C. (2009). Theory of Nanostructures, New York: Wiley, 141.

30. Edelstein, A. S., & Cammaratra, R. C. (1998). Nanomaterials: Synthesis, Properties and Applications, CRC Press.

31. Grzelczak, M. et al. (2010). Directed Self-assembly of Nanoparticles, ACS Nano, *4(7)*, 3591–3605.

32. Hung, C. et al. (1999). Strain Directed Assembly of Nanoparticle Arrays within a Semiconductor. Journal of Nanoparticle Research, *1(3)*, 329–347.

33. Wang, L., & Hong, R. (2001). Synthesis, Surface Modification and Characterisation of Nanoparticles, Polymer Composites, *2*, 13–51.

34. Lai, W. et al. (2012). Synthesis of Nanostructured Materials by Hot and Cold Plasma, in International Plasma Chemistry Society, Orleans, France, *5*.

35. Petermann, N. et al. (2011). Plasma Synthesis of Nanostructures for Improved Thermoelectric Properties Journal of Physics D: Applied Physics, *44(17)*, 174034.

36. Ye, Y. et al. (2001). RF Plasma Method, Google Patents: USA.
37. Hyeon, T. (2003). Chemical Synthesis of Magnetic Nanoparticles, Chemical Communications, *8*, 927–934.
38. Galvez, A. et al. (2002). Carbon Nanoparticles from Laser Pyrolysis Carbon, *40(15)*, 2775–2789.
39. Porterat, D. (2012). Synthesis of Nanoparticles by Laser Pyrolysis, Google Patents: USA.
40. Tiwari, J. N., Tiwari, R. N., & Kim, K. S. (2012). Zero-Dimensional, One-Dimensional, Two-Dimensional and Three-Dimensional, Nanostructured Materials for Advanced Electrochemical Energy Devices, Progress in Materials Science, *57(4)*, 724–803.
41. Murray, P. T. et al. (2006). Nanomaterials Produced by Laser Ablation Techniques Part I: Synthesis and Passivation of Nanoparticles, in Nondestructive Evaluation for Health Monitoring and Diagnostics, International Society for Optics and Photonics 61750–61750.
42. Dolgaev, S. I. et al. (2002). Nanoparticles Produced by Laser Ablation of Solids in Liquid Environment, Applied Surface Science, *186(1)*, 546–551.
43. Becker, M. F. et al. (1998). Metal Nanoparticles Generated by Laser Ablation. Nanostructured Materials, *10(5)*, 853–863.
44. Bonneau, F. et al. (2004). Numerical Simulations for Description of UV Laser Interaction with Gold Nanoparticles Embedded in Silica. Applied Physics B, *78(3–4)*, 447–452.
45. Chen, Y. H., & Yeh, C. S. (2002). Laser Ablation Method: Use of Surfactants to Form the Dispersed Ag Nanoparticles. Colloids and Surfaces A: Physicochemical and Engineering Aspects, *197(1)*, 133–139.
46. Andrady, A. L. (2008). Science and Technology of Polymer Nanofibers, Hoboken: John Wiley & Sons, Inc. 404.
47. Wang, H. S., Fu, G. D., & Li, X. S. (2009). Functional Polymeric Nanofibers from Electrospinning, Recent Patents on Nanotechnology, *3(1)*, 21–31.
48. Ramakrishna, S. (2005). An Introduction to Electrospinning and Nanofibers: World Scientific Publishing Company, 396.
49. Reneker, D. H., & Chun, I. (1996). Nanometer Diameter Fibers of Polymer produced by Electrospinning, Nanotechnology, *7(3)*, 216.
50. Doshi, J., & Reneker, D. H. (1995). Electrospinning Process and Applications of Electrospun Fibers, Journal of Electrostatics, *35(2)*, 151–160.
51. Burger, C., Hsiao, B., & Chu, B. (2006). Nanofibrous Materials and their Applications, Annual Reviews Material Researches, *36*, 333–368.
52. Fang, J. et al. (2008). Applications of Electro spun Nanofibers, Chinese Science Bulletin, *53(15)*, 2265–2286.
53. Ondarcuhu, T., & Joachim, C. (1998). Drawing a Single Nanofibers over Hundreds of Microns, EPL (Euro physics Letters), *42(2)*, 215.
54. Nain, A. S. et al. (2006). Drawing Suspended Polymer Micro Nanofibers Using Glass Micropipettes. Applied Physics Letters, *89(18)*, 183105–183105–3.
55. Bajakova, J. et al. (2011). Drawing the Production of Individual Nanofibers by Experimental Method in Nanoconference, Brno, Czech Republic, EU.
56. Feng, L. et al. (2002). Super Hydrophobic Surface of Aligned Polyacrylonitrile Nanofibers, Angewandte Chemie, *114(7)*, 1269–1271.
57. Delvaux, M. et al. (2000). Chemical and Electrochemical Synthesis of Polyaniline Micro and Nanotubules, Synthetic Metals, *113(3)*, 275–280.
58. Barnes, C. P. et al. (2007). Nanofiber Technology: Designing the Next Generation of Tissue Engineering Scaffolds. Advanced Drug Delivery Reviews, *59(14)*, 1413–1433.
59. Palmer, L. C., & Stupp, S. I. (2008). Molecular Self-Assembly into One-dimensional Nanostructures, Accounts of Chemical Research, *41(12)*, 1674–1684.

60. Hohman, M. M. et al. (2001). Electrospinning and Electrically Forced Jets. I. Stability Theory. Physics of Fluids, *13*, 2201–2220.

61. Hohman, M. M. et al. (2001). Electrospinning and Electrically Forced Jets. II. Applications, Physics of Fluids, *13*, 2221.

62. Shin, Y. M. et al. (2001). Experimental Characterization of Electrospinning: The Electrically Forced Jet and Instabilities. Polymer, *42(25)*, 9955–9967.

63. Fridrikh, S. V. et al. (2003). Controlling the Fiber Diameter during Electrospinning, Physical Review Letters, *90(14)*, 144502–144502.

64. Yarin, A. L., Koombhongse, S., & Reneker, D. H. (2001). Taylor Cone and Jetting from Liquid droplets in Electrospinning of Nanofibers, Journal of Applied Physics, *90(9)*, 4836–4846.

65. Zeleny, J. (1914). The Electrical discharge from Liquid Points and a Hydrostatic Method of Measuring the Electric Intensity at their Surfaces, Physical Review, *3(2)*, 69–91.

66. Reneker, D. H. et al. (2000). Bending Instability of Electrically Charged Liquid Jets of Polymer Solutions in Electrospinning, Journal of Applied Physics, *87*, 4531–4547.

67. Frenot, A., & Chronakis, I. S. (2003). Polymer Nanofibers Assembled by Electro spinning Current Opinion in Colloid & Interface Science, *8(1)*, 64–75.

68. Gilbert, W., De Magnete Transl. Mottelay. P. F. (1958). Dover, UK. New York: Dover Publications, Inc. 366.

69. Tucker, N. et al. (2012). The History of the Science and Technology of Electrospinning from 1600 to 1995, Journal of Engineered Fibers and Fabrics, *7*, 63–73.

70. Hassounah, I. (2012). Melt Electro spinning of Thermoplastic Polymers: Aachen: Hochschulbibliothek Rheinisch Westfälische Technischen Hochschule Aachen. 650.

71. Taylor, G. I. (1971). The Scientific Papers of Sir Geoffrey Ingram Taylor, Mechanics of Fluids, *4*.

72. Yeo, L. Y., & Friend, J. R. (2006). Electrospinning Carbon Nanotube Polymer Composite Nanofibers, Journal of Experimental Nanoscience, *1(2)*, 177–209.

73. Bhardwaj, N., & Kundu, S. C. (2010). Electro spinning: a Fascinating Fiber Fabrication Technique, Biotechnology Advances, *28(3)*, 325–347.

74. Huang, Z. M. et al. (2003). A Review on Polymer Nanofibers by Electrospinning and their Applications in Nanocomposites, Composites Science and Technology, *63(15)*, 2223–2253.

75. Haghi, A. K. (2011). Electrospinning of Nanofibers in Textiles, North Calorina: Apple Academic PressInc. 132.

76. Bhattacharjee, P., Clayton, V., & Rutledge, A. G. (2011). Electrospinning and Polymer Nanofibers Process Fundamentals in Comprehensive Biomaterials Elsevier, 497–512.

77. Garg, K., & Bowlin, G. L. (2011). Electrospinning Jets and Nanofibrous Structures, Biomicrofluidics, *5*, 13403–13421.

78. Angammana, C. J., & Jayaram, S. H. (2011). A Theoretical Understanding of the Physical Mechanisms of Electrospinning, in Proc. ESA Annual Meeting on Electrostatics: Case Western Reserve University, Cleveland OH, 1–9.

79. Reneker, D. H., & Yarin, A. L. (2008). Electrospinning Jets and Polymer Nanofibers, Polymer, *49(10)*, 2387–2425.

80. Deitzel, J. et al. (2001). The Effect of Processing Variables on the Morphology of Electro spun Nanofibers and Textiles, Polymer, *42(1)*, 261–272.

81. Rutledge, G. C., & Fridrikh, S. V. (2007). Formation of Fibers by Electrospinning Advanced Drug Delivery Reviews, *59(14)*, 1384–1391.

82. DeVrieze, S. et al. (2009). The Effect of Temperature and Humidity on Electrospinning, Journal of Materials Science, *44(5)*, 1357–1362.

83. Kumar, P. (2012). Effect of Collector on Electrospinning to Fabricate Aligned Nanofiber, in Department of Biotechnology and Medical Engineering, National Institute of Technology Rourkela: Rourkela, 88.

84. Sanchez, C., Arribart, H., & Guille, M. (2005). Biomimetism and Bioinspiration as Tools for the Design of Innovative Materials and Systems, Nature Materials, *4(4)*, 277–288.

85. Ko, F. et al. (2003). Electrospinning of Continuous Carbon Nanotube Filled Nanofiber Yarns, Advanced Materials, *15(14)*, 1161–1165.

86. Stuart, M. et al. (2010). Emerging Applications of Stimuli Responsive Polymer Materials, Nature Materials, *9(2)*, 101–113.

87. Gao, W., Chan, J., & Farokhzad, O. (2010). PH-Responsive Nanoparticles for Drug Delivery, Molecular Pharmaceutics, *7(6)*, 1913–1920.

88. Li, Y. et al. (2010). Stimulus Responsive Polymeric Nanoparticles for Biomedical Applications, Science China Chemistry, *53(3)*, 447–457.

89. Tirelli, N. (2006). (Bio) Responsive Nanoparticles, Current Opinion in Colloid and Interface Science, *11(4)*, 210–216.

90. Bonini, M. et al. (2002). A New Way to Prepare Nanostructured Materials: Flame Spraying of Microemulsions. The Journal of Physical Chemistry B, *106(24)*, 6178–6183.

91. Thierry, B. et al. (2003). Nanocoatings onto Arteries via Layer-By-Layer Deposition: Toward the in Vivo Repair of Damaged Blood Vessels. Journal of the American Chemical Society, *125(25)*, 7494–7495.

92. Andrady, A. (2008). Science and Technology of Polymer Nanofibers, Wiley com

93. Carroll, C. P. et al. (2008). Nanofibers from Electrically Driven Viscoelastic Jets: Modeling and Experiments. Korea-Australia Rheology Journal, *20(3)*, 153–164.

94. Zhao, Y., & Jiang, L. (2009). Hollow Micro/Nanomaterials with Multilevel Interior Structures, Advanced Materials, *21(36)*, 3621–3638.

95. Carroll, C. P. (2009). The Development of a Comprehensive Simulation Model for Electrospinning, *70*, Cornell University 300.

96. Song, Y. S., & Youn, J. R. (2004). Modeling of Rheological Behavior of Nanocomposites by Brownian Dynamics Simulation, Korea-Australia Rheology Journal, *16(4)*, 201–212.

97. Dror, Y. et al. (2003). Carbon Nanotubes Embedded in Oriented Polymer Nanofibers by Electrospinning. Langmuir, *19(17)*, 7012–7020.

98. Gates, T. et al. (2005). Computational Materials: Multi-Scale Modeling and Simulation of Nanostructured Materials. Composites Science and Technology, *65(15)*, 2416–2434.

99. Agic, A. (2008). Multiscale Mechanical Phenomena in Electrospun Carbon Nanotube Composites, Journal of Applied Polymer Science, *108(2)*, 1191–1200.

100. Teo, W., & Ramakrishna, S. (2009). Electrospun Nanofibers as a Platform for Multifunctional, Hierarchically Organized Nanocomposite, Composites Science and Technology, *69(11)*, 1804–1817.

101. Silling, S., & Bobaru, F. (2005). Peridynamic Modeling of Membranes and Fibers, International Journal of Non-Linear Mechanics, *40(2)*, 395–409.

102. Berhan, L. et al. (2004). Mechanical Properties of Nanotube Sheets: Alterations in Joint Morphology and Achievable Modulii in Manufacturable Materials, Journal of Applied Physics, *95(8)*, 4335–4345.

103. Heyden, S. (2000). Network Modeling for the evaluation of Mechanical Properties of Cellulose Fiber Fluff, Lund University.

104. Collins, A. J. et al. (2011). The Value of Modeling and Simulation Standards, Virginia Modeling, Analysis and Simulation Center, Old Dominion University: Virginia, 1–8.

105. Kuwabara, S. (1959). The Forces Experienced by Randomly Distributed Parallel Circular Cylinders or Spheres In a Viscous Flow at Small Reynolds Numbers, Journal of the Physical Society of Japan, *14*, 527.

106. Brown, R. (1993). Air Filtration: An Integrated Approach to the Theory and Applications of Fibrous Filters, New York: Pergamon Press.

107. Buysse, W. M. et al. (2008). A 2D model for the Electrospinning Process, in Department of Mechanical Engineering, Eindhoven University of Technology: Eindhoven 75.
108. Ante, A., & Budimir, M. (2010). Design Multifunctional Product by Nanostructures, Sciyo. com. 27.
109. Jackson, G., & James, D. (1986). The Permeability of Fibrous Porous Media, the Canadian Journal of Chemical Engineering, *64(3)*, 364–374.
110. Sundmacher, K. (2010). Fuel Cell Engineering: Toward the Design of Efficient Electrochemical Power Plants, Industrial & Engineering Chemistry Research, *49(21)*, 10159–10182.
111. Kim, Y. et al. (2008). Electrospun Bimetallic Nanowires of PtRh and PtRu with Compositional Variation for Methanol Electrooxidation Electrochemistry Communications, *10(7)*, 1016–1019.
112. Kim, H. et al. (2009). Pt and PtRh Nanowire Electrocatalysts for Cyclohexane Fueled Polymer Electrolyte Membrane Fuel Cell. Electrochemistry Communications, *11(2)*, 446–449.
113. Formo, E. et al. (2008). Functionalization of Electrospun TiO_2 Nanofibers with Pt Nanoparticles and Nanowires for Catalytic Applications, Nano letters, *8(2)*, 668–672.
114. Xuyen, N. et al. (2009). Hydrolysis-Induced Immobilization of Pt (Acac) 2 on Polyimide-Based Carbon Nanofiber Mat and Formation of Pt Nanoparticles, Journal of Materials Chemistry, *19(9)*, 1283–1288.
115. Lee, K. et al. (2009). Nafion Nanofiber Membranes, ECS Transactions, *25(1)*, 1451–1458.
116. Qu, H., Wei, S., & Guo, Z. (2013). Coaxial Electrospun Nanostructures and their Applications, Journal of Material Chemistry A, *1(38)*, 11513–11528.
117. Thavasi, V., Singh, G., & Ramakrishna, S. (2008). Electrospun Nanofibers in Energy and Environmental Applications, Energy & Environmental Science, *1(2)*, 205–221.
118. Dersch, R. et al. (2005). Nanoprocessing of Polymers: Applications in Medicine, Sensors, Catalysis, Photonics. Polymers for Advanced Technologies, *16(2–3)*, 276–282.
119. Yih, T., & Al-Fandi, M. (2006). Engineered Nanoparticles As Precise Drug Delivery Systems, Journal of Cellular Biochemistry, *97(6)*, 1184–1190.
120. Kenawy, E. et al. (2002). Release of Tetracycline Hydrochloride from Electrospun Poly (Ethylene-Co-Vinylacetate), Poly (Lactic Acid), and a Blend, Journal of Controlled Release, *81(1)*, 57–64.
121. Verreck, G. et al. (2003). Incorporation of Drugs in an Amorphous State into Electrospun Nanofibers Composed of a Water-Insoluble, Nonbiodegradable Polymer. Journal of Controlled Release, *92(3)*, 349–360.
122. Zeng, J. et al. (2003). Biodegradable Electrospun Fibers for Drug Delivery, Journal of Controlled Release, *92(3)*, 227–231.
123. Luu, Y. et al. (2003). Development of a Nanostructured DNA Delivery Scaffold via Electrospinning of PLGA and PLA–PEG Block Copolymers, Journal of Controlled Release, *89(2)*, 341–353.
124. Zong, X. et al. (2002). Structure and Process Relationship of Electrospun Bioabsorbable Nanofiber Membranes, Polymer, *43(16)*, 4403–4412.
125. Yu, D. et al. (2012). PVP Nanofibers Prepared using Co-Axial Electrospinning with Salt Solution as Sheath Fluid. Materials Letters, *67(1)*, 78–80.
126. Verreck, G. et al. (2003). Preparation and Characterization of Nanofibers containing Amorphous Drug Dispersions Generated by Electrostatic Spinning, Pharmaceutical Research, *20(5)*, 810–817.
127. Jiang, H. et al. (2004). Preparation and Characterization of Ibuprofen-Loaded Poly (Lactide-Co-Glycolide) Poly (Ethylene Glycol)-G-Chitosan Electrospun Membranes, Journal of Biomaterials Science, Polymer Edition, *15(3)*, 279–296.
128. Yang, D., Li, Y., & Nie, J. (2007). Preparation of Gelatin/PVA Nanofibers and their Potential Application in Controlled Release of Drugs, Carbohydrate Polymers, *69(3)*, 538–543.

129. Kim, K. et al. (2004). Incorporation and Controlled Release of a Hydrophilic Antibiotic Using Poly (Lactide-Co-Glycolide)-Based Electrospun Nanofibrous Scaffolds. Journal of Controlled Release, *98(1)*, 47–56.

130. Xu, X. et al. (2005). Ultrafine Medicated Fibers Electrospun from W O emulsions, Journal of Controlled Release, *108(1)*, 33–42.

131. Zeng, J. et al. (2005). Poly (vinyl alcohol) Nanofibers by Electrospinning as a Protein Delivery System and the Retardation of Enzyme Release by Additional Polymer Coatings, Biomacromolecules, *6(3)*, 1484–1488.

132. Yun, J. et al. (2010). Effects of Oxyfluorination on Electromagnetic Interference Shielding behavior of MWCNT/PVA/PAAc composite Microcapsules, European Polymer Journal, *46(5)*, 900–909.

133. Jiang, H. et al. (2005). A Facile Technique to Prepare Biodegradable Coaxial Electrospun Nanofibers for Controlled Release of Bioactive Agents, Journal of Controlled Release, *108(2)*, 237–243.

134. He, C., Huang, Z., & Han, X. (2009). Fabrication of Drug-Loaded Electrospun Aligned Fibrous Threads for Suture Applications. Journal of Biomedical Materials Research Part A, *89(1)*, 80–95.

135. Qi, R. et al. (2010). Electrospun Poly (lactic-co-glycolic acid) Halloysite Nanotube Composite Nanofibers for Drug Encapsulation and Sustained Release. Journal of Materials Chemistry, *20(47)*, 10622–10629.

136. Reneker, D. H. et al. (2007). Electro spinning of Nanofibers from Polymer Solutions and Melts, Advances in Applied Mechanics, *41*, 343–346.

137. Haghi, A. K., & Zaikov, G. (2012). Advance in Nanofibre Research, Smithers Rapra Technology, 194.

138. Maghsoodloo, S. et al. (2012). A Detailed Review on Mathematical Modeling of Electrospun Nanofibers, Polymers Research Journal *6*, 361–379.

139. Fritzson, P. (2010). Principles of Object Oriented Modeling and Simulation with Modelica 2.1, Wiley-IEEE Press.

140. Robinson, S. (2004). Simulation: The Practice of Model Development and Use: Wiley. 722.

141. Carson, I. I., & John, S. (2004). Introduction to Modeling and Simulation, in Proceedings of the 36th Conference on Winter Simulation, Winter Simulation Conference: Washington, DC 9–16.

142. Banks, J. (1998). Handbook of Simulation: Wiley Online Library, 342.

143. Pritsker, A. B., & Alan, B. (1998). Principles of Simulation Modeling, New York: Wiley, 426.

144. Yu, J. H., Fridrikh, S. V., & Rutledge, G. C. (2006). The Role of Elasticity in the Formation of Electrospun Fibers, Polymer, *47(13)*, 4789–4797.

145. Han, T., Yarin, A. L., & Reneker, D. H. (2008). Viscoelastic Electrospun Jets: Initial Stresses and Elongational Rheometry. Polymer, *49(6)*, 1651–1658.

146. Bhattacharjee, P. K. et al. (2003). Extensional Stress Growth and Stress Relaxation in Entangled Polymer Solutions, Journal of Rheology, *47*, 269–290.

147. Paruchuri, S., & Brenner, M. P. (2007). Splitting of a Liquid Jet, Physical Review Letters, *98(13)*, 134502–134504.

148. Ganan-Calvo, A. M. (1997). On the Theory of Electrohydrodynamically Driven Capillary Jets. Journal of Fluid Mechanics, *335*, 165–188.

149. Liu, L., & Dzenis, Y. A. (2011). Simulation of Electrospun Nanofibre Deposition on Stationary and Moving Substrates, Micro & Nano Letters, *6(6)*, 408–411.

150. Spivak, A. F., & Dzenis, Y. A. (1998). Asymptotic Decay of Radius of a Weakly Conductive Viscous Jet in an External Electric Field, Applied Physics Letters, *73(21)*, 3067–3069.

151. Jaworek, A., & Krupa, A. (1999). Classification of the Modes of EHD Spraying, Journal of Aerosol Science, *30(7)*, 873–893.

152. Senador, A. E., Shaw, M. T., & Mather, P. T. (2000). Electrospinning of Polymeric Nanofibers: Analysis of Jet Formation, in Material Research Society, Cambridge University Press: California, USA 11.
153. Feng, J. J. (2002). The Stretching of an Electrified Non-Newtonian Jet: A Model for Electrospinning. Physics of Fluids, *14(11)*, 3912–3927.
154. Feng, J. J. (2003). Stretching of a Straight Electrically charged Viscoelastic Jet, Journal of Non-Newtonian Fluid Mechanics, *116(1)*, 55–70.
155. Spivak, A. F., Dzenis, Y. A., & Reneker, D. H. (2000). A Model of Steady State Jet in the Electrospinning Process, Mechanics Research Communications, *27(1)*, 37–42.
156. Yarin, A. L., Koombhongse, S., & Reneker, D. H. (2001). Bending instability in Electrospinning of Nanofibers, Journal of Applied Physics, *89*, 3018.
157. Gradoń, L. Principles of Momentum, Mass and Energy Balances. Chemical Engineering and Chemical Process Technology, *1*, 1–6.
158. Bird, R. B., Stewart, W. E., & Lightfoot, E. N. (1960). Transport Phenomena, *2*, New York: Wiley & Sons, Incorporated, John 808.
159. Peters, G. W. M., Hulsen, M. A., & Solberg, R. H. M. (2007). A Model for Electrospinning Viscoelastic Fluids, in Department of Mechanical Engineering, Eindhoven University of Technology: Eindhoven 26.
160. Whitaker, R. D. (1975). An Historical note on the Conservation of Mass, Journal of Chemical Education, *52(10)*, 658.
161. He, J. H. et al. (2007). Mathematical Models for Continuous Electrospun Nanofibers and Electrospun Nanoporous Microspheres, Polymer International, *56(11)*, 1323–1329.
162. Xu, L., Liu, F., & Faraz, N. (2012). Theoretical Model for the Electrospinning Nanoporous Materials Process, Computers and Mathematics with Applications, *64(5)*, 1017–1021.
163. Heilbron, J. L. (1979). Electricity in the 17th and 18th Century: A Study of Early Modern Physics, University of California Press. 437.
164. Orito, S., & Yoshimura, M. (1985). Can the Universe be charged? Physical Review Letters, *54(22)*, 2457–2460.
165. Karra, S. (2007). Modeling Electrospinning Process and a Numerical Scheme Using Lattice Boltzmann Method to Simulate Viscoelastic Fluid Flows, In Indian Institute of Technology, Texas A&M University: Chennai 60.
166. Hou, S. H., & Chan, C. K. (2011). Momentum Equation for Straight Electrically Charged Jet. Applied Mathematics and Mechanics, *32(12)*, 1515–1524.
167. Maxwell, J. C. (1878). Electrical Research of the Honorable Henry Cavendish, 426, in Cambridge University Press, Cambridge, Editor., Cambridge University Press, Cambridge, UK.
168. Vught, R. V. (2010). Simulating the Dynamical behavior of Electrospinning processes, in Department of Mechanical Engineering, Eindhoven University of Technology, Eindhoven 68.
169. Jeans, J. H. (1927). The Mathematical Theory of Electricity and Magnetism, London, Cambridge University Press, 536.
170. Truesdell, C., & Noll, W. (2004). The Nonlinear Field Theories of Mechanics, Springer, 579.
171. Roylance, D. (2000). Constitutive Equations, in Lecture Notes, Department of Materials Science and Engineering, Massachusetts Institute of Technology: Cambridge, 10.
172. He, J. H., Wu, Y., & Pang, N. (2005). A Mathematical Model for Preparation by AC-Electrospinning Process, International Journal of Nonlinear Sciences and Numerical Simulation, *6(3)*, 243–248.
173. Little, R. W. (1999). Elasticity: Courier Dover Publications, 431.
174. Clauset, A., Shalizi, C. R., & Newman, M. E. J. (2009). Power-Law Distributions in Empirical Data, SIAM Review *51(4)*, 661–703.

175. Wan, Y., Guo, Q., & Pan, N. (2004). Thermo-Electro-Hydrodynamic Model for Electrospinning Process, International Journal of Nonlinear Sciences and Numerical Simulation, *5(1)*, 5–8.

176. Giesekus, H. (1966). Die elastizität von flüssigkeiten. Rheologica Acta, *5(1)*, 29–35.

177. Giesekus, H. (1973). The Physical Meaning of Weissenberg's Hypothesis with Regard to The Second Normal-Stress Difference, in the Karl Weissenberg 80th Birthday Celebration Essays, J. Harris and Weissenberg, K., Editors., East African Literature Bureau 103–112.

178. Wiest, J. M. (1989). A Differential Constitutive Equation for Polymer Melts. Rheologica Acta, *28(1)*, 4–12.

179. Bird, R. B., & Wiest, J. M. (1995). Constitutive Equations for Polymeric Liquids, Annual Review of Fluid Mechanics, *27(1)*, 169–193.

180. Giesekus, H. (1982). A Simple Constitutive Equation for Polymer Fluids Based on the Concept of Deformation-Dependent Tensorial Mobility, Journal of Non-Newtonian Fluid Mechanics, *11(1)*, 69–109.

181. Oliveira, P. J. (2001). On the Numerical Implementation of Nonlinear Viscoelastic Models in a Finite Volume Method, Numerical Heat Transfer: Part B: Fundamentals, *40(4)*, 283–301.

182. Simhambhatla, M., & Leonov, A. I. (1995). On the Rheological modeling of Viscoelastic Polymer Liquids with Stable Constitutive Equations, Rheological Act, *34(3)*, 259–273.

183. Giesekus, H. (1982). A Unified Approach to a Variety of Constitutive Models for Polymer Fluids Based on the Concept of Configuration Dependent Molecular Mobility. Rheologica Acta, *21(4–5)*, 366–375.

184. Eringen, A. C., & Maugin, G. A. (1990). Electrohydrodynamics, in Electrodynamics of Continua II, Springer, 551–573.

185. Hutter, K. (1991). Electrodynamics of Continua (A Cemal Eringen and Gerard A. Maugin), SIAM Review, *33(2)*, 315–320.

186. Kröger, M. (2004). Simple Models for Complex Nonequilibrium Fluids, Physics Reports, *390(6)*, 453–551.

187. Denn, M. M. (1990). Issues in Viscoelastic Fluid Mechanics, Annual Review of Fluid Mechanics, *22(1)*, 13–32.

188. Rossky, P. J., Doll, J. D., & Friedman, H. L. (1978). Brownian Dynamics as Smart Monte Carlo Simulation, the Journal of Chemical Physics, *69*, 4628–4633.

189. Chen, J. C., & Kim, A. S. (2004). Brownian Dynamics, Molecular Dynamics, and Monte Carlo Modeling of Colloidal Systems, Advances in Colloid and Interface Science, *112(1)*, 159–173.

190. Pasini, P., & Zannoni, C. (2005). Computer Simulations of Liquid Crystals and Polymers *177*, Erice: Springer 380.

191. Zhang, H., & Zhang, P. (2006). Local Existence for the FENE-Dumbbell Model of Polymeric Fluids, Archive for Rational Mechanics and Analysis, *181(2)*, 373–400.

192. Isihara, A. (1951). Theory of High Polymer Solutions (The Dumbbell Model), the Journal of Chemical Physics, *19*, 343-397.

193. Masmoudi, N. (2008). Well-Posedness for the FENE Dumbbell Model of Polymeric Flows, Communications on Pure and Applied Mathematics, *61(12)*, 1685–1714.

194. Stockmayer, W. H. et al. (1970). Dynamic Properties of Solutions, Models for Chain Molecule Dynamics in Dilute Solution, Discussions of the Faraday Society, *49*, 182–192.

195. Graham, R. S. et al. (2003). Microscopic Theory of Linear, Entangled Polymer Chains under Rapid Deformation Including Chain Stretch and Convective Constraint Release, Journal of Rheology, *47*, 1171–1200.

196. Gupta, R. K., Kennel, E., & Kim, K. S. (2010). Polymer Nanocomposites Handbook: CRC Press.

197. Marrucci, G. (1972). The Free Energy Constitutive Equation for Polymer Solutions from the Dumbbell Model, Journal of Rheology, *16*, 321–331.
198. Reneker, D. H. et al. (2007). Electrospinning of Nanofibers from Polymer Solutions and Melts, Advances in Applied Mechanics, *41*, 43–195.
199. Kowalewski, T. A., Barral, S., & Kowalczyk, T. (2009). Modeling Electrospinning of Nanofibers in IUTAM Symposium on Modeling Nanomaterials and Nanosystems, Springer, Aalborg, Denmark, 279–292.
200. Macosko, C. W. (1994). Rheology: Principles, Measurements, and Applications. Poughkeepsie, Newyork: Wiley-VCH 578.
201. Kowalewski, T. A., Blonski, S., & Barral, S. (2005). Experiments and Modeling of Electrospinning Process, Technical Sciences, *53(4)*, 385–394.
202. Ma, W. K. A. et al. (2008). Rheological Modeling of Carbon Nanotube Aggregate Suspensions, Journal of Rheology, *52*, 1311–1330.
203. Buysse, W. M. (2008). A 2D Model for the Electrospinning Process, in Department of Mechanical Engineering, Eindhoven University of Technology: Eindhoven, *71*.
204. Silling, S. A., & Bobaru, F. (2005). Peridynamic Modeling of Membranes and Fibers, International Journal of Non-Linear Mechanics, *40(2)*, 395–409.
205. Teo, W. E., & Ramakrishna, S. (2009). Electrospun Nanofibers as a Platform for Multifunctional, Hierarchically Organized Nanocomposite, Composites Science and Technology, *69(11)*, 1804–1817.
206. Wu, X., & Dzenis, Y. A. (2005). Elasticity of Planar Fiber Network, Journal of Applied Physics, *98(9)*, 93501.
207. Tatlier, M., & Berhan, L. (2009). Modeling the Negative Poisson's Ratio of Compressed Fused Fiber Networks Physical Status Solid (b), *246(9)*, 2018–2024.
208. Kuipers, B. (1994). Qualitative Reasoning: Modeling and Simulation with Incomplete Knowledge, the MIT press 554.
209. West, B. J. (2004). Comments on the Renormalization Group, Scaling and Measures of Complexity. Chaos, Solitons and Fractals, *20(1)*, 33–44.
210. DeGennes, P. G., & Witten, T. A. (1980). Scaling Concepts in Polymer Physics, Vol. Cornell University Press. 324.
211. He, J. H., & Liu, H. M. (2005). Variational Approach to Nonlinear Problems and a Review on Mathematical Model of Electrospinning, Nonlinear Analysis, *63*, e919–e929.
212. He, J. H., Wan, Y. Q., & Yu, J. Y. (2004). Allometric Scaling and Instability in Electrospinning, International Journal of Nonlinear Sciences and Numerical Simulation *5(3)*, 243–252.
213. He, J. H., Wan, Y. Q., & Yu, J. Y. (2004). Allometric Scaling and Instability in Electrospinning, International Journal of Nonlinear Sciences and Numerical Simulation, *5*, 243–252.
214. He, J. H., & Wan, Y. Q. (2004). Allometric Scaling for Voltage and Current in Electrospinning, Polymer, *45*, 6731–6734.
215. He, J. H., Wan, Y. Q., & Yu, J. Y. (2005). Scaling Law in Electrospinning, Relationship between Electric Current and Solution Flow Rate, Polymer, *46*, 2799–2801.
216. He, J. H., Wanc, Y. Q., & Yuc, J. Y. (2004). Application of Vibration Technology to Polymer Electrospinning, International Journal of Nonlinear Sciences and Numerical Simulation, *5(3)*, 253–262.
217. Kessick, R., Fenn, J., & Tepper, G. (2004). The use of AC Potentials in Electrospraying and Electrospinning Processes, Polymer, *45(9)*, 2981–2984.
218. Boucher, D. F., & Alves, G. E. (1959). Dimensionless Numbers, Parts 1 and 2.
219. Ipsen, D. C. (1960). Units Dimensions and Dimensionless Numbers, New York: McGraw Hill Book Company Inc. 466.
220. Langhaar, H. L. (1951). Dimensional Analysis and Theory of Models, *2*, New York: Wiley, 166.

221. McKinley, G. H. (2005). Dimensionless Groups for Understanding Free Surface Flows of Complex Fluids, Bulletin of the Society of Rheology, 6–9.
222. Carroll, C. P. et al. (2008). Nanofibers from Electrically Driven Viscoelastic Jets: Modeling and Experiments. Korea-Australia Rheology Journal, *20(3)*, 153–164.
223. Saville, D. (1997). Electrohydrodynamics: the Taylor-Melcher Leaky Dielectric Model. Annual Review of Fluid Mechanics, *29(1)*, 27–64.
224. Ramos, J. I. (1996). Force Fields on Inviscid, Slender, Annular Liquid, International Journal for Numerical Methods in Fluids, *23*, 221–239.
225. Saville, D. A. (1997). Electrohydrodynamics: the Taylor-Melcher leaky Dielectric Model. Annual Review of Fluid Mechanics, *29(1)*, 27–64.
226. Senador, A. E., Shaw, M. T., & Mather, P. T. (2000). Electrospinning of Polymeric Nanofibers, Analysis of Jet Formation in MRS Proceedings: Cambridge University Press.
227. Reneker, D. H. et al. (2000). Bending Instability of Electrically Charged Liquid Jets of Polymer Solutions in Electrospinning, Journal of Applied Physics, *87*, 4531.
228. Peters, G., Hulsen, M., & Solberg, R. A. Model for Electrospinning Viscoelastic Fluids.
229. Wan, Y. et al. (2012) Modeling and Simulation of the Electrospinning Jet with Archimedean Spiral, Advanced Science Letters, *10(1)*, 590–592.
230. Dasri, T. (2012). Mathematical Models of Bead-Spring Jets during Electrospinning for Fabrication of Nanofibers, Walailak Journal of Science and Technology, *9*.
231. Solberg, R. H. M. (2007). Position Controlled Deposition for Electrospinning, Eindhoven University of Technology: Eindhoven *75*.
232. Holzmeister, A., Yarin, A. L., & Wendorff, J. H. (2010). Barb Formation in Electrospinning: Experimental and Theoretical Investigations. Polymer, *51(12)*, 2769–2778.
233. Karra, S. (2012). Modeling Electrospinning Process and a Numerical Scheme Using Lattice Boltzmann Method to Simulate Viscoelastic Fluid Flows.
234. Arinstein, A. et al. (2007). Effect of Supramolecular Structure on Polymer Nanofibre Elasticity, Nature Nanotechnology, *2(1)*, 59–62.
235. Lu, C. et al. (2006). Computer Simulation of Electrospinning, Part I. Effect of Solvent in Electrospinning. Polymer, *47(3)*, 915–921.
236. Greenfeld, I. et al. (2011). Polymer Dynamics in Semidilute Solution during Electrospinning: A Simple Model and Experimental Observations. Physical Review *84(4)*, 41806–41815.
237. Ly, H. V., & Tran, H. T. (2001). Modeling and Control of Physical processes Using Proper Orthogonal Decomposition, Mathematical and Computer Modeling, *33(1)*, 223–236.
238. Peiró, J., & Sherwin, S. (2005). Finite Difference, Finite Element and Finite Volume Methods for Partial Differential Equations, in Handbook of Materials Modeling, Springer: London. 2415–2446.
239. Kitano, H. (2002). Computational Systems Biology, Nature, *420(6912)*, 206–210.
240. Gerald, C. F., & Wheatley, P. O. (2007). Applied Numerical Analysis, ed. 7th, Addison-Wesley, 624.
241. Burden, R. L., & Faires, J. D. (2005). Numerical Analysis, *8*, Thomson Brooks/Cole, 850.
242. Lawrence, C. E. (2010). Partial Differential Equations: American Mathematical Society 749.
243. Quarteroni, A., Quarteroni, A. M., & Valli, A. (2008). Numerical Approximation of Partial Differential Equations, *23*, Springer, 544.
244. Butcher, J. C. (1996). A History of Runge-Kutta Methods, Applied Numerical Mathematics, *20(3)*, 247–260.
245. Cartwright, J. H. E., & Piro, O. (1992). The Dynamics of Runge-Kutta Methods, International Journal of Bifurcation and Chaos, *2(03)*, 427–449.
246. Zingg, D. W., & Chisholm, T. T. (1999). Runge–Kutta Methods for Linear Ordinary Differential Equations, Applied Numerical Mathematics, *31(2)*, 227–238.

247. Butcher, J. C. (1987). The Numerical Analysis of Ordinary Differential Equations: Runge-Kutta and General Linear Methods: Wiley-Interscience, 512.
248. Reznik, S. N. et al. (2006). Evolution of a Compound Droplet Attached to a Core-shell Nozzle under the Action of a Strong Electric Field, Physics of Fluids, *18(6)*, 062101–062101–13.
249. Reznik, S. N. et al. (2004). Transient and Steady Shapes of Droplets Attached to a Surface in a Strong Electric Field. Journal of Fluid Mechanics, *516*, 349–377.
250. Donea, J., & Huerta, A. (2003). Finite Element Methods for Flow Problems: Wiley.com. 362.
251. Zienkiewicz, O. C., & Taylor, R. L. (2000). The Finite Element Method: Solid Mechanics, *2*, Butterworth-Heinemann. 459.
252. Brenner, S. C., & Scott, L. R. (2008). The Mathematical Theory of Finite Element Methods, 15, Springer, 397.
253. Bathe, K. J. (1996). Finite Element Procedures, 2, Prentice Hall Englewood Cliffs, 1037.
254. Reddy, J. N. (2006). An Introduction to the Finite Element Method 2, McGraw-Hill New York, 912.
255. Ferziger, J. H., & Perić, M. (1996). Computational Methods for Fluid Dynamics 3, Springer Berlin, 423.
256. Baaijens, P. T. F. (1998). Mixed Finite Element Methods for Viscoelastic Flow Analysis: A Review, Journal of Non-Newtonian Fluid Mechanics, 79(2), 361–385.
257. Angammana, C. J., & Jayaram, S. H. (2011). A Theoretical Understanding of the Physical Mechanisms of Electro spinning in Proceedings of the ESA Annual Meeting on Electrostatics.
258. Costabel, M. (1987). Principles of Boundary Element Methods, Computer Physics Reports, *6(1)*, 243–274.
259. Kurz, S., Fetzer, J., & Lehner, G. (1995). An Improved Algorithm for the BEM-FEM-coupling Method Using Domain Decomposition. IEEE Transactions on Magnetics, *31(3)*, 1737–1740.
260. Mushtaq, M., Shah, N. A., & Muhammad, G. (2010). Advantages and Disadvantages of Boundary Element Methods for Compressible Fluid Flow Problems, Journal of American Science, *6(1)*, 162–165.
261. Gaul, L., Kögl, M., & Wagner, M. (2003). Boundary Element Methods for Engineers and Scientists Springer, 488.
262. Kowalewski, T. A., Barral, S., & Kowalczyk, T. (2009). Modelling Electrospinning of Nano-fibers, in IUTAM Symposium on Modeling Nanomaterials and Nanosystems, Springer.
263. Toro, E. F. (2009). Riemann Solvers and Numerical Methods for Fluid Dynamics: A Practical Introduction, Springer, 724.
264. Tonti, E. (2001). A Direct Discrete Formulation of Field Laws: The Cell Method CMES Computer Modeling in Engineering and Sciences, *2(2)*, 237–258.
265. Thomas, P. D., & Lombard, C. K. (1979). Geometric Conservation Law and its Application to Flow Computations on moving Grids, American Institute of Aeronautics and Astronautics Journal, *17(10)* 1030–1037.
266. Lyrintzis, A. S. (2003). Surface Integral Methods in Computational Aeroacoustics from the (CFD) Near Field to the (Acoustic) Far-Field, International Journal of Aeroacoustics, 2(2), 95–128.
267. Škerget, L., Hriberšek, M., & Kuhn, G. (1999). Computational Fluid Dynamics by Boundary-Domain Integral Method, International Journal for Numerical Methods in Engineering, *46(8)*, 1291–1311.
268. Rüberg, T., & Cirak, F. (2011). An Immersed Finite Element Method with Integral Equation Correction, International Journal for Numerical Methods in Engineering, *86(1)*, 93–114.
269. Feng, J. J. (2002). The Stretching of an Electrified Non-Newtonian Jet: A Model for Electro-spinning, Physics of Fluids, *14*, 3912–3926.

270. Varga, R. S. (2009). Matrix Iteratives Analysis *27*, Springer, 358.
271. Stoer, J., & Bulirsch, R. (2002). Introduction to Numerical Analysis *12*, Springer, 744.
272. Bazaraa, M. S., Sherali, H. D., & Shetty, C. M. (2006). Nonlinear Programming: Theory and Algorithms, John Wiley & Sons, 872.
273. Fox, L. (1947). Some Improvements in the Use of Relaxation Methods for the Solution of Ordinary and Partial Differential Equations, Proceedings of the Royal Society of London. Series A. Mathematical and Physical Sciences, *190(1020)*, 31–59.
274. Zauderer, E. (2011). Partial Differential Equations of Applied Mathematics, *71*, Wiley.com, 968.
275. Fisher, M. L., (2004). The Lagrangian Relaxation Method for Solving Integer Programming Problems. Management Science, *50(12 Supplement)*, 1861–1871.
276. Steger, T. M. (2005). Multi-Dimensional Transitional Dynamics: A Simple Numerical Procedure. Macroeconomic Dynamics, *12(3)*, 301–319.
277. Roozemond, P. C. (2007). A Model for Electrospinning Viscoelastic Fluids, in Department of Mechanical Engineering, Eindhoven University of Technology, 25.
278. Succi, S. (2001). The Lattice Boltzmann Equation: For Fluid Dynamics and Beyond, Oxford University Press. 288.
279. Chen, S., & Doolen, G. D. (1998). Lattice Boltzmann Method for Fluid Flows. Annual Review of Fluid Mechanics, *30(1)*, 329–364.
280. Aidun, C. K., & Clausen, J. R. (2010). Lattice-Boltzmann Method for Complex Flows, Annual Review of Fluid Mechanics, *42*, 439–472.
281. Guo, Z., & Zhao, T. S. (2003). Explicit Finite-difference Lattice Boltzmann Method for Curvilinear Coordinates. Physical Review E, *67(6)*, 066709–066712.
282. Albuquerque, P. et al. (2006). A Hybrid Lattice Boltzmann Finite Difference Scheme for the Diffusion Equation, International Journal for Multiscale Computational Engineering, *4(2)*, 209–219.
283. Tsutahara, M. (2012). The Finite-Difference Lattice Boltzmann Method and its Application in Computational Aero-Acoustics, Fluid Dynamics Research, *44(4)*, 045507–045518.
284. Junk, M. (2001). A Finite Difference Interpretation of the Lattice Boltzmann Method, Numerical Methods for Partial Differential Equations, *17(4)*, 383–402.
285. So, R. M.,C., Fu, S. C., & Leung, R. C. K. (2010). Finite Difference Lattice Boltzmann Method for Compressible Thermal Fluids, American Institute of Aeronautics and Astronautics Journal, *48(6)*, 1059–1071.
286. Karra, S. (2007). Modelling Electrospinning Process and a Numerical Scheme Using Lattice Boltzmann Method to Simulate Viscoelastic Fluid Flows in Mechanical Engineering, Indian Institute of Technology Madras, 60.
287. De Pascalis, R. (2010). The Semi-Inverse Method in Solid Mechanics: Theoretical Underpinnings and Novel Applications, in Mathematics. (2010), Universite Pierre et Marie Curie and Universita del Salento, 140.
288. Nemenyi, P. F. (1951). Recent Developments in Inverse and Semi-inverse Methods in the Mechanics of Continua, Advances in Applied Mechanics, *2(11)*, 123–151.
289. Chen, J. T., Lee, Y. T., & Shieh, S. C. (2009). Revisit of Two Classical Elasticity Problems by using the Trefftz Method, Engineering Analysis with Boundary Elements, *33(6)*, 890–895.
290. Zhou, X. W. (2013). A Note on the Semi Inverse Method and a Variational Principles for the Generalized KdV-mKdV Equation, in Abstract and Applied Analysis, Hindawi Publishing Corporation.
291. Narayan, A. S. P., & Rajagopal, K. R. (2013). Unsteady Flows of a Class of Novel Generalizations of the Navier–Stokes Fluid. Applied Mathematics and Computation, *219(19)*, 9935–9946.

292. He, J. H. (2004). Variational Principles for Some Nonlinear Partial Differential Equations with Variable Coefficients, Chaos, Solitons & Fractals, *19(4)*, 847–851.

293. Tarantola, A. (2002). Inverse Problem Theory: Methods for Data Fitting and Model Parameter Estimation. Elsevier Science, 613.

294. He, J. H., Liu, H. M., & Pan, N. (2003). Variational Model for Ionomeric Polymer Metal Composite, Polymer, *44(26)*, 8195–8199.

295. He, J. H. (2001). Coupled Variational Principles of Piezoelectricity, International Journal of Engineering Science, *39(3)*, 323–341.

296. Starovoitov, É., & Nağıyev, F. (2012). Foundations of the Theory of Elasticity, Plasticity, and Viscoelasticity, CRC Press, 320.

297. Bertero, M. (1986). Regularization Methods for Linear Inverse Problems, in Inverse Problems, Springer, 52–112.

298. Wan, Y. Q., Guo, Q., & Pan, N. (2004). Thermo Electro-Hydrodynamic Model for Eectro-spinning Process, International Journal of Nonlinear Sciences and Numerical Simulation, *5(1)*, 5–8.

299. Mccann, J., Li, D., & Xia, Y. (2005). Electrospinning of Nanofibers with Core-sheath, Hollow, or Porous Structures, Journal of Materials Chemistry, *15(7)*, 735–738.

300. Srivastava, Y. et al. (2008). Electrospinning of Hollow and Core Sheath Nanofibers using a Microfluidic manifold, Microfluidics and Nanofluidics, *4(3)*, 245–250.

301. Sun, Z. et al. (2003). Compound Core Shell Polymer Nanofibers by Co-Electrospinning, Advanced Materials, *15(22)*, 1929–1932.

302. Li, D., & Xia, Y. (2004). Direct Fabrication of Composite and Ceramic Hollow Nanofibers by Electrospinning, Nano Letters, *4(5)*, 933–938.

303. Moghe, A., & Gupta, B. (2008). Co-axial Electrospinning for Nanofiber Structures, Preparation and Applications, Polymer Reviews, *48(2)*, 353–377.

304. Sakuldao, S., Yoovidhya, T., & Wongsasulak, S. (2011). Coaxial Electrospinning and Sustained Release Properties of Gelatin Cellulose Acetate Core Shell Ultrafine Fibres, Science Asia, *37(4)*, 335–343.

305. Yarin, A. et al. (2007). Material Encapsulation and Transport in Core–Shell Micro Nanofibers, Polymer and Carbon Nanotubes and Micro Nanochannels, Journal of Materials Chemistry, *17(25)*, 2585–2599.

306. Yarin, A. (2011). Coaxial Electrospinning and Emulsion Electrospinning of Core Shell Fibers, Polymers for Advanced Technologies, *22(3)*, 310–317.

307. Hufenus, R. (2008). Electrospun Core-Sheath Fibers for Soft Tissue Engineering, National Textile Center Annual Report, 18–31.

308. Mccann, J., Marquez, M., & Xia, Y. (2006). Melt Coaxial Electrospinning: A Versatile Method For The Encapsulation of Solid Materials and Fabrication of Phase Change Nanofibers. Nano Letters, *6(12)*, 2868–2872.

309. Loscertales, G. et al. (2004). Electrically Forced Coaxial Nanojets for One-step Hollow Nanofiber Design, Journal of the American Chemical Society, *126(17)*, 5376–5377.

310. Tan, S., Huang, X., & Wu, B. (2007). Some Fascinating Phenomena in Electrospinning Processes and Applications of Electrospun Nanofibers, Polymer International, *56(11)*: 1330–1339.

311. Ganan-Calvo, A. M. (1997). Cone-Jet Analytical extension of Taylor's Electrostatic Solution and the Asymptotic Universal Scaling Laws in Electrospraying. Physical Review Letters, *79(2)*, 217–220.

312. Chen, X. et al. (2005). Spraying Modes in Coaxial Jet Electrospray with outer Driving Liquid, Physics of Fluids, *17*, 032101.

313. Gañán-Calvo, A. M. (1998). Generation of Steady Liquid Microthreads and Micron Sized Monodisperse Sprays in Gas Streams, Physical Review Letters, *80(2)*, 285.

314. Hwang, Y., Jeong, U., & Cho, E. (2008). Production of Uniform Sized Polymer Core Shell Microcapsules by Coaxial Electrospraying, Langmuir, *24(6)*, 2446–2451.
315. Lopez-Herrera, J. et al. (2003). Coaxial Jets Generated from Electrified Taylor Cones, Scaling Laws, Journal of Aerosol Science, *34(5)*, 535–552.
316. Loscertales, I. et al. (2002). Micro Nano Encapsulation via Electrified Coaxial Liquid Jets, Science, *295(5560)*, 1695–1698.
317. Reznik, S. et al. (2006). Evolution of a Compound Droplet Attached to a Core-Shell Nozzle Under The Action of a Strong Electric Field. Physics of Fluids, *18(6)*, 62101–62113.
318. Zussman, E. et al. (2006). Electrospun Polyaniline Poly (methyl methacrylate)-Derived Turbostratic Carbon Micro Nanotubes, Advanced Materials, *18(3)*, 348–353.
319. La Mora, D., & Fernandez, J. (1994). The Current Emitted by Highly Conducting Taylor Cones. Journal of Fluid Mechanics, *260(1)*, 155–184.
320. Gamero-Castano, M., & Hruby, V. (2002). Electric Measurements of Charged Sprays emitted by Cone Jets, Journal of Fluid Mechanics, *459(1)*, 245–276.
321. Higuera, F. (2003). Flow Rate and Electric Current Emitted by a Taylor Cone. Journal of Fluid Mechanics, *484*, 303–327.
322. Artana, G., Romat, H., & Touchard, G. (1998). Theoretical Analysis of Linear stability of Electrified Jets Flowing at High Velocity inside a Coaxial Electrode, Journal of Electrostatics, *43(2)*, 83–100.
323. Yu, J., Fridrikh, S., & Rutledge, G. C. (2004). Production of Sub Micrometer Diameter Fibers by Two Fluid Electrospinning, Advanced Materials, *16(17)*, 1562–1566.
324. Li, F., Yin, X., & Yin, X. (2005). Linear Instability Analysis of an Electrified Coaxial Jet, Physics of Fluids, *17*, 77104.
325. Boubaker, K. (2012). A Confirmed Model to Polymer Core-Shell Structured Nanofibers Deposited via Coaxial Electrospinning, ISRN Polymer Science.
326. Zhang, L. et al. (2012). Coaxial Electrospray of Microparticles and Nanoparticles for Biomedical Applications, Expert Review of Medical Devices, *9(6)*, 595–612.
327. Mei, F., & Chen, D. (2007). Investigation of Compound Jet Electrospray: Particle Encapsulation, Physics of Fluids, *19*, 103303.
328. Batchelor, G. (2000). An Introduction to Fluid Dynamics, Cambridge University Press.
329. Castellanos, A. (1998). Electrohydrodynamics, Springer.
330. Gibbs, J., Bumstead, H., & Longley, W. (1928). The Collected works of J. Willard Gibbs, *1*, Longmans, Green and Company.
331. Slattery, J. (1980). Interfacial Transport Phenomena Invited Review. Chemical Engineering Communications, *4(1–3)*, 149–166.
332. Castellanos, A., & Gonzalez, A. (1996). Nonlinear waves and instabilities on electrified free surfaces, in Conduction and Breakdown in Dielectric Liquids, ICDL'96, 12th International Conference on IEEE.
333. Drazin, P., & Reid, W. (2004). Hydrodynamic Stability, Cambridge University Press.
334. Levich, V., & Spalding, D. (1962). Physicochemical Hydrodynamics *689*, Prentice Hall Englewood Cliffs, NJ.
335. Taylor, G. (1969). Electrically Driven Jets, Proceedings of the Royal Society of London, A. Mathematical and Physical Sciences, *313(1515)*, 453–475.
336. Melcher, J., & Waves, F. (1963). A. Comparative Study of Surface Coupled Electro Hydrodynamic and Magneto Hydrodynamic Systems, MIT Press, Cambridge, MA, 433.
337. Bailey, A. (1988). Electrostatic spraying of Liquids, Research Studies Press Somerset, England.
338. Lin, S., & Kang, D. (1987). Atomization of a Liquid Jet, Physics of Fluids, *30*, 2000.
339. Reznik, S. N. et al. (2006). Evolution of a Compound Droplet Attached to a Core-shell Nozzle under the Action of a Strong Electric Field, Physics of Fluids, *18(6)*, 62101–62113.

340. Hu, Y., & Huang, Z. (2007). Numerical study on Two-Phase Flow Patterns in Coaxial Electrospinning, Journal of Applied Physics, *101(8)*, 084307–084377.
341. Sugiyama, H. et al. (2013). Simulations of Electro Hydrodynamic Jet flow in Dielectric Fluids, Journal of Applied Fluid Mechanics, *6(3)*.
342. Li, J. et al. (1991). Numerical Study of Laminar Flow Past one and Two Circular Cylinders, Computers & Fluids, *19(2)*, 155–170.
343. Theron, S., Zussman, E., & Yarin, A. (2004). Experimental Investigation of the Governing Parameters in the Electrospinning of Polymer Solutions, Polymer, *45(6)*, 2017–2030.
344. Marín, Á. et al. (2007). Simple and Double Emulsions via Coaxial jet Electrosprays, Physical Review Letters, *98(1)*, 14502.
345. Chen, X. et al. (2005). Spraying Modes in Coaxial jet Electrospray with Outer Driving Liquid, Physics of Fluids, *17*, 32101.
346. Li, F., Yin, X., & Yin, X. (2005). Linear Instability Analysis of an Electrified Coaxial Jet, Physics of Fluids, *17*, 077104.
347. Li, F., Yin, X., & Yin, X. (2006). Linear Instability of a Coflowing Jet under an Axial Electric Field, Physical Review E, *74(3)*, 036304.
348. Higuera, F. (2007). Stationary Coaxial Electrified Jet of a Dielectric Liquid Surrounded by a Conductive Liquid, Physics of Fluids, *19*, 012102.
349. Chandrasekhar, S. (1961). Hydrodynamic and Hydromagnetic Stability, International Series of Monographs on Physics, *1*.
350. Shen, J., & Li, X. (1996). Instability of an Annular Viscous Liquid Jet, Acta Mechanica, *114*, 167–183.
351. Chen, J., & Lin, S. (2002). Instability of an Annular Jet surrounded by a Viscous Gas in a Pipe, Journal of Fluid Mechanics, *450*, 235–258.
352. Turnbull, R. (1992). On the instability of an Electrostatically Sprayed Liquid Jet, Industry Applications, IEEE Transactions, *28(6)*, 1432–1438.
353. Mestel, A. (1994). Electrohydrodynamic Stability of a Slightly Viscous Jet, Journal of Fluid Mechanics, *274*, 93–114.
354. González, H., García, F., & Castellanos, A. (2003). Stability Analysis of Conducting Jets Under Ac Radial Electric Fields for Arbitrary Viscosity, Physics of Fluids, *15*, 395.
355. Huebner, A., & Chu, H. (1971). Instability and Breakup of Charged Liquid Jets, J. Fluid Mech, *49(2)*, 361–372.
356. Son, P., & Ohba, K. (1998). Theoretical and Experimental Investigations on Instability of an Electrically Charged Liquid Jet, International Journal of Multiphase Flow, *24(4)*, 605–615.
357. Lopez-Herrera, J., & Ganan-Calvo, A. (2004). A Note on Charged Capillary Jet Breakup of Conducting Liquids: Experimental Validation of a Viscous One-Dimensional Model. Journal of Fluid Mechanics, *501*, 303–326.
358. Zakaria, K. (2000). Nonlinear Instability of a Liquid Jet in the Presence of a Uniform Electric Field, Fluid Dynamics Research, *26(6)*, 405.
359. Elhefnawy, A., Agoor, B., & Elcoot, A. (2001). Nonlinear Electrohydrodynamic Stability of a Finitely Conducting Jet under an Axial Electric Field, Physica A: Statistical Mechanics and Its Applications, *297(3)*, 368–388.
360. Elhefnawy, A., Moatimid, G., & Elcoot, A. (2004). Nonlinear Electrohydrodynamic Instability of a Finitely Conducting Cylinder: Effect of Interfacial Surface Charges, Zeitschrift für angewandte Mathematik und Physik ZAMP, *55(1)*, 63–91.
361. Moatimid, G. (2003). Non-Linear Electrorheological Instability of two Streaming Cylindrical Fluids, Journal of Physics A: Mathematical and General, *36(44)*, 11343.
362. Si, T. et al. (2009). Modes in Flow Focusing and Instability of Coaxial Liquid-Gas Jets, Journal of Fluid Mechanics, *629(1)*, 1–23.

363. Si, T. et al. (2010). Spatial Instability of Coflowing Liquid-Gas Jets in Capillary Flow Focusing, Physics of Fluids, *22*, 112105.
364. Lin, S. P. (2003). Break up of Liquid Sheets and jets, Cambridge University Press, New York.
365. Li, F., Yin, X., & Yin, X. (2009). Axisymmetric and Non-Axisymmetric Instability of an Electrified Viscous Coaxial Jet, Journal of Fluid Mechanics, *632(1)*, 199–225.
366. Li, F., Yin, X., & Yin, X. (2006). Linear Instability of a Coflowing Jet under an Axial Electric Field, Physical Review E, *74(3)*, 36304.
367. Si, T., Zhang, L., & Li, G. (2012). Co-Axial Electrohydrodynamic Atomization for Multimodal Imaging and Image-Guided Therapy, in Proc. SPIE, Multimodal Biomedical Imaging VII, San Francisco, California, USA .
368. López-Herrera, J., Riesco-Chueca, P., & Gañán-Calvo, A. (2005). Linear Stability Analysis of Axisymmetric Perturbations in Imperfectly Conducting Liquid Jets, Physics of Fluids, *17*, 034106.
369. Lim, D., & Redekopp, L. (1998). Absolute Instability conditions for Variable Density, Swirling Jet Flows, European Journal of Mechanics B Fluids, *17(2)*, 165–185.
370. Meyer, J., & Weihs, D. (1987). Capillary Instability of an Annular Liquid Jet, Journal of Fluid Mechanics, *179*, 531–545.
371. Chen, F. et al. (2003). On the Axisymmetry of Annular Jet Instabilities, Journal of Fluid Mechanics, *488*, 355–367.
372. Liu, J., & Xue, D. (2010). Hollow Nanostructured Anode Materials for Li-Ion Batteries, Nanoscale Research Letters, *5(10)*, 1525–1534.
373. Chen, J., Archer, L., & Lou, X. (2011). SnO2 Hollow Structures and TiO2 Nanosheets for Lithium-Ion Batteries, Journal of Materials Chemistry, *21(27)*, 9912–9924.
374. Lee, B. et al. (2012). Fabrication of Si Core C Shell Nanofibers and their Electrochemical Performances as a Lithium-Ion Battery Anode. Journal of Power Sources, *206*, 267–273.
375. Liu, B. et al. (2011). An Enhanced Stable-Structure Core-Shell Coaxial Carbon Nanofiber Web as a Direct Anode Material for Lithium-Based Batteries, Electrochemistry Communications, *13(6)*, 558–561.
376. Hwang, T. et al. (2012). Electrospun Core-Shell Fibers for Robust Silicon Nanoparticle-Based Lithium Ion Battery Anodes, Nano letters, *12(2)*, 802–807.
377. Han, H. et al. (2011). Nitridated TiO2 Hollow Nanofibers as an Anode Material for High Power Lithium Ion Batteries, Energy & Environmental Science, *4(11)*, 4532–4536.
378. Ueno, S., & Fujihara, S. (2011). Effect of an Nb_2O_5 Nanolayer Coating on ZnO electrodes in Dye-Sensitized Solar Cells, Electrochimical Acta, *56(7)*, 2906–2913.
379. Greene, L. et al. (2007). ZnO-TiO2 Core-Shell Nanorod P3HT Solar Cells, the Journal of Physical Chemistry C, *111(50)*, 18451–18456.
380. Yang, H., Lightner, C., & Dong, L. (2011). Light-Emitting Coaxial Nanofibers, ACS nano, *6(1)*, 622–628.
381. Qin, X., & Wang, S. (2006). Filtration Properties of Electrospinning Nanofibers, Journal of Applied Polymer Science, *102(2)*, 1285–1290.
382. Anka, F., & Jr. Balkus, K. (2013). Novel Nanofiltration Hollow Fiber Membrane Produced via Electrospinning, Industrial & Engineering Chemistry Research, *52(9)*, 3473–3480.
383. Sill, T., & Von Recum, H. (2008). Electrospinning: Applications in Drug Delivery and Tissue Engineering Biomaterials, *29(13)*, 1989–2006.
384. Kenawy, E. et al. (2009). Processing of Polymer Nanofibers through Electrospinning as Drug Delivery Systems, Materials Chemistry and Physics, *113(1)*, 296–302.
385. Huang, Z. et al. (2006). Encapsulating Drugs in biodegradable Ultrafine Fibers through Co-axial Electrospinning, Journal of Bio-medical Materials Research PartA, *77(1)*, 169–179.
386. Chakraborty, S. et al. (2009). Electrohydrodynamics: A Facile Technique to Fabricate Drug Delivery Systems. Advanced Drug Delivery Reviews, *61(12)*, 1043–1054.

387. Zhang, Y. et al. (2006). Coaxial Electrospinning of (Fluorescein Isothiocyanate-Conjugated Bovine Serum Albumin)-Encapsulated Poly (E-Caprolactone) Nanofibers for Sustained Release. Biomacromolecules, *7(4)*, 1049–1057.
388. Srikar, R. et al. (2008). Desorption-Limited Mechanism of Release from Polymer Nanofibers, Langmuir, *24(3)*, 965–974.
389. Bognitzki, M. et al. (2000). Polymer, Metal, and Hybrid Nano and Mesotubes by Coating Degradable Polymer Template Fibers (TUFT Process), Advanced Materials, *12(9)*, 637–640.

CHAPTER 15

A STUDY ON POLYMER/ ORGANOCLAY NANOCOMPOSITES

K. S. DIBIROVA, G. V. KOZLOV, and G. M. MAGOMEDOV

Dagestan State Pedagogical University, Makhachkala 367003, Yaragskii st., 57, Russian Federation

CONTENTS

Abstract..352
15.1 Introduction...352
15.2 Experimental Part..352
15.3 Results and Discussion 353
15.4 Conclusions...357
Keywords 358
References..358

ABSTRACT

It has been shown that crystalline phase morphology in nanocomposites polymer/ organoclay with semicrystalline matrix defines the dimension of fractal space, in which the indicated nanocomposites structure is formed. In its turn, this dimension influences strongly on both deformational behavior and mechanical characteristics of nanocomposites.

15.1 INTRODUCTION

It has been shown earlier [1, 2], that particles (aggregates of particles) of filler (nanofiller) form network in polymer matrix, possessing fractal (in the general case – multifractal) properties and characterized by fractal (Hausdorff) dimension D_n. Hence, polymer matrix structure formation in nanocomposites can be described not in Euclidean space, but in fractal one. This circumstance tells to a considerable degree on both structure and properties of nanocomposites. As it has been shown in work [2], polymer nanocomposites properties change is defined by polymer matrix structure change, which is due to nanofiller introduction. So, the authors [3] demonstrated that the introduction of organoclay in high-density polyethylene (HDPE) resulted in matrix polymer crystalline morphology change, that is, in spherolites size increasing about twice. Therefore, the present work purpose is the study of polymer matrix crystalline morphology influence on structure and properties of nanocomposites high-density polyethylene/Na⁺-montmorillonite (HDPE/MMT). This study was performed within the framework of fractal analysis [4].

15.2 EXPERIMENTAL PART

As a matrix polymer HDPE with melt flow index of ~1.0 g/10 min and crystallinity degree K of 0.72, determined by samples density, manufactured by firm Huntsman LLC, was used. As nanofiller organoclay Na⁺-montmorillonite of industrial production of mark Cloisite 15, supplied by firm Southern Clay (USA), was used. A maleine anhydride (MA) was applied as a coupling agent. Conventional signs and composition of nanocomposites HDPE/MMT are listed in Table 15.1 [3].

TABLE 15.1 Composition and Spherolites Average Diameter of Nanocomposites HDPE/ MMT

Sample conventional sign	MA contents, mass. %	MMT contents, mass. %	Spherolites average diameter D_{sp}, mcm
A	-	-	5.70
B	1.0	-	5.80

TABLE 15.1 *(Continued)*

Sample conventional sign	MA contents, mass. %	MMT contents, mass. %	Spherolites average diameter D_{sp}, mcm
C	-	1.0	11.62
D	-	2.5	10.68
E	-	5.0	10.75
F	1.0	1.0	11.68
G	2.5	2.5	11.12
H	5.0	5.0	11.0
I	5.0	2.5	11.30

Compositions HDPE/MA/MMT were prepared by components mixing on twin-screw extruder of mark Haake TW 100 at temperatures 390–410 K. Samples with thickness of 1 mm were obtained on one-screw extruder Killion [3].

Uniaxial tension mechanical tests have been performed on apparatus Rheometric Scientific Instrument (RSA III) according to ASTM D882–02 at temperature 293° and strain rate 2×10^{-3} s^{-1}. The morphology of nanocomposites HDPE/MMT was studied with the help of polarized optical microscope Zeiss with magnification 40′ and thus obtained spherolites average diameter D_{sp} in this way is also adduced in Table 15.1.

15.3 RESULTS AND DISCUSSION

As it is known [5], the fractal dimension of an object is a function of space dimension, in which it is formed. In the computer model experiment this situation is considered as fractals behavior on fractal (but not Euclidean) lattices [6]. The space (or fractal lattice) dimension D_n can be determined with the aid of the following equation [1]:

$$v_F = \frac{2.5}{2 + D_n},$$ (1)

where n_F is Flory exponent, connected with macromolecular coil dimension D_f by the following relationship [1]:

$$v_F = \frac{1}{D_f}.$$ (2)

In its turn, the dimension D_f value for linear polymers the following simple equation gives [7]:

$$D_f = \frac{2d_f}{3}, \qquad (3)$$

where d_f is the fractal dimension of polymeric material structure, determined as follows [8]:

$$d_f = (d-1)(1+v), \qquad (4)$$

where d is the dimension of Euclidean space, in which a fractal is considered (it is obvious, that in our case d=3), n is Poisson's ratio, estimated according to the results of mechanical tests with the help of the relationship [9]:

$$\frac{\sigma_Y}{E} = \frac{1-2v}{6(1+v)}, \qquad (5)$$

where σ_Y is yield stress, E is elasticity modulus.

In Fig. 15.1, the dependence of fractal space (fractal lattice) dimension D_n, in which the studied nanocomposites structure is formed, on spherolites mean diameter D_{sp} is adduced. As one can see, the correlation $D_n(D_{sp})$ is linear and described analytically by the following empirical equation:

$$D_n = 2.1 \times 10^{-2} D_{sp} + 2.5, \qquad (6)$$

where value D_{sp} is given in mcm.

From the Eq. (6) it follows, that the minimum value D_n is achieved at $D_{sp} = 0$ and is equal to 2.5, that according to the Eqs. (1)–(3) corresponds to structure fractal dimension $d_f \approx 2.17$. Since the greatest value of any fractal dimension for real objects, including D_n at that, cannot exceed 2.95 [8], then from the Eq. (6) the limiting value D_{sp} for the indicated matrix polymer can be evaluated, which is equal to ~21.5 mcm. Let us also note that the large scatter of the data in Fig. 15.1 is due to the difficulty of the value D_{sp} precise determination.

As it is known [7], in semicrystalline polymers deformation process partial melting-recrystallization (mechanical disordering) of a crystalline phase can be realized, which is described quantitatively within the framework of a plasticity fractal theory [10]. According to the indicated theory Poisson's ratio value v_Y at a yield point can be evaluated as follows:

$$v_Y = \chi v + 0.5(1-\chi), \qquad (7)$$

where c is a relative fraction of elastically deformed polymer, n is Poisson's ration in elastic strains region, determined according to the Eq. (5), and v_Y value is accepted equal to 0.45 [7].

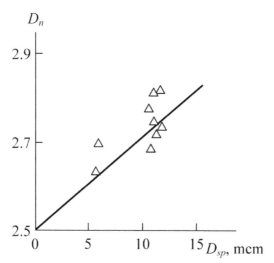

FIGURE 15.1   The dependence of space dimension D_n, in which nanocomposite structure is formed, on spherolites average diameter D_{sp} for nanocomposites HDPE/MMT.

The calculation of a relative fraction of crystalline phase χ_{cr}, subjecting to mechanical disordering, can be performed according to the equation [7]:

$$\chi_{cr} = \chi - \alpha_{am} - \phi_{cl},$$ (8)

where α_{am} is an amorphous phase relative fraction, which is equal to $(1-K)$, ϕ_{cl} is a relative fraction of local order domains (nanoclusters), which can be determined with the aid of the following fractal relationship [7]:

$$d_f = 3 - 6 \times 10^{-10} \left(\frac{\phi_{cl}}{SC_\infty} \right)^{1/2},$$ (9)

where S is cross-sectional area of macromolecule, which is equal to 14.4 Å² for HDPE [7], C_∞ is characteristic ratio, which is an indicator of polymer chain statistical flexibility [11], and connected with the dimension d_f by the following relationship [7]:

$$C_\infty = \frac{2d_f}{d(d-1)(d-d_f)} + \frac{4}{3}$$ (10)

As it is known [7], parameter χ_{cr} effects essentially on deformational behavior and mechanical properties of semicrystalline polymers. In Fig. 15.2, the dependence

$\chi_{cr}(D_n)$ is adduced for the studied nanocomposites, which turns out to be linear, that allows to describe it analytically as follows:

$$\chi_{cr} = 1.88(D_n - 2.55)$$

(11)

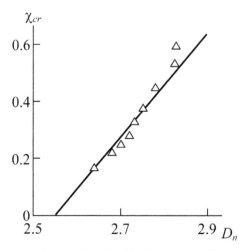

FIGURE 15.2   The dependence of relative fraction of crystalline phase χ_{cr}, subjecting to mechanical disordering, on space dimension D_n for nanocomposites HDPE/MMT.

From the Eq. (8) it follows that the greatest value χ_{cr} (χ_{cr}^{max}) is achieved at the following conditions: c=1.0 and φ_{cl}=0. In this case the condition χ_{cr}^{max} = K is realized, that was to be expected from the most common considerations. For the studied nanocomposites the value χ_{cr}^{max} =K=0.72 is achieved according to the Eq. (11) at $D_n \approx 2.933$, that is close to the indicated above limiting value D_n=2.95 [8]. The minimum value χ_{cr}=0 according to the Eq. (11) is achieved at D_n=2.55 or, according to the Eqs. (1)–(3), at d_f=2.73. As it is known [7], the value d_f can be determined alternatively as follows:

$$d_f \approx 2 + K$$

(12)

From the Eqs. (11) and (12) it follows, that a common variation χ_{cr}=0–0.72 is realized at the constant value K=0.72, that is, this parameter does not depend on crystallinity degree and it is defined only by polymer matrix crystalline morphology change.

As it is known [4], the parameter χ_{cr} influences essentially on nanocomposites polymer/organoclay properties. One from the most important mechanical character-istics of polymeric materials, namely, elasticity modulus E depends on the value χ_{cr} as follows:

$$E = \left(40 + 54.9\chi_{cr}\right)\sigma_Y \tag{13}$$

In Fig. 15.3, the comparison of experimental E and calculated according to the Eq. (13) E^T elasticity modulus values for the considered nanocomposites is adduced. In this case the value χ_{cr} was determined according to the Eq. (11). As it follows from the data of Fig. 15.3, a theory and experiment good correspondence is obtained (the average discrepancy between E^T and E makes up ~ 14%), that will be enough for preliminary estimations performance.

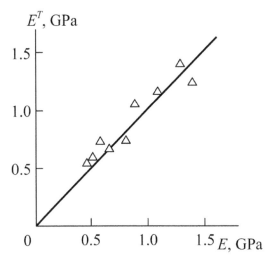

FIGURE 15.3   The relation between experimental E and calculated according to the Eqs. (11) and (13) E^T elasticity modulus values for nanocomposites HDPE/MMT.

15.4   CONCLUSIONS

Thus, the present work results have shown that the fractal dimension of space, in which nanocomposites structure is formed, is defined by their crystalline polymer phase morphology and does not depend on crystallinity degree. The indicated di-mension defines unequivocally partial melting-recrystallization process at nano-composites deformation and influences strongly on their mechanical characteristics.

KEYWORDS

- **elasticity modulus**
- **fractal space**
- **morphology**
- **nanocomposite**
- **organoclay**
- **semicrystalline polymer**

REFERENCES

1. Kozlov, G. V., Yanovskii, Yu. G., & Zaikov, G. E. (2010). Structure and Properties of Partic-ulate-Filled Polymer Composites: The Fractal Analysis. Nova Science Publishers, Inc., New York, 282.
2. Miktaev, A. K., Kozlov, G. V., & Zaikov, G. E. (2008). Polymer Nanocomposites, *Variety of Structural Forms and Applications*. Nova Science Publishers, Inc. New York, 319.
3. Ranade, A., Nayak, K., Fairbrother, D., & D'Souza, N. A. (2005). *Polymer, 46(23)*, 7323–7333.
4. Kozlov, G. V., & Miktaev, A. K. (2013). Structure and Properties of Nanocomposites Polymer Organoclay. Lap Lambert Academic Publishing GmbH, Saarbrücken, 318.
5. Aharony, A., & Harris, A. B. (1989). *J. Stat. Phys., 54(3/4)*, 1091–1097.
6. Vannimenus, J. (1989). Physica D, *38(2)*, 351–355.
7. Kozlov, G. V., & Zaikov, G. E. (2004). Structure of the Polymer Amorphous State. Brill Academic Publishers, Utrecht, Boston, 465.
8. Balankin, A. S. (1991). Synergetic of Deformable Body. Publishers of Ministry Defense of SSSR, Moscow, 404.
9. Kozlov, G. V., & Sanditov, D. S. (1994). An Harmonic Effects and Physical-Mechanical Properties of Polymers, Nauka, Novosibirsk, 261.
10. Balankin, A. S., & Bugrimov, A. L. (1992). Vysokomolek. Soed. A, *34(3)*, 129–132.
11. Budtov, V. P. (1992). Physical Chemistry of Polymer Solutions. Khimiya, Sankt-Peterburg, 384.

CHAPTER 16

A VERY DETAILED REVIEW ON APPLICATION OF NANOFIBERS IN ENERGY AND ENVIRONMENTAL

SAEEDEH RAFIEI, BABAK NOROOZI, and A. K. HAGHI

CONTENTS

Abstract 360
16.1   Introduction..360
16.2   Carbon Fibers...362
16.3   Polyacrylonitrile (PAN) 364
16.4   Activated Carbon Fibers 366
16.5   Activated Carbon 369
16.6   Carbon Nanostructures 370
16.7   Polyacrylonitrile Based Carbon Nanofibers 373
16.8   Electrospinning Methodology 375
16.9   Electrospinning of PAN Solution 377
16.10   Stabilization 380
16.11   Carbonization...380
16.12   Comparison of Stabilization and Carbonization.....................383
16.13   FT-IR Study of PAN and Stabilized PAN Fiber.....................384
16.14   CNF from Other Type of Precursors.......................................385
16.15   Activation 386
16.16   Applications...392
16.17   Summary and Outlook..409
Keywords 410
References..410

ABSTRACT

Carbon nanofibers (sometimes known as carbon nanofilaments or CNF) can be produced in a relative large scale by electrospinning of Polyacrylonitrile solution in dimethyl formaldehyde, stabilization and carbonization process. The porosity, pore volume and surface area of carbon nanofiber enhanced during the chemical or physical activation. This chapter is a review of electrospinning method of CNF production and heat pretreatments and activations process to make activated carbon nanofiber. Attention is also given to some of the possible applications of this nano-structures which center around the unique blend of properties exhibited by the material, which include: hydrogen adsorption properties, energy storage media, catalyst support, and regenerative medicine.

16.1   INTRODUCTION

In recent years, porous materials have been of immense interest because of their potential for applications in various fields, ranging from chemistry to physics, and to biotechnology [1]. Introducing nanometer-sized porosity has been shown as an effective strategy to achieve desired properties while maintaining micro structural feature sizes commensurate with the decreasing length-scales of devices and membranes [2]. On the other hand, fibrous materials have filled many needs in many areas due to their intrinsically high surfaces, interfiber pores and engineering versatility. In concept, coupling nano porosity in the ultra-fine nanofibers should lead to the highest possible specific surface and fibrous materials. Hence, the synthesis of organic polymer or inorganic nanofibers with nanoporous structures might be very useful in a widely areas, such as membranes technology [3], tissue engineering, drug delivery [4], adsorption materials [5–8], filtration and separation [9], sensors [10], catalyst supports [11, 12] and electrode materials [13–17] and so forth.

Activated carbon fibers (ACF) and activated carbon nanofibers (ACNF) are a relatively modern form of porous carbon material with a number of significant advantages over the more traditional powder or granular forms. Advantages include high adsorption and desorption rates, thanks to the smaller fiber diameter and hence very low diffusion limitations, great adsorption capacities at low concentrations of adsorbates, and excellent flexibility [18, 19].

The adsorption and the isolation technique is an efficient method in environmental problems, the toxic materials in sorts of industrial wastewater and exhaust fumes can be absorbed by using various adsorbents, so that the fumes and the liquid are up to standard of environmental protection. The key problem of the adsorption and the isolation technique lies in the adsorbents; the commonly used adsorbents are activated carbon, silica gel, acid terra alba and zeolite molecular sieve, etc. [20–22]. But, not only the adsorption characterization of these materials but also the operating characterization and the reproducing ability of these materials are all very weak.

So searching for a high quality adsorption material has become a subject concerned by experts all over the world.

Although several methods have been proposed for nanofiber manufacturing so far, an efficient and cost effective procedure of production is still a challenge and is debated by many experts [23]. Some of different methods for nanofiber manufacturing are drawing [24], template synthesis [25], phase separation [26], self assembly [27] vapor growths, arc discharge, laser ablation and chemical vapor deposition [28]. One of the most versatile methods in recent decades is electrospinning which is a broadly used technology for electrostatic fiber formation which uses electrical forces to produce polymer fibers with diameters ranging from several nm to several micrometers using polymer solutions of both natural and synthetic polymers has seen a tremendous increase in research and commercial attention over the past decade [29–31]. Electrospinning is a remarkably simple and versatile technique to prepare polymers or composite materials nanofibers. As many research showed that, by changing appropriate electrospinning parameters, it can be possible to obtain the fibers with nanoporous structures, for which the basic principle based on away that phase separation processes take place during electrospinning [1, 30, 32]. Nanomaterials are traditionally defined to be those with at least one of the three dimensions equal to or less than 100 nm, 8 whereas an upper limit of 1000 nm is usually adopted to define nanofibers. A large number of special properties of polymer nanofibers have been reported due to the high specific surface area and surface area to volume ratio. In contrast to the normal PAN fibers with diameters of about 20 μm [33, 34], electrospun PAN nanofibers with diameters in the submicron range result in ACNF with very much higher specific surface area [1, 35].

ACNF is an excellent adsorbent and has found usage in applications such as gas-phase and liquid phase adsorption [5, 20, 36] as well as electrodes for super capacitors and batteries [14, 37]. The literature review concerning ACNF is summarized as follows:

Kim and Yang [14] prepared ACNF from electrospun PAN nanofibers activated in steam at 700–800°C and found that the specific surface area of the ACNF activated at 700°C was the highest but the Mesopore volume fraction was the lowest. However, the work by Lee et al. [38] showed an opposite result. Song et al. [21] investigated the effect of activation time on the formation of ACNF (ultra-thin PAN fiber based) activated in steam at 1000°C. Ji et al. [39] made mesoporous ACNF produced from electrospun PAN nanofibers through physical activation with silica and conducted chemical activations by potassium hydroxide and zinc chloride to increase specific surface area and pore volume of ACNF [40]. It must be pointed out that different methods and conditions of activation lead to very different physical properties and adsorption capacities for ACNF.

This review provides an overview of the most preferable production methods of carbon fiber and nanofiber and activated form of them. Because of the extraordinary combination of physical and chemical properties exhibited by carbon nanofibers and activated form of it, which blends two properties that rarely coexist: high surface area and high electrical conductivity, which are the result of the unique stacking and crystalline order present within the structure, there are tremendous opportunities to exploit the potential of this form of carbon in a number of areas, some of which are discussed in this chapter.

16.2   CARBON FIBERS

The existence of carbon fiber (CFs) came into being in 1879 when Thomas Edison recorded the use of carbon fiber as a filament element in electric lamp. Fibers were first prepared from rayon fibers by the US Union Carbide Corporation and the US Air Force Materials Laboratory in 1959 [41]. In 1960, it was realized that carbon fiber is very useful as reinforcement material in many applications. Since then a great deal of improvement has been made in the process and product through research work carried out in USA, Japan and UK. In 1960's, High strength Polyacrylonitrile (PAN) based carbon fiber was first produced in Japan and UK and pitch based carbon fiber in Japan and USA.

Carbon fibers can be produced from a wide variety of precursors in the range from natural materials to various thermoplastic and thermosetting precursors Materials, such as PAN, mesophase pitch, petroleum, coal pitches, phenolic resins, polyvinylidene chloride (PVDC) and rayon (viscose) and etc. [42–43]. About 90% of world's total carbon fiber productions are PAN-based. To make carbon fibers from PAN precursor, PAN-based fibers are generally subjected to four pyrolysis processes, namely oxidation stabilization, carbonization and graphitization or activation; they will be explained in following sections later [43].

Among all kind of carbon fibers, PAN-based carbon fibers are the preferred reinforcement for structural composites with the result of their excellent specific strength and stiffness combined with their light weight as well as lower cost, but in general, PAN-based carbon fibers have lower carbonization yield than aromatic structure based precursors, such as pitch, phenol, polybenzimidal, polyimide, etc. The carbon yield strongly depends on the chemical and morphological structures of the precursor fibers [44].

Carbon fibers are expected to be in the increasing demand for composite materials in automobile, housing, sport, and leisure industries as well as airplane and space applications [45]. They requires high strength and high elasticity, the high strength type is produced from PAN and the high elasticity type is manufactured from coat tar pitch. In order to meet expanded use in some high-tech sectors, many novel approaches, such as dry-wet spinning [46], steam drawing [34], increasing the molecular weight of precursors polymer [47], modifying the precursors prior to stabilization [43–47], etc., have been performed to increase the tensile strength of PAN-based carbon fibers.

The quality of the high performance carbon fibers depends mainly on the composition and quality of the precursor fibers. In order to obtain high performance PAN-based carbon fibers, the combination of both physical mechanical properties and chemical composition should be optimized [48]. Producing a high performance PAN-based carbon fiber and activated carbon fiber is not an easy task, since it involves many steps that must be carefully controlled and optimized. Such steps are the dope formulation, spinning and post spinning processes as well as the pyrolysis process. At the same time, there are several factors that need to be considered in order to ensure the success of each step. However, among all steps, the pyrolysis process is the most important step and can be regarded as the heart of the carbon fiber production [43].

Carbon fibers possess high mechanical strengths and module, superior stiffness, excellent electrical and thermal conductivities, as well as strong fatigue and corrosion resistance; therefore, they have been widely used for numerous applications particularly for the development of large load-bearing composites. Conventional carbon fibers are prepared from precursors such as PAN [49, 50]. Ch. Kim et al. [44] produced two-phase carbon fibers from electrospinning by pitch and PAN precursor. He proved that the fiber diameter, the carbon yield and the electrical conductivity are increased with increasing Pitch component.

Carbon fibers have various applications because of their porous structure [51]. The preparation of drinkable, high quality water for the electronics and pharmaceutical industries, treatment of secondary effluent from sewage processing plants, gas separation for industrial application, hemo dialyzers, and the controlled release of drugs to mention only a few applications [49–52].

16.2.1 CLASSIFICATIONS OF CARBON FIBERS

The manufacturing technology for carbon fibers is based on the high-temperature pyrolysis of organic compounds, conducted in an inert atmosphere. Some carbon fiber classifications are based on the magnitude of the final heat-treatment temperature (HTT) during production of the carbon fibers through pyrolysis, and also on the carbon content of the final product. Accordingly, carbon fibers may be subdivided into three classes: partially carbonized fibers (HTT 500°C, carbon content up to 90 wt.%), carbonized fibers (HTT 500–1500°C, carbon content 91–99 wt.%), and graphitized fibers (HTT 2000–3000°C, carbon content over 99 wt.%). Carbon fibers can be classified by raw materials as well, for example, rayon-based fibers, PAN-based fibers, pitch-based mesophase fibers, lignin based fibers and gas phase production fibers [52].

16.2.2   PROPERTIES OF CARBON FIBERS

The characteristics of carbon fiber material are influenced by choices of the initial polymer raw material, conditions of carbonization and heat treatment, and also by introduction of certain additives.

16.2.2.1   MECHANICAL PROPERTIES

PAN-based carbon fibers demonstrate 200–400 GPa Young's modulus upon longitudinal stretching of the fiber, while upon transverse stretching the Young's modulus is 5–25 GPa, and the compressive strength is 6 GPa. It has been suggested that the Young's modulus of carbon fibers depends on the orientation of graphite crystallites in the carbon fiber, while the strength is determined by the intrafiber bonding [52].

16.2.2.2   CHEMICAL STABILITY

An important property of carbon fibers that largely determines their prospective uses in many fields is their stability with respect to aggressive agents. This property is related to some structural features of carbon fibers and primarily depends on the type of initial raw material, the heat-treatment temperature, and the presence of element in the fiber. It is evident that the acid stability of carbon fibers increases with increase in heat treatment temperatures as the proportion of the more stable bonds increases, while the more perfect carbon structure excludes reagent diffusion into the fiber matrix. While at room temperature there has been little change observed in carbon fibers even after prolonged periods of exposure to corrosive liquids. The stability of carbon fibers at elevated temperatures decreases, especially if the reagents are oxidizing (i.e., nitric acid, sodium hypo chloride) [52].

16.2.2.3   APPLICATIONS OF CARBON FIBERS

The expansion of the areas of application for carbon fibers is stimulated by their attractive properties, not found in other materials, such as strength, electrical conductivity, stability on exposure to reactive media, low density, low-to-negative coefficient of thermal expansion, and resistance to shock heating. The most representative applications of carbon fibers and element carbon fibers are as sorption materials, electrostatic discharge materials, catalysts, and reinforcement materials in composites.

16.3   POLYACRYLONITRILE (PAN)

It is well know that PAN is made from acrylonitrile, which was prepared by Moureu in 1893. The chemical structure of PAN is illustrated in Fig. 16.1. It is a resinous, fibrous, or rubbery organic polymer and can be used to make acrylic fibers. PAN is sometimes used to make plastic bottles, and as a starting material for making carbon fibers. It is chemically modified to make the carbon fibers found in plenty of both high-tech and common daily applications such as civil and military aircraft primary and secondary structures, missiles, solid propellant rocket motors, pressure vessels, fishing rods, tennis rackets, badminton rackets & high-tech bicycles. Homopolymer of PAN has been used as fibers in hot gas filtration systems, outdoor awnings, sails for yachts, and even fiber reinforced concrete. The homopolymer was developed for the manufacture of fibers in 1940, after a suitable solvent had been discovered by DuPont in the USA, while Bayer developed an aqueous based solution [52]. Almost all polyacrylonitrile resins are copolymers made from mixtures of monomers; with acrylonitrile as the main component. It is a component repeat unit in several important copolymers, such as styrene-acrylonitrile or SAN and ABS plastic. Copolymers containing polyacrylonitrile are often used as fibers to make knitted clothing, like socks and sweaters, as well as outdoor products like tents and similar items. If the label of a piece of clothing says "acrylic," then it is made out of some copolymer of polyacrylonitrile. It was made into spun fiber at DuPont in 1941 and marketed under the name of Orlon. Acrylonitrile is commonly employed as a comonomer with styrene (e.g., SAN, ABS, and ASA plastics).

PAN fibers are used in weaving (blanket, carpet and clothes) and in engineering- housing (instead of asbestos) and most importantly for producing carbon fibers [53]. In recent decade PAN fibers are considered as main material for production of carbon fibers. PAN fibers manufactured presently are composed of at least 85% by weight of acrylonitrile (AN) units. The remaining 15% consists of neutral and/or ionic comonomers, which are added to improve the properties of the fibers. Neutral comonomers like methyl acrylate (MA), vinyl acetate (VA), or methyl methacrylate (MMA) are used to modify the solubility of the PAN copolymers in spinning solvents, to modify the PAN fiber morphology, and to improve the rate of diffusion of dyes into the PAN fiber. Ionic and acidic comonomers including the sulfonate groups like SMS, SAMPS, sodium p-styrene sulfonate (SSS), sodium p- sulfophenyl methallyl ether (SMPE), and IA also can be used to provide dye sites apart from end groups and to increase hydrophilicity [54].

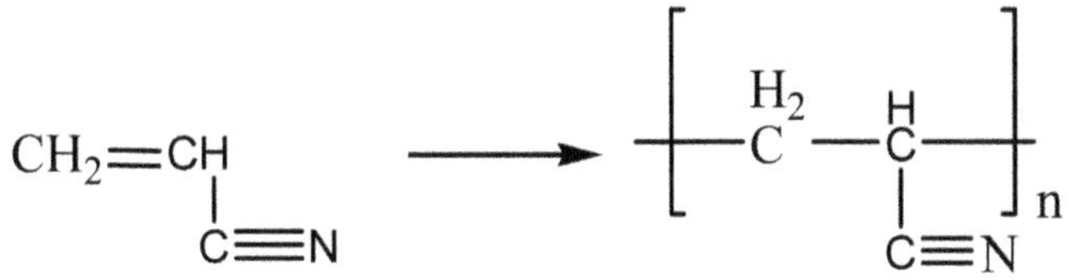

FIGURE 16.1 Chemical structure of PAN.

16.4   ACTIVATED CARBON FIBERS

One of the disadvantages of carbon fibers is its low surface area This fact restrict the applications of these materials like hydrogen (or energy) storage or treatment of water and waste water; therefore is necessary increase the surface area to improve the yield in these materials [55]. Activation treatment greatly increases the number of micropores and mesopores [18].

Chemical activation with KOH or NaOH is an effective method to prepare activated carbon materials [55, 56]. Chemical agent activation soaks the carbon material using chemical agent, during the process of heating and activating, carbon element will liberate with tiny molecule such as CO or CO_2. $ZnCl_2$, KOH, H_3PO_4 [57] are the commonly used chemical agents [22].

Chemical activation presents several advantages and disadvantages compared to physical activation. The main advantages are the higher yield, lower temperature of activation, less activation time and generally, higher development of porosity. Among the disadvantages, the activating agents are more expensive (KOH and NaOH vs. CO_2 and H_2O) and it is also necessary an additional washing stage. Because the character of ACF manufactured by this method is unstable, we seldom use it [22]. Moreover, these hydroxides are very corrosive. Physical activation with carbon dioxide or steam is the usual procedure to obtain ACF. Chemical activation of carbon fibers by $ZnCl_2$, $AlCl_3$, H_3PO_4, $H3BO_3$, ..., has been reported [58].

ACF is one member of carbon family with multiple holes, which has its unique performance. It has the characters such as big specific surface area, developed microcellular structure, large adsorptive capacity, fast absorbed and desorbed velocity and easy regeneration capacity, etc. They have begun to find use in fluid filtration applications as an alternative to activated carbon granules. Their properties differ from granules with respect to porosity and physical form, which can confer certain advantages, e.g., their smaller dimensions give improved access of the adsorptive and the micropores are accessible from the surface. Adsorption rates of organic vapors are therefore faster than in granules where diffusion through macropores and mesopores must occur first. A further advantage is their ability to be formed into both woven and nonwoven mats, where problems due to channeling when granules are used are avoided. Materials of this type can act as aerosol and particulate filters, as well as microporous adsorbents [59]. According to the aperture classification standard by IUPAC, the pores can be classified into three categories, namely micropore (pore size < 2 nm), mesopore (2 nm < pore size < 50 nm) and macropore (pore size >50 nm). It is proved that mesopores or macropores in ACF can thus improve the adsorption effectivity for larger molecules or macromolecules such as protein and virus [19].

The PAN based activated carbon hollow fibers (ACHF) have brought on many investigators' interest, since PAN-based ACHF shows large adsorption capacity. PAN hollow fibers are pretreated with ammonium dibasic phosphate and then further oxidized in air, carbonized in nitrogen, and activated with carbon dioxide.

One of the important applications of ACF is to remove formaldehyde from atmosphere. It has been investigated for many years. It was found that porous carbons derived from PAN showed larger formaldehyde adsorption capacity, which originated from its abundant nitrogen functionalities on the surface [20].

16.4.1 METHODS OF ACTIVATED CARBON FIBER PREPARATION

Fibers are formed from above-mentioned precursors, which are then subject to various heat treatments in controlled atmospheres to yield carbon fibers, often with specific mechanical properties. A final step in this procedure can be either chemical or physical activation of the carbon fibers. In general, nongraphitizable carbons, which are more disordered can be activated physically in steam or CO whereas the more ordered graphitizable carbons require chemical activation for the generation of porosity [60].

As a result activated carbon fiber is produced through a series of process consisting of stabilization, carbonization, and activation of precursor fibers. It is important to improve the efficiency of the production process as well as to select low cost precursors [42]. The stabilization process, air oxidation of precursor fibers at 200–300°C, is the process required to prevent the precursor fibers from melting during the subsequent carbonization process [61, 62]. It is essential for PAN and pitches, but is not essential for phenolic resin and cellulose, because the latter precursors are thermosetting resins. Phenolic resin is known to produce higher surface area ACF as compared with other precursors. It is, therefore, very advantageous if we could improve the production efficiency of phenol resin based ACF by, possibly, simple and cost effective methods.

16.4.2 PROPERTIES OF ACF

Some of the most important properties of ACF are listed in Table 16.1.

TABLE 16.1   Activated Carbon Fiber Properties

Most important properties of ACF	References
Fast speed of adsorption and desorption as a result of large surface area and average aperture of micropore	[63]
ACF also has excellent adsorption to low concentration substance, because its adsorption forces and unit adsorptive volume	[64, 65]
The capability of turning into different patterns, like paper, nonwoven, a honeycomb structure and corrugated cardboard because of good tensile strength	[63]

TABLE 16.1   *(Continued)*

Most important properties of ACF	References
ACF itself is not easy to become powder so it will not cause second pollution	[63]
ACF's thin and light adsorption layer allow it be used in the small treatment with high efficiency	[66]
ACFs adsorption can reach to the expectable efficiency in short time because it's Low density and small loss of press	[66]
ACF can also be applied as fuel cell electrodes material	[67]
Without reducing its adsorption function, it's easy to be regenerated. And it can be used for very long time	[67]
ACF can be a deoxidizer to recycle the precious metal	[66]
ACFs have acid-proof and alkali-proof properties	[68, 69]

16.4.3   APPLICATION OF ACF

ACFs have been successfully applied in many fields, such as the treatment of organic and inorganic waste gases, the recovery of organic solvent, air cleaning and deodorization, treatment of wastewater and drinking water, separation and recovery of precious metals, as medical adsorbents and protective articles, and in electrodes [57]. Some of the most common applications of ACF are listed in Table 16.2.

TABLE 16.2   Activated Carbon Fiber Applications

ACF applications	Explanation	Reference
Recovery of Organic Compounds and Solvents	ACF can be used in gas/air separation, recovery of organic compounds and solvents, especially for caustic nitrides, and low boiling point solvent.	[64, 70]
Air purification	ACF can eliminate malodorous substance in the air, especially for aryl substance which will generate carcinogens	[57]
Wastewater treatment	ACF is suitable in organic waste-water treatment, like substance content phenol, medical waste, etc. which are hard to decompose by organism. With large quantity of adsorption volume, fast speed of adsorption, excellent desorption function, and easy to regenerate. ACF can be used in a small, continuous, simple design condition which cost low and do not make second pollution.	[9]

TABLE 16.2 *(Continued)*

ACF applications	Explanation	Reference
Water purification	They can be applied in deodorizing and de-coloring Applications in different industrial field, like food and beverage, pharmaceutical, sugar-making, wine-making and also for super-pure water treatment system of electronic industry and aqueous filtration treatment.	[9, 71]
Domestic products	ACF can be a refrigerator deodorizer and keep food fresh.	[57]

16.5 ACTIVATED CARBON

Needs for porous materials in industrial application and in our daily life are increasing. Porous carbon materials, especially those containing micropores or mesopores, are being used in various applications such as adsorbent and catalyst supports. Activated carbon has played a major role in adsorption technology over the last few years [72]. Porous carbons have highly developed porosity and an extended surface area. Their preparation involves two methods: a physical method that initiates the activation at high temperature with CO_2 or a water steam stream, and a chemical method that initiates the activation using the micro explosion behaviors of chemical agents [73]. Normally, all carbonaceous materials can be converted into porous carbons, although the properties of the final products will differ according to the nature of the raw materials used.

Toxic organic compounds, such as aromatics and chlorinated hydrocarbons in the gaseous or liquid effluent streams from the production of chemicals and pharmaceuticals are normally removed by adsorption with activated carbons (ACs) in a fixed-bed purifier. However, both contaminant and water molecules are adsorbed at the same time owing to the hydrophilic character acquired during the production process of AC, thus limiting the capacity of the adsorbent which is directly related to its highly micro porous structure, which unfortunately leads to weak mechanical strength resulting in decrepitation into fines in practical application [74]. Most commercially available ACs are extremely micro porous and ACs are of high surface area, and consequently they have high efficiency for the adsorption or removal of low-molecular weight compounds. The mesoporous activated carbons are expected to be excellent adsorbents for the removal and recovery of mesomolecular weight compounds.

16.6   CARBON NANOSTRUCTURES

16.6.1   *CARBON NANOTUBES (CNTS)*

Nanotubes of carbon and other materials are arguably the most fascinating materials playing an important role in nanotechnology today. One can think of carbon nanotubes as a sheet of graphite rolled into a tube with bonds at the end of the sheet forming the bonds that close the tube. Due to their small size and their extraordinary physicochemical properties, much attention has been paid to the interesting sp2-based fibrous carbons, including carbon nanotubes. It is generally accepted that carbon nanotubes consist of single or multiple grapheme sheets rolled into concentric cylinders: thus giving rise to single wall carbon nanotubes (SWNTs) or multiwall carbon nanotubes (MWNTs). A SWNT can have a diameter of 2 nm and a length of 100 pm, making it effectively a one-dimensional structure called a nanowires [75].

Their unique mechanical, electronic, and other properties are expected to result in revolutionary new materials and devices. These nanomaterials, produced mostly by synthetic bottom up methods, are discontinuous objects, and this leads to difficulties with their alignment, assembly, and processing into applications. Partly because of this, and despite considerable effort, a viable carbon nanotubes reinforced super nanocomposite is yet to be demonstrated. Advanced continuous fibers produced a revolution in the field of structural materials and composites in the last few decades as a result of their high strength, stiffness, and continuity, which, in turn, meant processing and alignment that were economically feasible [76]. Fiber mechanical properties are known to substantially improve with a decrease in the fiber diameter. Hence, there is a considerable interest in the development of advanced continuous fibers with nanoscale diameters. Electrospinning technology enables production of these continuous polymer nanofibers from polymer solutions or melts in high electric fields [23, 77].

16.6.2   *CARBON NANOFIBERS (CNFS)*

Carbon nanofibers, like other one-dimensional (1D) nanostructures such as nanowires, nanotubes, and molecular wires, are receiving increasing attention because of their large length to diameter ratio. With the development of nanotechnology in fiber fields, CNFs gradually attracted much attention after the discovery of carbon nanotubes by Iijima in 1991 [78]. CNF cost significantly less to produce than CNT and therefore offer significant advantages over nanotubes for certain applications, providing a high performance to cost ratio [28, 79].

CNFs are currently widely used in many fields, such as reinforcement materials, gas adsorption/desorption, rechargeable batteries, templates for nanotubes, high-temperature filters, supports for high-temperature catalysis, nanoelectronics and supercapacitors, due to their high aspect ratio, large specific surface area, high-temperature resistance and good electrical/thermal conductivities. The conventional preparation methods for CNFs, including the substrate method, spraying method, vapor growth method and plasma-enhanced chemical vapor deposition method, are known to be very complicated and costly. Therefore, a simple and inexpensive electrospinning process, first patented by Cooley in 1902, second has been increasingly used as the optimum method to fabricate continuous CNFs during the last decade [41]. The primary characteristic that distinguishes CNF from CNT resides in grapheme plane alignment: if the graphene plane and fiber axis do not align, the structure is defined as a CNF, but when parallel, the structure is considered as a CNT [79].

Carbon nanofibers with controllable nanoporous structures can be prepared via different ways. An oldest method is the catalytic decomposition of certain hydrocarbons on small metal particles such as iron, cobalt, nickel, and some of their alloys [80]. The mechanism includes hydrocarbon adsorption on a metal surface, conversion of the adsorbed hydrocarbon to adsorbed surface carbon via surface reactions, subsequent segregation of surface carbon into the layers near the surface, diffusion of carbon through metal particles, and then precipitation on the rare side of the particle [81]. The size of the catalyst nanoparticles seems to be the determining factor for the diameter of the carbon nanostructures grown on it. Small nanoparticles catalyze this grown better than the big ones due to that exhibit peculiar electronic properties (and thus catalytic properties) as consequence of their unusual ratio surface atom/bulk atom [80, 82, 83].

In a typical operation, about 100 mg of the powdered catalyst is placed in a ceramic boat, which is positioned in a quartz tube located in a horizontal tube furnace. The sample is initially reduced in a 10% hydrogen helium stream at 600°C and then quickly brought to the desired reaction temperature. Following this step, a predetermined mixture of hydrocarbon, hydrogen, and inert gas is introduced into the system and the reaction allowed proceeding for periods of about 2 h [6].

Carbon nanofibers were successfully prepared via electrospinning of PAN solution [78, 84, 85]. Electrospinning is a unique method for producing nanofibers or ultrafine fibers, and uses an electromagnetic field with a voltage sufficient to overcome the surface tension. Electrospun fibers have the characteristics of high specific surface area, high aspect ratio, dimensional stability, etc. [20].

Barranco et al. [86] divided carbon nanofiber into two groups, (i) highly graphitic CNFs and (ii) lowly graphitic ones. He investigated that the lowly graphitic CNFs have been obtained by certain procedures:

§ from blends of polymers in which a polymer acts as the carbon precursor and the other gives way to porosity, the latter being removed along the carbonization process;

§ by electrospinning of precursors, this gives way to webs of CNFs after carbonization;

§ by using an anodic alumina as a template, which leads to CNFs with marked mesoporosity;

§   from electrochemical decomposition of chloroform; and

§   from flames of ethanol.

The two last procedures lead to nanofibers with high oxygen contents. In all cases, the low crystallinity of the CNFs could be an advantage for subsequent activation, and thus, activated CNFs with a larger specific surface area and a higher specific capacitance can be obtained.

16.6.2.1   STRUCTURE OF CARBON NANOFIBER

The structural diversity of CNFs occurs by the anisotropic alignment of grapheme layers like the graphite, providing several particular structures such as platelet, herringbone and tubular CNFs according to the alignment direction to the fiber axis [87].

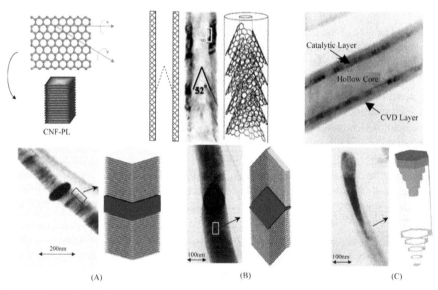

FIGURE 16.2   Different structures of carbon nanofibers: (a) platelet structure, (b) herringbone structure, and (c) tubular structure.

There are mainly three types of carbon nanofibers: the herringbone in which the graphene layers are stacked obliquely with respect to the fiber axis; the platelet in which the graphene layers are perpendicular to the fiber axis; and the tubular in which the graphene layers are parallel to the growth axis [55].

Figure 16.3 shows the SEM image of carbon nanofiber film. It can be seen from Fig. 16.2(a) that the film is composed of aggregated nanofibers. They have been called *platelet carbon nanofibers*. The fibers are of several tens μm long, zigzag, and most of them have a bright ellipsoidal particle on their tip. The particle has the nearly same width as that of the PCNFs [88].

Herringbone-type carbon nanofibers are a special kind of CNF with angles between the graphene plane direction and the axis of CNFs in the range of 0–90°. Theoretical calculations indicate that this type of nanofiber may exhibit some excellent properties, such as enhanced field emission, caused by the open edge sites, Aharonov–Bohm magnetic effects and magneto conductance, and localized states at the Fermi level may give rise to materials with novel electronic and magnetic properties Moreover, they can be expected to be used as absorbent materials, catalyst supports, gas storage materials and composite fillers, due to their special structural characteristics. Herringbone-type CNFs with large diameter and a very small or completely hollow core have been synthesized through a catalytic chemical vapor deposition (CVD) method [38].

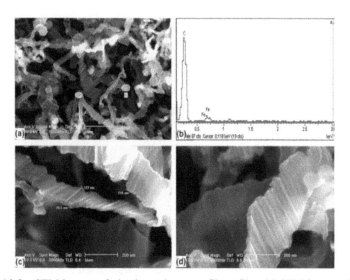

FIGURE 16.3 SEM images of platelet carbon nanofibers film. (a) SEM image of PCNFs film (b) Energy dispersive X-ray spectrometry (EDXS) of a light particles (indicated by *A* in (a)) on the top of a fiber. (c) The size of a typical PCNFs, which shows the width (100–300 nm) and thick (30 nm) of the fiber (d) surface morphology of a PCNF.

16.7 POLYACRYLONITRILE BASED CARBON NANOFIBERS

As we discuss there are various methods to produce carbon nanofibers or carbon nanotubes, for example, vapor growth [89], arc discharge, laser ablation and chemical vapor deposition [28, 89]. However, these are very expensive processes owing to the low product yield and expensive equipment. Preparation carbon nanofiber by electrospinning of proper precursors is preferred because of its lower cost and more output [90].

Polyacrylonitrile is a common precursor of general carbon nanofibers. As described in the previous section, PAN is the most common Polymer for the preparation of CNFs due mainly to its relatively high melting point and carbon yield and the ease of obtaining stabilized products by forming a thermally stable structure. Additionally, the surface of PAN-based CNFs can be modified and functionalized using a coating or activation process [18]. Moreover, PAN can be blended with other polymers (miscible or immiscible) to carry out coelectrospinning or be embedded with nanoscale components (e.g., nanoparticles, nanowires, nanotubes or catalysts) to obtain multiphase precursors and subsequently to make composite CNFs through high-temperature treatment.

PAN nanofibers were produced using electrospinning by dissolving polyacrylonitrile in N,N-Dimethylformamide (DMF) solution. The spinning can be carried out for 8, 10, 13, 15 and 20% (by weight) concentrations at different voltages, flow rates, varying distance between needle and collector and at different needle diameters [91]. Kurban et al. have synthesized CNF by heat treatment of electrospun Polyacrylonitrile in dimethylsulphoxide, offering a new solution route of low toxicity to manufacture sub-60 nm diameter CNFs [28, 92].

After that PAN nanofiber would be stabilized and carbonized which will be discussed later. During the stabilization, PAN was stretched to improve the molecular orientation and the degree of crystallinity to enhance the mechanical properties of the fibers [85].

Chun et al. [93] produced carbon nanofibers with diameter in the range from 100 nm to a few microns from electrospun polyacrylonitrile and mesophase pitch precursor fibers. Wang et al. [94, 95] produced carbon nanofibers from carbonizing of electrospun PAN nanofibers and studied their structure and conductivity. Hou et al. [96] reported a method to use the carbonized electrospun PAN nanofibers as substrates for the formation of multiwall carbon Nanotubes. Kim et al. [14, 97] produced carbon nanofibers from PAN-based or pitch-based electrospun fibers and studied the electrochemical properties of carbon nanofibers web as an electrode for supercapacitor.

Sizhu Wu et al. [85] electrospun PAN nanofibers from PAN solution in DMF (Table 16.3) and hot-stretched them by weighing metal in a temperature controlled oven to improve its crystallinity and molecular orientation, and then were converted to CNFs by stabilization and carbonization. PAN was stretched to improve the molecular orientation and the degree of crystallinity to enhance the mechanical properties of the fibers. The hot-stretched nanofiber sheet can be used as a promising precursor to produce high-performance carbon nanofiber composites.

Several research efforts have been attempted to prepare the electrospun PAN precursor nanofibers in the form of an aligned nanofiber bundle [50, 98]. The bundle was then tightly wrapped onto a glass rod, so that tension existed in a certain degree during the oxidative stabilization in air.

TABLE 16.3 Polyacrylonitrile/DMF Solution Properties

Viscosity	Temperature	Conductivity	Surface tension	Solution composition		Ref.
(Cp)	(°c)	(µs)	(m N/m)	Polymer	DMF	
8%	333.3	20.8	39.0	72.9	8.0	[99]
10%	1723.0	22.0	43.2	77.0	10.0	[100]
13%	2800.0	21.4	50.0	97.6	13.0	[101]
15%	3240.4	23.3	56.7	105.0	15.0	[102]
20%	3500	21.1	58.0	122.5	20.0	[102]

The mechanical resonance method and Weibull statistical distribution can be used to analyze the mechanical properties (i.e., bending modulus and fracture strength) of CNFs. Using these methods, Zussman et al. [103] found that the stiffness and strength of CNFs were inferior to those of commercial PAN-based carbon fibers. The average bending modulus of individual CNFs was 63 GPa, lower than the lowest tensile modulus (230 GPa) for commercial High-strength PAN-based carbon fibers. Nevertheless, Zussman et al. [103] pointed out that the mechanical properties had the potential to be significantly improved if the microstructure of precursor nanofibers and thermal treatment process were optimized.

16.8 ELECTROSPINNING METHODOLOGY

Electrospinning is a process involving polymer science, applied physics, fluid mechanics, electrical, mechanical, chemical, material engineering and rheology [104]. It has been recognized as an efficient technique for the fabrication of fibers in nanometer to micron diameter range from polymer solutions or melts [105]. In a typical process, an electrical potential is applied between a droplet of polymer solution or melt held at the end of a capillary and a grounded collector. When the applied electric field overcomes the surface tension of the droplet, a charged jet of polymer solution or melt is ejected. The jet grows longer and thinner due to bending instability or splitting [83] until it solidifies or collects on the collector.

The fiber morphology is controlled by the experimental parameters and is dependent upon solution conductivity, concentration, viscosity, polymer molecular weight, applied voltage, etc. [34, 106] much work has been done on the effect of parameters on the electrospinning process and morphology of fibers.

Sinan Yördem et al. [107] reported that the fiber diameter increased with increasing polymer concentration according to a power law relationship. Filtering application [108] is also affected by the fiber size. Therefore, it is important to have control over the fiber diameter, which is a function of material and process parameters.

Deitzel et al. [109] reported a bimodal distribution of fiber diameter for fibers spun from higher concentration solution. Boland et al. [110] obtained a strong linear relationship between fiber diameter and concentration in electrospun poly(glycolic acid) (PGA). Ryu et al. [111] and Katti et al. [112] also reported a significant relationship between fiber diameter and concentration in electrospinning process. For the effect of applied voltage, Reneker et al. [113] obtained a result that fiber diameter did not change much with electric field when they studied the electrospinning behavior of polyethylene oxide. Mo et al. and Katti et al. [112] reported that fiber diameter tended to decrease with increasing electrospinning voltage, although the influence was not as great as that of polymer concentration. But Demir et al. [114] reported that fiber diameter increased with increasing electrospinning voltage when they electrospun polyurethane fibers.

Sukigara [115] studied the effect of electrospinning parameters (electric field, tip-to-collector distance and concentration) on the morphology and fiber diameter of regenerated silk from Bombyx mori using response surface methodology and concluded that the silk concentration was the most important parameter in producing uniform cylindrical fibers less than 100 nm in diameter.

In order to optimize and predict the morphology and average fiber diameter of electrospun PAN precursor, design of experiment was employed. Morphology of fibers and distribution of fiber diameter of PAN precursor were investigated varying concentration and applied voltage [116].

Another advantage of the electrospinning is that it can be used to produce a web structure [117]. When used as an electrode, it does not need a second processing step adding a binder and an electric conductor such as carbon black. Therefore, the webs from electrospinning have important advantages such as an ease of handling, an increase in the energy density due to large specific surface area, an improvement of the conductivity due to increased density of the contact points, and low cost of preparations of the electrodes [14].

During the electrospinning process, the droplet of solution at the capillary tip gradually elongates from a hemispherical shape to a conical shape or Taylor cone as the electric field is increased. A further increase in the electric field results in the ejection of a jet from the apex of the cone. The jet grows longer and thinner until it solidifies and collects on the collector. The fiber morphology is controlled by process parameters. In this study the effect of solution concentration and applied voltage on fiber morphology and average fiber diameter were investigated. Polymer concentration was found to be the most significant factor controlling the fiber diameter in the electrospinning process [78] (Table 16.4).

TABLE 16.4 The Electrospinning Conditions That Gave the Best Alignment of the Nanofibers [116]

Concentration (wt %)	Tip to target Distance (cm)	Voltage (kv)	Width of gap (cm)	Fiber diameter (nm)
10	15	9	1	165±25
11	15	9	1.2	205±25
12	20	9.5	1.5	245±20
13	20	10	2.2	255±30
14	20	11	2.5	300±30
15	20	11	3	400±40

16.9 ELECTROSPINNING OF PAN SOLUTION

The proper solvent has to be selected for dissolving a polymer source completely before carrying out the electrospinning. N, N-dimethyl formamide (DMF) is considered to be the best solvent among various organic solvents due to its proper boiling point (426 K) and enough electrical conductivity (electrical conductivity = 10.90 mS/cm, dipole moment = 3.82 Debye) for electrospinning [1, 41, 95, 98]. This mixture should be vigorously stirred by an electromagnetically driven magnet at around 60°C until it becomes a homogeneous polymer solution. Different concentrations of PAN solution (8–20 wt.%) can be used [91].

The following parameters affect the PAN solution electrospinning:

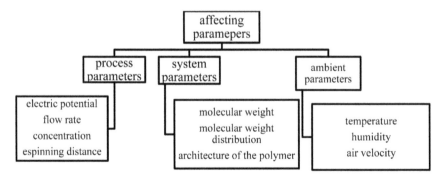

FIGURE 16.4 Classifications of parameters affect PAN solution electrospinning.

The electrospinning equipment for PAN solution can act with two different collectors, stationary and rotating drum. One important physical aspect of the electrospun PAN nanofibers collected on the grounded collectors is their dryness from the solvent used to dissolve the polymer (i.e., DMF). At the distances of tip to target of

5 and 7.5 cm, the structures of nanofibers were not completely stabilized and consequently the cross-sections of spun nanofibers became more flat and some nanofibers shuck together and bundles of nanofibers were collected. At the distances
of tip to target of 15 cm and longer, the nanofibers exhibited a straight, cylindrical
morphology indicating that the nanofibers are mostly dry when they have reached
the target [118]. Another important factor is electrospinning distance, as we know
a certain minimum value of the solution volume suspended at the end of the needle
should be maintained in order to form an equilibrium Taylor cone. Therefore, different morphologies of electrospun nanofibers can be obtained with the change in
feeding rates at a given voltage. Jalili et al. [118] proved that at lower feeding rate
of 2 mL/h, a droplet of solution remains suspended at the end of the syringe needle
and the electrospinning jet originates from a cone at the bottom of the droplet. The
nanofibers produced under this condition have a uniform morphology and no bead
defect was present. At lower feeding rate of 1 mL/h, the solution was removed from
the needle tip by the electric forces, faster than the feeding rate of the solution onto
the needle tip. This shift in the mass balance resulted in sustained but unstable jet
and nanofibers with beads were formed.

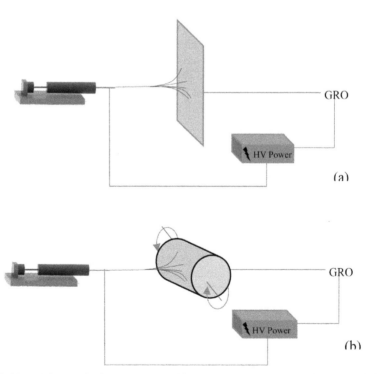

FIGURE 16.5   Electrospinning apparatus with (a) a stationary, grounded target and (b) a
rotating, grounded target.

Studies confirmed that PAN nanofibers were formed by varying the solution concentration, voltage, solution flow rate and distance between needle and collector. The morphology of the PAN fibers for 10, 15, and 20% concentrations were studied using SEM and is shown in Figs. 16.6a, 16b, and 16c, respectively [91]. As the concentration of the solution was increases, the diameter of the fiber was also increases.

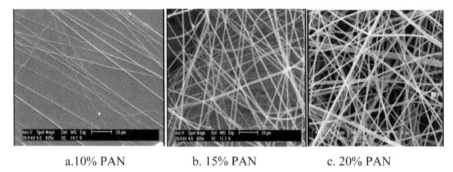

    a.10% PAN   b. 15% PAN   c. 20% PAN

FIGURE 16.6   SEM analysis of PAN fibers for 10, 15, and 20% concentration.

Ma et al. [119] demonstrated a facile method for the preparation of porous ultrafine nanofibers of PAN as a precursor for carbon nanofiber. They prepared the PAN/NaHCO3 composite nanofibers by electrospinning, and then NaHCO3 was removed by a selective dissolution and reaction with the solution of hydrochloric acid (10 wt%). The obtained PAN fibers showed highly porous surfaces after the extraction of $NaHCO_3$.

For many applications, it is necessary to control the spatial orientation of 1D nanostructure. In the fabrication of electronic and photonic devices, for example, well aligned and highly ordered architectures are often required [120, 121]. Even for application as fiber-based reinforcement, it is also critical to control the alignment of fibers [50]. Because of the bending instability associated with a spinning jet, electrospun fibers are often deposited on the surface of collector as randomly oriented, nonwoven mats. The whipping instability is mainly caused by the electrostatic interactions between the external electric field and the surface charges on the jet. The formation of fibers with fine diameters is mainly achieved by the stretching and acceleration of the fluid filament in the instability region [98]. In the past several years, a number of approaches have been demonstrated to directly collect electrospun nanofibers as uniaxially aligned arrays [50, 98, 120–122]. The most popular method of obtaining aligned fibers is by using rotating drum target.

The aligned fiber mechanism behind the rotating drum technique is as follows:

When a linear speed of the rotating drum surface, which serves as a fiber take-up device, matches that of evaporated jet depositions, the fibers are taken up on the surface of the drum tightly in a circumferential manner, resulting in a fair alignment.

Such a speed can be called as an *alignment speed*. If the surface speed of the cylinder is slower than the alignment speed, randomly deposited fibers will be collected, as it is the fast chaos motions of jets determine the final deposition manner. On the other hand, there must be a limit rotating speed above which continuous fibers cannot be collected since the over fast take-up speed will break the fiber jet. The reason why a perfect alignment is difficult to achieve can be attributed to the fact that the chaos motions of polymer jets are not likely to be consistent and are less controllable [91].

16.10 STABILIZATION

The next step in preparing carbon nanofiber is stabilization. In stabilization, in order to prevent precursor fibers fusing together during carbonization, the thermoplastic precursor nanofibers are converted to highly condense thermosetting fibers by complex chemical and physical reactions, such as dehydrogenation, cyclization and polymerization [54]. If the temperature is raised too rapidly during polymerization, a very large amount of heat will be released leading to a loss of the orientation and melting of the polymer. Therefore, the heating rate during stabilization is usually controlled at a relatively low value (e.g., $1°C$ min^{-1}). It is noted that external tension is necessary during stabilization to avoid shrinkage of the fibers and to maintain the preferential orientation of the molecules along the fiber axis [41].

Stabilization process which is carried out in air (oxidative stabilization) constitutes the first and very important operation of the conversion of the PAN fiber precursor to carbon as well as activated carbon fiber. During stabilization, the precursor fiber is heated to a temperature in the range of 180–$300°C$ for over an hour. Because of the chemical reactions involved, cyclization, dehydrogenation, aromatization, oxidation and crosslinking occur and as a result of the conversion of C≡N bonds to C=N bonds a fully aromatic cyclized ladder type structure forms [36, 49].

This new structure is thermally stable (infusible). Also, it has been reported that during stabilization, CH_2 and CN groups disappear while C =C, C= N and = C–H groups form. At the same time the color of precursor fiber changes gradually and finally turns black when carbonized [36]. Research shows that optimum stabilization conditions lead to high modulus carbon fibers. Too low temperatures lead to slow reactions and incomplete stabilization, whereas too high temperatures can fuse or even burn the fibers [49, 93, 117, 123].

16.11 CARBONIZATION

Carbonization is the last step for producing carbon nanofiber [49, 124]. It involves crosslinking, reorganization and the coalescence of cyclized sections accompanying the structural transformation from a ladder structure to a graphite-like one and a morphological change from smooth to wrinkle. This process needs to be carried out in a dynamic inert gas atmosphere (e.g., nitrogen and argon), which can prevent oxidation, remove the pyrolysis products (i.e., volatile molecules such as H_2O, H_2, HCN, N_2, NH_3 and CO_2) and transfer energy. Vacuum carbonization can also be employed, but the degree of carbonization is lower than that in nitrogen or argon. Furthermore, fiber shrinkage occurs during carbonization. Fiber shrinkage may further lead to the formation of a large number of pits on the surface of CNFs having a deleterious effect on their appearance [41].

During this process, the noncarbon elements remove in the form of different gases and the fibers shrink in diameter and lose approximately 50% of its weight [54].

Wang et al. [125] proved that the conductivity of PAN-based carbon nanofibers produced by electrospinning increases sharply with the pyrolysis temperature, and also increases considerably with pyrolysis time at lower pyrolysis temperatures.

The stabilized nanofiber would be subsequently carbonized at a temperature around 1000°C in an inert (high purity nitrogen gas) environment with the heating rate 1–2°C/min [34, 50, 116]. Zh. Zhou, et al. reported that the carbonized PAN nanofiber bundles can be further carbonized in vacuum at relatively high temperatures between 1400°C to 2200°C. A Lindberg high temperature reactor with inside diameter and depth of 12 cm and 25 cm, respectively, can be used for conducting the high-temperature carbonization; and the heating rate must be set at 5°C/min. All of the carbonized PAN nanofiber bundles must be held at the respective final temperatures for 1 h to allow the carbonization to complete [50].

The optimum processing conditions (i.e., temperature and time) for stabilization and carbonization can be investigated using TGA. Micro structural parameters for CNFs and ACNFs, such as integrated intensity ratio of D and G peaks (RI), interlayer spacing ($d002$), crystallite size, molecular orientation and pore characteristics, which are of great significance for final applications, can be analyzed with Fourier transform infrared (FTIR) spectroscopy, Raman spectroscopy, XRD, electron energy loss spectroscopy (EELS), transmission electron microscopy (TEM) and isothermal nitrogen adsorption/desorption [35, 49, 123]. The disappearance of peaks of functional groups in FTIR spectra and a very broad diffraction peak at a 2θ value of about 24° in XRD can demonstrate the conversion from the as-spun nanofibers to CNFs. The disappearance of peaks of functional groups in FTIR spectra and a very broad diffraction peak at a 2θ value of about 24° in XRD can demonstrate the conversion from the as-spun nanofibers to CNFs. The degree of carbonization can be evaluated with Raman spectroscopy, EELS and XRD. In Raman spectra, the lower the RI value, the higher the degree of the transformation from disordered carbon to graphitic carbon and the fewer the number of defects in the CNFs. The in plane graphitic crystallite size (La) can be further calculated from RI according to the equation $La = 4.4/RI$ (nm) [90, 126, 127]. In EELS, the degree of carbonization can be monitored by analyzing the contents of sp2 bonds (structural order) and sp3 bonds (structural disorder). In XRD spectra, the average crystallite dimensions (e.g. Lc (002) and La (110)) and $d002$ can be calculated using the Scherrer and Bragg equations. It is noticed that micro structural parameters measured from XRD and Raman spectra are usually inconsistent due to the different measurement mechanisms; that is, XRD provides average bulk structural information, whereas Raman spectroscopy gives structural information merely within the surface layer (*ca* 10 nm). The commonly observed core–shell structure of CNFs can be clearly inspected using high-resolution TEM. The specific surface area and average pore size of CNFs can be measured based on the Brunner–Emmett–Teller (BET) method. Mesopore and micropore size distribution can be obtained using the Barrett–Joiner–Halenda (BJH) and Horvath–Kawazoe (HK) methods, respectively [41].

The morphology of as spun, stabilized and carbonized electrospun 15% PAN fibers are studied using scanning electron microscopy (Figs. 8a–8c) [91]. All these figures show the formation of random fibers. The diameter of the fiber decreases with increase in stabilization and carbonization temperatures.

A problem during the heat treatment of the web of these nanofibers is that crack and wrinkle occur on a sheet only when holding on a metal plate because the nanofiber sheet cannot be shrunk. Using a fixing frame, which holds the sheet and prevents it from shrinkage can keep the structure safe [128].

a   b   c   5x

As spun (1-1.3μ)   Stabilized @ 280°C (0.9-1.1μ)   Carbonized @ 900°C

FIGURE 16.7   SEM analysis of as spun, stabilized and carbonized PAN fibers (magnification of 5000x).

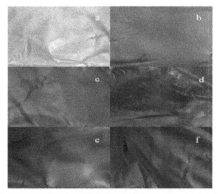

FIGURE 16.8   Stabilized samples in different temperatures: (a) 180, (b) 200, (c) 220, (d) 250, (e) 270°C.

IGURE 16.9 Process of PAN stabilization and subsequent carbonization.

16.12 COMPARISON OF STABILIZATION AND CARBONIZATION

The average diameter of the stabilized PAN nanofibers appeared to be almost the same as that of the as-electrospun nanofibers, while the average diameters of the carbonized PAN nanofibers were significantly reduced [14]. With increase of the final carbonization temperature, the carbon nanofibers became more graphitic and structurally ordered. the microstructure of the low-temperature (1000°C) carbonized nanofibers was primarily turbostratic and the sheets of carbon atoms were f folded and/or crumpled together, while the microstructure of the high-temperature (2200°C) carbonized nanofibers was graphitic and the grapheme sheets stacked together to form ribbon-shaped structures [62].

Both electrical conductivities and mechanical properties of the carbon nanofiber increased with the increase of the final carbonization temperature. It is noteworthy that the electrical conductivities and mechanical properties of the carbon nanofiber cannot be directly interpreted as those of individual nanofibers in the bundles [35, 123].

In order to develop carbon nanofibers with superior mechanical properties particularly tensile strength, the electrospun PAN precursor nanofibers have to be extensively stretched; the stabilization (and probably carbonization as well) has to be conducted under optimal tension; and the PAN copolymer instead of homopolymer has to be used as the precursor because the electrospun carbon nanofibers with superior mechanical and electrical properties are expected to be an innovative type of nanomaterials with many potential applications [50]. Commercial PAN Precursor fibers are usually drawn prior to stabilization in order to reduce the probability of encountering a critical flaw during thermal treatment. However, few studies have applied stretching to electrospun precursor nanofibers before stabilization [129]. Additionally, although aligned nanofibers with various degrees of alignment have been obtained by using various modified collecting devices, tension is rarely applied to the nanofiber assembly during stabilization to prevent shrinkage of the fibers and to ensure molecular orientations along the fiber axis to a large extent. Therefore, there is still considerable room for further improvement of the micro structural, electrical and mechanical properties of the final CNFs. Considering the increased interest in electrospinning and the wide range of potential applications for electrospun CNFs, commercialized products can be expected in the future [41, 129].

16.13   FT-IR STUDY OF PAN AND STABILIZED PAN FIBER

IR spectra are considered as tool for determination the chemical interaction during heat treatment on PAN fibers. By using of these spectra, it is possible to study the relation between chemical changes and strength, aromatic index and fiber contraction during fabrication process. But analyzing these relations is so difficult because the intensity of bonds used for analysis depends on samples type, form and the way of preparation. The study of FT-IR spectra of PAN fibers sample with different comonomers shows that during stabilization of PAN fibers, the peaks related to $C\equiv N$ bonds and CH_2 are reduced sharply. These reductions are related to cyclization of nitrile groups and stabilization procedure [54].

FT-IR spectra of PAN fibers have many peaks which related to existence of CH_2, $C\equiv N$, $C=O$, $C-O$ and $C-H$ bonds. The absorption peaks are in range of 2926–2935 cm^{-1} are related to $C-H$ bonds in CH, CH_2 and CH_3 but in this range the second weak peak is observed which is related to C-H bonds also. Another peak is observed in the range of 2243–2246 cm^{-1}, which is related to presence nitrile ($C\equiv N$) bonds and indicates the nitrile group exists in polyacrylonitrile chain. The absorption peaks in the ranges of 1730–1737 cm-1 and 1170 cm^{-1} are related to $C=O$ or $C-O$ bonds and are resulted from presence of comonomers like MA. Absorption in the range of 1593–1628 cm^{-1} is related to resonance $C-O$ bonds. The peaks in the range of 1455–1460 cm^{-1} is related to tensile vibration and peaks in the range of 1362–1382 cm^{-1} and range of 1219 cm^{-1} are related to vibration in different situation [54, 130].

After stabilization, links in the range of 2926–2935 cm^{-1} are reduced which are related to CH$_2$ bond. These links are weakened and with some displacement are observed in the range of 2921–2923 cm^{-1}. Additionally, these bonds in the range of 1455–1460 cm^{-1} are mainly omitted or would be reduced which is related to CH$_2$ bonds too. Also, the main reduction is observed in links in the range of 2243–2246 cm^{-1}, which is resulted from the change C≡N bonds and their conversion to C=N. C≡N peak is weakened in stabilized samples or completely would be removed [54, 131]. New peaks in the range of 793–796 cm^{-1} are created as the result of =C–H bond formation. Increasing the intensity of =C–H groups and reduction of intensity of CH$_2$ groups shows that =C–H is created during aromatization of structure in the presence of oxygen.

During the stabilization step, the absorption peaks in the ranges of 1730–1737 cm^{-1} and 1170 cm^{-1} (related to C=O or C–O bonds) and two groups of peaks in the ranges of 1362–1382 cm^{-1} and 1219 cm^{-1} (related to C–H in different situation) mainly are removed in types of fibers. In the range of 670 cm^{-1}, some changes are observed which are not related to chemical changes of PAN fibers during oxidized stabilization. It is believe that this issue is may be related to different modes of CN and C–CN bonds appearance in FT-IR [130].

16.14 CNF FROM OTHER TYPE OF PRECURSORS

Many investigations shave dealt with PAN-based CNFs, as presented in Section 16.13. However, other precursors (e.g., pitch, PVA, PI and PBI) have also been used successfully to prepare electrospun CNFs. Pitch is generally obtained from petroleum asphalt, coal tar and poly(vinyl chloride) with a lower cost and a higher carbon yield compared with PAN [41]. Pitch is generally obtained from petroleum asphalt, coal tar and poly (vinyl chloride) with a lower cost and a higher carbon yield compared with PAN. However, impurities in the pitch are required to be fully removed to obtain high-performance carbon fibers leading to a great increase in the cost. Therefore, there have been limited studies on electrospun pitch-based carbon microfibers [93].

The diameter of the electrospun pitch fibers was in the micrometer range and difficult to become thinner due to the low boiling point (65–67°C) of the solvent tetrahydrofuran (THF). Electrical conductivity was increased on increasing the carbonization temperature from 0 (700°C) to 83 S cm^{-1} (1200°C).

PVA, another thermoplastic precursor, allows the preparation of CNFs through thermal treatment processes [132]. Unlike thermoplastic precursors, electrospun thermosetting nanofibers can directly undergo carbonization for preparing CNFs without the need of the costly stabilization process. Therefore, they have gained increasing attention in recent years. For example, electrospun PI-based CNFs were prepared by Kim et al [90]. The electrical conductivity of the CNFs carbonized at 1000°C was measured to be 2.5 S cm^{-1}, higher than that (1.96 S cm^{-1}) of PAN-based

CNFs treated at the same carbonization temperature. PBI and PXTC are other precursors for CNF preparation [41].

Another thermosetting polymer precursor (PXTC) in the form of aligned nanofiber yarn (parallel with the electric field lines), which was several centimeters long, was obtained by electrospinning merely using a conventional flat aluminum plate as the collector. The formation of the yarn might result from the ionic conduction of PXTC. The presence of D and G peaks in the Raman spectra demonstrated a successful conversion from electrospun PXTC yarn to CNFs in the temperature range from 600 to 1000°C, while the yarn carbonized at 500°C did not show these two peaks in the Raman spectra [41]. The mole fraction of graphite for the carbonized nanofibers was determined to be 0.21–0.24, less than that (0.25–0.37) obtained by Wang et al. [126] for PAN-based CNFs treated from 873 to 1473 K, showing a lower degree of carbonization. Depending on the properties of the precursor nanofibers and subsequent high-temperature treatment processes, diverse microstructural and electrical properties have been achieved [41].

16.15   ACTIVATION

The most principal disadvantage of CNFs is its relatively low surface area and porosity (around 10–200 m^2/g), which limit the applications of these materials like hydrogen storage or catalyst support; therefore is necessary increase the surface area to improve the yield in these materials. It is conceivable that CNF webs consisting of CNFs with porous surfaces, which are obtained through the activation of electrospun PAN-based CNFs, can lead to a significant expansion of applications of CNFs, such as in electrode materials, high-temperature filtration and removal of toxic gases.

The surface area can be modifying by means of activation process in which a part of structural carbon atoms are eliminated (mainly, the most reactive) by an activate agent. As consequence, the porosity and surface area increase and so, their applications as hydrogen storage or catalyst support improve [55].

Barranco et al. [86] expressed that Activation does not produce any important change in the shape, surface roughness, diameter, graphene sheet size, and electrical conductivity of starting nanofibers; it leads to new micropores and larger surface areas as well as a higher content of basic oxygen groups.

During the activation treatment, high porosity is formed within the material through the interaction of activating agents (usually oxidizing medium like steam, carbon dioxide, etc.) with carbon structures. By this interaction, the surface chemical properties are altered or transformed to some extent [133].

It is known there are still some amounts of nitrogen left in the structures of carbon fibers, which exist in various types of nitrogen functionalities after the carbonization treatment. These functionalities were bounded or attached to the edge parts of the carbon structures. During activation treatment, these edge parts would be preferentially attacked by the oxidizing agents, which probably led to the removal of these edge structures. Due to this effect, the surface nitrogen level decreased upon activation treatment. Wang et al. suggested that the activation treatment in steam helps the elimination of nitrogen from the carbon structure [133].

Comparatively to nanotubes, nanofibers present a nanostructure made of grapheme layer stacking which is favorable to activation. Two activation systems can be used for activated carbon nanofibers: physical activation by CO_2 or heat, and chemical activation by KOH or RbOH [134, 135]. A range of potential adsorbents was thus prepared by varying the temperature and time of activation. The structure of the CNF proved more suitable to activation by KOH than by CO_2, with the former yielding higher surface area carbons (up to 1000 m^2 g^{-1}).

Depending on the final application of the activated materials, it is possible to control their pore structure by choosing the suitable activation conditions. The increased surface area, however, did not correspond directly with a proportional increase in hydrogen adsorption capacity. Although high surface areas are important for hydrogen storage by adsorption on solids, it would appear that it is essential that not only the physical, but also the chemical, properties of the adsorbents have to be considered in the quest for carbon based materials, with high hydrogen storage capacities [106, 136].

16.15.1 PHYSICAL ACTIVATION

Physical activation involves carbonization of a carbonaceous precursor followed by gasification of the resulting char or direct activation of the starting material in the presence of an activating agent such as CO_2, steam or a combination of both, that among them steam activation is more common. This gasification or activation process eliminates selectively the most reactive carbon atoms of the structure generating the porosity [56, 129].

Steam is widely used as an activating agent for fabricating ACNFs because of its low cost and environmentally friendly character. Kim and Yang [14] reported that the specific surface area of steam-activated carbon nanofibers (steam-ACNFs) decreased with increasing activation temperature (from 700 to 850°C) due to the unification of micropores at elevated temperatures, whereas the electrical conductivity of ACNF electrodes and the accessibility of ions were increased according to cyclic voltammetry curves and impedance Nyquist plots.

For physical activation, about 2 g of CNF are placed in the center of a quartz tube in a tube furnace. Then the CNF are heated to the required reaction temperature (800–1000°C) under neutral environment, for a reaction time of 15–45 min. The sample is then allowed to cool under argon [106]. The ACNFs activated at 800°C afforded the highest specific surface area but low mesopore volume [137]. After activation, the average diameters of the fibers decreased about 100 nm compared with the CNFs without activation process. The reduction in diameter may be a consequence of the subsequent carbonization and burn-off through activation at elevated temperatures. However, no severe shrinkage is found in the ACNFs. Therefore, activation does not create defects on the surfaces; although the roughness of the ACNF surfaces is somewhat increased and nano-sized fibers are created [137].

A representative case of physical activation is activation with silica, reported by Im et al. [138]. The activation agent was embedded into the fibers and then removed by physically removing the agents. The process of activation by silica is presented in Fig. 16.11. The silica-activated carbons nanofibers are shown in Fig. 16.12. The pores generated by physical activation are clearly observed.

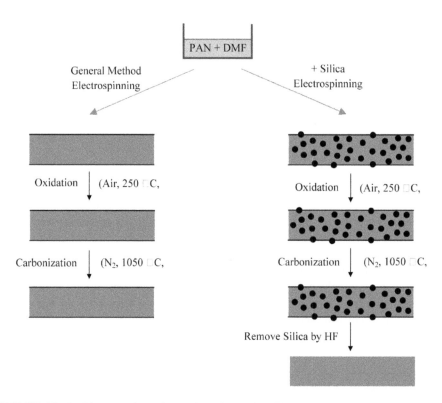

FIGURE 16.10   The procedure of manufacturing carbon fibers and silica-activated carbon fibers.

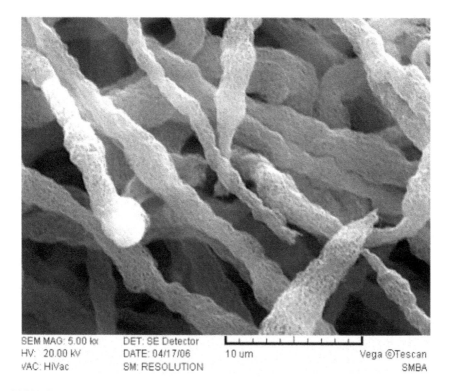

SEM MAG: 5.00 kx   DET: SE Detector
HV: 20.00 kV   DATE: 04/17/06   10 um   Vega ©Tescan
VAC: HiVac   SM: RESOLUTION   SMBA

FIGURE 16.11   Silica-activated carbon nanofibers [138].

16.15.2   CHEMICAL ACTIVATION

In addition to the steam activation, activating agents (e.g., KOH, $ZnCl_2$, NaOH, Na_2CO_3, K_2CO_3, SiO_2) can also be used for activation of CNFs [41]. In chemical activation the precursor is impregnated with a given chemical agent and, after that, is pyrolyzed. As a result of the pyrolysis process, a much richer carbon content material with a much more ordered structure is produced, and once the chemical agent is eliminated after the heat treatment, the porosity is so much developed. Several activating agents have been reported for the chemical activation process: phosphoric acid, zinc chloride and alkaline metal compounds [42, 55, 56, 60, 106, 134–136, 139] (Table 16.5).

Phosphoric acid and zinc chloride are activating agents usually used for the activation of lignocellulosic materials which of coal precursors or chars have not been previously carbonized. Contrarily, alkaline metal compounds, usually KOH, are used for the activation.

TABLE 16.5   Surface Area and Yield to Different Activating Agents [55]

Activating agents	Initial CNFS	KOH	NaOH	K$_2$CO^3	KHCO$_3$	Mg(OH)$_2$
Surface area (m²/g)	127.0	407.5	177.0	261.0	233.0	173.8
Yield (%)	-	55.1	76.2	62.7	67.2	77.5

An important advantage of chemical activation is that the process normally takes place at a lower temperature and shorter time than those used in physical activation. In addition, it allows us to obtain very high surface area activated carbons. Moreover, the yields of carbon in chemical activation are usually higher than in physical activation because the chemical agents used are substances with dehydrogenation properties that inhibit formation of tar and reduce the production of other volatile products [42, 134].

However, the general mechanism for the chemical activation is not so well understood as for the physical activation. Other disadvantages of chemical activation process are the need of an important washing step because of the incorporation of impurities coming from the activating agent, which may affect the chemical properties of the activated carbon and the corrosiveness of the chemical activation process [56].

Im et al. [36]studied the hydrogen adsorption capacities of ACNFs activated using KOH (chemical activation; applied after carbonization) and ZnCl$_2$ (physical activation; applied along with carbonization) with the volumetric method. Although KOH-activated CNFs had higher specific surface area and total volume than ZnCl$_2$-activated CNFs, the ultra micropore (0.6–0.7 nm) volume of ZC-W4 (PAN: DMF: ZnCl2 = 3:50: 4 weight ratio) or ZC-W6 (PAN: DMF: ZnCl2 = 3:60: 6 weight ratio) was much larger than that of KOH-activated CNFs. ZC-W6 with the largest ultra micropore volume (0.084 cm^3 g^{-1}) exhibited the highest hydrogen adsorption capacity. Therefore, it was concluded that hydrogen adsorption capacity is mainly determined by ultra micropore (0.6–0.7 nm) volume. Beaded nanofibers are generally unfavorable during the electrospinning process.

It is known that the traditional activation processes are relatively complex and costly. To resolve this issue, it is possible to fabricate porous CNFs by precreating pores in the as-spun precursor nanofibers, such as by choosing a particular solvent system, by changing the environmental humidity or by using polymer mixtures, and then using a thermal treatment.

Cheng et al. produced chemically activated carbon nanofibers based on a novel solvent-free coextrusion and melt-spinning of polypropylene-based core/ sheath polymer blends and their morphological and microstructure characteristics analyzed by scanning electron microscopy, atomic force microscopy (AFM), Raman spectroscopy, and X-ray diffractometry [139].

Kim et al. used the third strategy via the removal of a poly(methyl methacrylate) (PMMA) component during the carbonization of electrospun immiscible polymers (PAN and PMMA). The higher the PAN content, the finer the electrospun composite nanofibers. The carbon surface was burned off by generating carbon monoxide and carbon dioxide from outside to inside by the following reactions (2–6) (here, M = Na or K) [140]:

$$6 \text{ MOH} + C \leftrightarrow 2 \text{ M} + 3H2 + 2 \text{ M2CO3–2}$$
$$\text{M2CO3} + C \leftrightarrow \text{M2O} + 2\text{CO } 3$$
$$\text{M2CO3} \leftrightarrow \text{M2O} + CO_2 - 4$$
$$2 \text{ M} + CO_2 \leftrightarrow \text{M2O} + CO \text{ } 5$$
$$\text{M2O} + C \leftrightarrow 2 \text{ M} + CO \text{ } 6$$

TABLE 16.6 Comparisons of Preparation Processes for Electrospun CNFs and ACNFs Discussed in the Open Literature

Precursor	Stabilization	Carbonization	Activation	References
PAN/DMF-ZnCl$_2$	250°C (1 h)	1050°C (2 h)	-	[36]
PAN/DMF-KOH	250°C (1 h)	-	750°C (3 h)	[36]
PAN/DMF-si	280°C (5 h)	700°C (1 h)	-	[141]
PAN/DMF	350°C (0.5 h)	750°C (1 h), 1100°C (1 h)	-	[117]
PAN/DMF	280°C (1 h)	1000°C (1 h)	-	[142]
PAN/DMF	280°C (1 h)	700, 800, 900 and 1000°C (1 h)	-	[90]
PAN/DMF-CNT	285°C (2, 4, 8, 16 h)	700°C (5°C min^{-1}), 1 h	900°C (in CO$_2$ for 1 h)	[51]
PAN/DMF-Mg	280°C (1 h)	1000°C (at a rate of 5°C/min^{-1}, 1 h)	800°C (0.5 h)	[143]
PAN/DMF	280°C (1h)	900°C(1h)	-	[91]
PAN/DMF	240°C (2 h)	1200°C (10 min)	-	[54]
PAN-PVP/DMF	270°C (1 h)	1000°C 5°C/min^{-1}	-	[1]
PAN/DMF	Not given	1200°C (0.5 h)	-	[126]

TABLE 16.6   *(Continued)*

Precursor	Stabilization	Carbonization	Activation	References
PAN/DMF	280°C (3 h)	1000°C (2°C/min 1 h), 1400°C, 1800°C, 2200°C (1 h)	-	[50]
PAN/DMSO	125–283°C (50 min)	1003–1350°C	-	[49]
PAN/DMF	230°C (2 h)	600°C (0.5 h)	800, 850, and 900°C (in CO$_2$ for 1 h)	[136]

16.16   APPLICATIONS

Many promising applications of electrospun activated carbon nanofibers can be expected if appropriate micro structural, mechanical and electrical properties become available.

16.16.1   *HYDROGEN STORAGE*

Due to the exhaustion of gasoline or diesel fuel, new energy sources are necessary to be developed as an assistant or alternative energy. Among them, hydrogen gas is an attractive possibility to provide new solutions for ecological and power problem [144]. Hydrogen is least polluting fuel. Since it is difficult to store hydrogen, its use as a fuel has been limited [145–147]. The possibility of developing hydrogen into an environmentally friendly, convenient fuel for transportation has lead to the search for suitable materials for its storage. The suitable media for hydrogen storage have to be light, industrial, and in compliance with national and international safety laws. In the last few years, researchers have paid much attention on hydrogen adsorption storage in nanostructured carbon materials [5], such as ACs [148], CNT [146], and CNF. Carbon nanomaterials due to their high porosity and large surface area have been suggested as a promising material for hydrogen storage [6] (Table 16.7).

TABLE 16.7   Summary of Reported Hydrogen Storage Capacity of Carbon Nanofiber

Sample	Purity	T(K)	P(Mpa)	H$_2$ (Wt %)	Ref.
GNF	-	77–300	0.8–1.8	0.08	Ahn et al. [145]
GNF herringbone		298	11.35	67.58	Browning et al. [149]
GNF platelet		298	11.35	53.68	Browning et al. [149]

TABLE 16.7   *(Continued)*

Sample	Purity	T(K)	P(Mpa)	H$_2$ (Wt %)	Ref.
Vapor grown carbon fiber		298	3.6	<0.1	Tibbetts et al. [146]
CNF		77	12	12.38	Rzepka et al. [150]
ACNF With CO$_2$		300	11	0.33	Blackman et al. [106]
ACNF With KOH		300	10	0.42	Blackman et al. [106]

Rodriguez and Baker investigated that the hydrogen storage capacities of CNF at room temperature and pressures up to 140 bars were quantified independently by gravimetric and volumetric methods, respectively [145, 149, 150]. Ji Sun Im and Soo-Jin Park [36] studied the relation between pore structure and the capacity of hydrogen adsorption, textural properties of activated CNFs with micropore size distribution, specific surface area, and total pore volume by using BET (Brunauer–Emmett–Teller) surface analyzer apparatus and the capacity of hydrogen adsorption was evaluated by PCT (pressure–composition–temperature) hydrogen adsorption analyzer apparatus with volumetric method.

They indicated that Even though specific surface area and total pore volume were important factors for increasing the capacity of hydrogen adsorption, the pore volume which has pore width (0.6–0.7 nm) was a much more effective factor than specific surface area and pore volume in PAN-based electrospun activated CNFs.

Chemically activated carbon nanofiber with NaOH and KOH, were evaluated by Figueroa-Torres [5] for hydrogen adsorption at 77 K and atmospheric pressure. Hydrogen adsorption reached values in the order of 2.7 wt.% for KOH activated carbon. The mechanism of formation of the porous nanostructures was found to be the key factor in controlling the hydrogen adsorption capacity of chemically activated carbon [134].

Loading of CNF with metallic particles can enhance the hydrogen storage capacity of it. The hydrogen storage behaviors of porous carbon nanofibers decorated by Pt nanoparticles were investigated, It was found that amount of hydrogen stored increased with increasing Pt content to 3.4 mass%, and then decreased [151].

Fuel cells are used to convert hydrogen, or hydrogen-containing fuels, directly into electrical energy plus heat through the electrochemical reaction of hydrogen and oxygen into water. The process is that of electrolysis in reverse. Overall reaction [152]:

$$2 \; H2 \; (gas) + O2 \; (gas) \rightarrow 2 \; H_2O + energy \; 1$$

Because hydrogen and oxygen gases are electrochemically converted into water, fuel cells have many advantages over heat engines. A simple example of a hydrogen fuel cell consists of an anode and a cathode with an electrolyte in-between that allows positive ions to pass through. Hydrogen fuel is fed to the anode and atmospheric oxygen is fed to the cathode. When activated by a catalyst, usually platinum on the cathode itself, the hydrogen atoms separate into electrons and protons, which take different paths to the cathode. The electrons take a path through an electrical circuit and load, while the protons take a path through the electrolyte. When the electrons and protons meet again at the cathode, they recombine along with the oxygen atoms to produce water and heat. This process is illustrated in Fig. 16.12 [153].

In order for a fuel cell to operate, it requires a constant supply of hydrogen. In transportation sector applications, this hydrogen must be stored locally in a safe and efficient manner. One method to do this that is proving to be viable and able to meet the D.O.E. benchmark is that of a pressurized tank containing hydrogen physically adsorbed on carbon [153]. Hydrogen has a kinetic diameter of 0.289 nm, which is slightly smaller than the ~0.335–0.342 nm interlayer spacing in carbon nanofibers (see Fig. 16.12) [149].

The principle application of a carbon nanofiber hydrogen storage medium is in a fuel tank for an integrated on-board fuel cell system with a polymer electrolyte membrane (PEM) fuel cell stack at its core and a hydrogen supply stored as adsorbed hydrogen in a pressurized tank containing carbon nanofibers.

In a PEM fuel cell, two half-cell reactions take place simultaneously, an oxidation reaction (loss of electrons) at the anode and a reduction reaction (gain of electrons) at the cathode. These two reactions make up the total oxidation-reduction (redox) reaction of the fuel cell, the formation of water from hydrogen and oxygen gases [152]. The D.O.E.'s target benchmarks for on-board hydrogen storage are based on a model hydrogen fuel cell powered vehicle to be able to travel 500 km without refueling and the metric that 3.1 kg of hydrogen would be required for a fuel cell powered car to travel those 500 km. Based on the 10–15 wt.% storage capability that has been demonstrated with properly prepared carbon nanofibers, as illustrated earlier in this paper, this would result in a 10 full tank of hydrogen adsorbed carbon nanofibers weighing between 21 and 31 kg with perhaps some additional weight required for the pressure valve and protective covering [154].

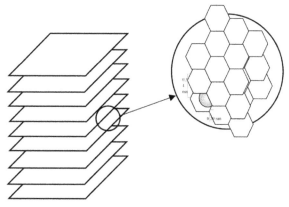

FIGURE 16.12 (a) Schematic representation of the structure of a carbon nanofiber; (b) enlarged section.

A new group of fuel cell is microbial fuel cells (MFCs), which is a novel technology that produces electricity using bacteria as electrocatalysts. The performance of MFCs is influenced by the type of electrode, the electrode distance, the type and surface area of their membrane, their substrate and their microorganisms. The most common catalyst used in cathodes is platinum (Pt). Ghasemi et al. applied chemically and physically activated carbon nanofibers as an alternative cathode catalyst to platinum in a two-chamber microbial fuel cell for the first time [155].

The main reason suggested for improved hydrogen adsorption was that electrospun activated carbon nanofibers might be expected to have an optimized pore structure with controlled pore size. This result may come from the fact that the diameters of electrospun fibers can be controlled easily, and optimized pore sizes can be obtained with a highly developed pore structure. To find the optimized activation conditions, carbon nanofibers were activated based on varying the chemical activation agents, reaction time, reaction temperature, and the rate of inert gas flow [156].

Porous carbon-nanofiber supported nickel nanoparticles also can be used as a promising material for hydrogen storage. It was found that the amount of hydrogen stored was enhanced by increasing nickel content [7] (Fig. 16.13).

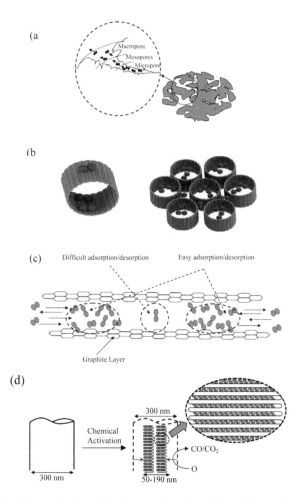

FIGURE 16.13   The mechanism of hydrogen adsorption using various carbon materials; (a) activated carbon, (b) single walled carbon nanotube, (c) graphite, (d) electrospun activated carbon nanofibers.

16.16.2   MEDICAL APPLICATIONS

CNTs and CNFs play an important role in nanomaterial research due to their mechanical, optical, electrical and structural properties. In the field of regenerative medicine, these nanofibers are becoming increasingly attractive as they can be modified to be integrated into human bodies for promoting tissue regeneration and treatment of various diseases [157]. Considering the excellent mechanical strength of CNTs and CNFs, it is natural that there are many studies focusing on using these carbon nanostructures as reinforcing agents in composite materials, especially in bone scaffolds. Since natural neural tissues have numerous nanostructured features (such as nanostructured extracellular matrices that neural cells interact with), CNFs/CNTs, which also have such nanofeatures and exceptional electrical, mechanical and biocompatible properties, are excellent candidates for neural tissue repair. CNFs have excellent properties comparable to CNTs but at a lower cost and are fabricated through an easier scale-up process thus, CNFs have generated much interest in regenerative neural tissue engineering applications [157, 158].

Carbon nanostructures specially carbon nanofibers provide a large surface area with high surface energy which can easily increase the interaction of nanoscaffolds and cells and improve the performance of implant, this fact make them appropriate for nerve regeneration [159]. CNF and ACNF can be applied in drug delivery system too. Drug delivery is an emerging field focused on targeting drugs or genes to a desirable group of cells. The goal of this targeted delivery is to transport a proper amount of drugs to the desirable sites (such as tumors, diseased tissues, etc.) while minimizing unwanted side effects of the drugs on other tissues. CNFs with the ability to cross cell membranes are good candidates to serve as drug delivery carriers to cells with high efficacy [157].

Some methods were used to modify CNFs to improve their biocompatibility properties and highlight some applications of these fibrous materials in creating regenerative scaffolds and drug and gene delivery vehicles [105, 110, 117]. Despite the tremendous potential CNFs can bring, toxicity of these materials is one of the issues that remain to be fully studied. The human health hazards associated with exposure to carbon nanoparticles have not been fully investigated, especially their potential for genotoxicity and carcinogenicity. Surprisingly, despite the current widespread use of carbon nanofibers, toxicological studies have mainly focused on carbon nanotubes, and only a few studies have evaluated different carbon nanofibers and their toxicity [79]. Importantly, the presence of unreacted catalysts in CNFs is a key factor promoting their toxicity so care should be taken when synthesizing CNFs. Clearly, toxicity effects of these fibers when implanted or injected need further investigation. With continued work from researchers, there is no doubt that these materials will become useful and safe to use for enhancing human health [108].

16.16.3 ENERGY STORAGE

Electrospun porous carbon nanofiber webs have attracted considerable attention as a promising electrode material in energy storage devices due to high electrical conductivity, high specific surface area and freestanding nature [160]. These materials possess some rather unique properties that may find use in a number of electro-chemical energy storage systems including primary and secondary batteries, fuel cells and electrochemical capacitors. The structure and properties of the nanofibers can be controlled at the nanometer level by manipulating the process variables [2, 161, 162]. The application of CNF and activated form of it in energy storage, divides into different parts such as lithium-ion batteries and electrochemical double-layer capacitors.

16.16.3.1   CNF AND ACNF APPLICATION IN THIN AND FLEXIBLE LITHIUM-ION BATTERIES (LIBS)

In most batteries, porous structure is an essential requirement. A sponge-like elec-trode will have high discharge current and capacity, and a porous separator between the electrodes can effectively stop the short circuit, but allow the exchange of ions freely. Solid electrolytes used in portable batteries, such as lithium ion battery (LIB), are typically composed of a gel or porous host to retain the liquid electrolyte inside. A porous membrane with well-interconnected pores, suitable mechanical strength and high electrochemical stability could be a potential candidate [163]. Lithium ion batteries offer very high energy densities and design flexibilities, thereby making them integral in modern day consumer devices such as cellular phones, camcorders and laptop computers. However, unlike electrochemical capacitors, lithium ion bat-teries are restricted to low achievable power densities. With the advent of electric vehicles (EV) and plug-in hybrid electric vehicles (PHEV) there has therefore been a growing need to build lithium-ion batteries that can not only provide high energy densities but also deliver high power densities in order to be considered as a po-tential replacement for conventional gasoline engines [164]. A lithium ion battery essentially comprises of three components cathode, anode and electrolyte. Cathodes are generally categorized into three types, namely (i) lithium based metal oxides such as $LiCoO_2$, (ii) transition metal phosphates such as Li3 V2 $(PO_4)_3$ and (iii) spi-nels such as $LiMn_2O_4$. Among anodes, carbon is the typical material used in lithium ion batteries [164]. A lithium ion battery and its charging and discharging processes are depicted in Fig. 16.14.

Nanostructured materials are attractive for lithium-ion batteries with their unique features that arise from their nanoscale structures. Due to the small size, the optimum transport of both electrons from the back contact to the front of the electrode and ions from the electrolyte to the electrode particles can lead to a rapid discharging and charging rate. Up to now, various nanomaterials such as metal oxides, carbonaceous materials, phosphate, and sulfides, etc., have been widely used as anode material for lithium-ion battery. Among the families of materials studied, nanostructured carbon materials such as carbon nanofibers have a lot of advantages such as availability, chemical stability, good cyclability, and low cost as anode materials for lithium-ion batteries [20].

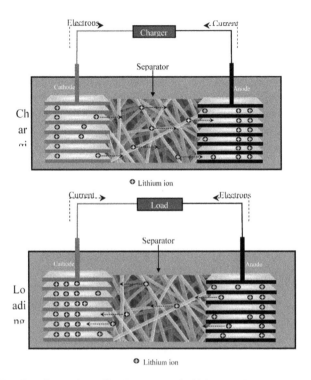

FIGURE 16.14   Charging and Loading Processes of Lithium Ion Battery.

These increases are attributed to the uniform distribution of network-like CNF and ACNF of high conductivity; CNF not only connects the surface of the active materials, its network penetrates into and connects each active material particle. CNF composite electrode also improves the electrochemical performance of thin and flexible lithium-ion batteries such as discharge capacity at high current densities, cycle-life stability, and low-temperature (at 20°C) discharge capacity [20]. These improved electrochemical properties are attributed to the well-distributed network-like carbon nanofibers, within the cathode. The addition of CNF reduces the electron conducting resistance and decreases the diffusion path for lithium ions, hence increases the utilization of active materials during high-current discharge and low-temperature discharge. In addition, network-like CNF forms a more uniform cathode structure so as to have a lower deterioration rate and correspondingly better life cycle stability.

In order to further improve the performance of carbon materials as anodes for LIBs, an effective porous structure in a controllable fashion is needed to provide desirable surface area and open pore structure, which can achieve larger energy conversion density, higher rate capability, and better cycle performance. Therefore, activated or porous CNFs with large specific surface area and controlled pore structure could be an ideal candidate to meet these requirements [97].

Mn-based oxide-loaded porous carbon nanofiber anodes, exhibiting large reversible capacity, excellent capacity retention, and good rate capability, are fabricated by carbonizing electrospun polymer/Mn $(CH_3COO)_2$ composite nanofibers without adding any polymer binder or electronic conductor. The excellent electrochemical performance of these organic/inorganic nanocomposites is a result of the unique combinative effects of nano-sized Mn-based oxides and carbon matrices as well as the highly developed porous composite nanofiber structure, which make them promising anode candidates for high-performance rechargeable lithium-ion batteries [93].

16.16.3.2 CNF AND ACNF APPLICATIONS IN ELECTROCHEMICAL DOUBLE-LAYER CAPACITORS (EDLCS)

In addition to lithium ion battery, electrospun carbon nanofibers could be used in electrochemical double-layer capacitors (EDLC). EDLCs or supercapacitors are promising high-power energy sources for many different applications where high-power density, high cycle efficiency and long cycle life are needed. In 1879, Helmholtz suggested that EDLCs accumulate electrical energy that is generated by the formation of an electrochemical double layer suggested, at the interface between electrode and electrolyte (non-Faradaic process), unlike secondary batteries such as the lithium ion battery or nickel metal hydride battery, which are based on a redox reaction (Faradaic process). This energy storage system based on a non-Faradaic process provides very fast charge and discharge making the EDLCs with the best candidates to meet the demand for high power and long durability [165].

However, because the energy density of EDLCs is small compared to that of rechargeable batteries, it is necessary to increase the capacitance of EDLCs. Recently, the relationships between the porous structures and electrochemical behavior have become increasingly important. Although the use of various materials as EDLCs has been investigated, the application is limited in terms of specific energy. To enhance the specific energy and power of EDLCs, several researchers have put much effort into the development and modification of carbonaceous materials, such as controlling the pore size distribution, introducing electro-active metallic particles or electro- conducting polymers, and fabricating hybrid type cells [166]. The double layer capacitance is correlated with the morphological properties of the porous electrodes, particularly the surface and the pore size distribution of the carbon materials. Thus, tailoring the porous structures of carbon materials is a major goal of EDLC optimization [17, 44]. Various forms of carbonaceous materials, that is, powder, fiber, paper or cloth (fabric or web), carbon nanotubes, carbon nanofibers, and related nanocomposites are candidates for electrodes of EDLCs. A paper type material particularly useful for application as electrodes as the addition of binder, which normally degrades the performance of capacitors, is not needed.

Electronic properties make the carbon nanostructures such as carbon nanofiber applicable inter alia in EDLC, batteries, catalyst supports, and field emission displays. By electrospinning of the PAN nanofiber and subsequent thermal treatments, CNFs as polarized electrodes in EDLC with a remarkable specific capacitance of ca. 297 F/g was obtained [167]. Seo et al. [137] showed that the ACNFs afforded good electronic conductivity, higher specific surface, suitable pore size, and higher content of surface oxygen functional groups. These unique properties of ACNFs were favorable for the diffusion of hydrated ions during charge/discharge within the electric double layers and more effective surface area was provided compared to CNFs, so On the bases of their high-power characteristics and excellent maintenance of specific capacitance, ACNFs could serve as useful electrode materials for supercapacitor applications.

16.16.4 REMOVAL OF POLLUTANTS

16.16.4.1 CNF AND ACNF FOR REMOVAL OF MICRO-ORGANISM FROM WATER

Treatment processes for wastewater reuse and water treatment usually have adopted process, such as, biological treatment, coagulation, sand filtration, membrane filtration and activated carbon adsorption [168]. Recently, membrane filtration in water treatment has been used worldwide for reduction of particle concentration and natural organic material in water. Among the membrane processes, nanofiltration (NF) is the most recent technology, having many applications, especially for drinking water and wastewater treatment [169].

These nanofilters are reusable filters that have controlled porosity at the nanoscale and at the same time, can be formed into macroscopic structures with controlled geometric shapes, density, and dimensions. A carbonaceous nanofilter was fabricated by carbon nanofiber, in the best condition was used for MS2 virus removal. The results showed that at pressures of 8–11 bar the MS2 viruses were removed with a high efficiency by using the fabricated nanofilter. The results showed that the fabricated nanofilter had good water permeability, filtrate flux and could be used for virus removal with high efficiency [170].

16.16.4.2   CNF AND ACNF FOR REMOVAL OF VOLATILE ORGANIC COMPOUNDS (VOCS)

The manufacturers of specialty chemicals and pharmaceuticals generate effluent streams that contain trace amounts of aromatic and chlorinated hydrocarbons. Careful handling, recovery, or disposal of these toxic organics is one of the major environmental issues that confront such industries. Methods for the elimination of such contaminants from gaseous and liquid effluent streams are normally based on fixed-bed adsorption on carbonaceous materials. Traditionally, when recovery steps prove to be uneconomical or difficult, destruction of the organic contaminants is carried out by incineration. Such procedures, however, require elevated temperatures with associated high fuel costs. Catalytic degradation of organic contaminants into less toxic products may be an alternative low-temperature option. An effective catalyst is one on which the contaminant is initially strongly adsorbed and on which reaction with atomic species generated by the interaction of metallic components with the aqueous environment takes place [171]. VOCs (volatile organic compounds), such as, toluene and benzene, are considered as pollutants [70]. Toluene is a hydrocarbon volatile organic compound with a low boiling point. This compound is quite harmful to human beings due to the easy conversion by other pollutants such as ozone and photochemical oxidants. Adsorption is a recommended method showing better control efficiency in removing toluene because it is emitted easily in low concentrations [20, 143].

There have been many attempts to reduce the level of pollution by VOCs using a variety of adsorption methods Activated carbon nanofibers have attracted considerable attention as potential effective adsorbents for low concentrations of organic compounds by adsorption yield of 20–36%. Activated carbon fibers with diameters <10 μm and pore sizes ranging from 8 to 20 Å can be prepared using electrospinning of PAN solution [34]. However, despite the high specific surface area, the toluene adsorption capacity was limited due to the underdeveloped narrow micropores [172].

16.16.4.3   CNF AND ACNF FOR REMOVAL OF TOXIC MATERIALS IN ENVIRONMENT

Formaldehyde is one of the main pollutants in the atmosphere. In indoor air, formaldehyde mainly comes from decorating materials, paint, furniture glue and chemical fiber carpets, and the concentration of formaldehyde is always relative low (<20 ppm). Even if the concentration of the formaldehyde is low, it can cause symptoms such as headache, nausea, coryza, pharyngitis, emphysema, lung cancer, and even death, so it is necessary to take effective measures for its removal. Adsorption by carbonaceous adsorbents is the most widely used method to purify the polluted air.

Carbon nanostructure such a carbon nanofiber and fiber especially in activated form can be good adsorbents for formaldehyde [65].

Dichlorodiphenyltrichloroethane (DDT), DDT is a potential endocrine disruptor even at ng·L-1 levels. It is forbidden as a kind of pesticides from 1980s. While DDT is found in a higher concentration from the lake, river and the atmosphere, water, sediment, soil. It has been detected in many aquatic systems, from the Arctic Antarctic marine mammals to the birds, in the people's milk for human consumption, fish and so on. This raises serious problems in aquatic organisms and animals. Due to their harmful effects on the environment and biological body and the difficulty to degradation by the common treatment methods, it s important to use a suitable adsorbent to remove it. activated carbon fiber and nanofiber are good adsorbent to eliminate it [173].

Another toxic material, which is harmful in environment is Arsenic. Arsenic contaminants in drinking water have been recognized as a serious environmental problem. Arsenic contamination in drinking water was found in the areas where water is extracted from groundwater with geological regions containing arsenic. But there are some cases of contamination from industries and mining as well. There are number of treatment methods for the removal of arsenic from water and wastewater. Chemical precipitation, ion exchange, ultra filtration, membrane techniques, lime softening and microbiological processes are the methods used for the treatment of water and wastewater containing arsenic. But the anaerobic process may be inhibited in chemical precipitation. Although reverse osmosis and ion exchange methods are effective in removing such pollutants, they are expensive in the operational procedure. These factors have limited the use of methods for the removal of arsenic and other toxic pollutants from water and wastewater especially in most of developing countries. Activated carbon nanofiber is successfully used to remove arsenic from wastewater [174].

16.16.5   CNF AND ACNF AS CATALYST SUPPORTS

In chemistry and biology, a carrier for catalyst is used to preserve high catalysis activity, increase the stability and life of the catalyst, and simplify the reaction process. An inert porous material with a large surface area and high permeability to reactants could be a promising candidate for efficient catalyst carriers [175]. Using an electrospun nanofiber mat as catalyst carrier, the extremely large surface could provide a huge number of active sites, thus enhancing the catalytic capability. The well-interconnected small pores in the nanofiber mat warrant effective interactions between the reactant and catalyst, which is valuable for continuous-flow chemical reactions or biological processes. Also, the catalyst can be grafted onto the electrospun nanofiber surface via surface coating or surface modification [163].

In the application of heterogeneous catalysts in liquid phase reactions, the rate of reaction as well as selectivity is often negatively influenced by mass transfer limitations in the stagnant liquid in the pores of the catalyst support [12]. Internal mass transfer limitations can be reduced by maximizing the porosity and lowering the tortuosity of the catalyst support. Particles and layers consisting of carbon nanofibers are promising catalyst supports because of the combination large pore volume ($0.5-2$ cm^3/g) and extremely open morphology, on one hand, and significant high surface area ($100-200$ m^2/g), on the other hand. The Scheme 6 compares the conventional and carbon nanofiber support catalysts (Fig. 16.15).

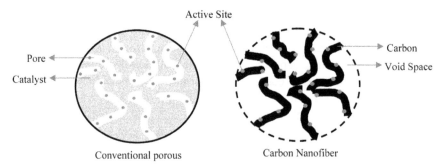

FIGURE 16.15   Carbon nanofiber as a catalyst support [11].

In order to maximize yield, in catalytic reactions, catalyst activity needs to be enhanced, this is typically achieved by developing catalytic sites having high intrinsic activities and by maximizing the number of active sites, for example, by using high surface area support materials [175]. Carbon filaments are formed catalytically in metallic catalysts, particularly in Ni, Fe, and Co based catalysts, used for the conversion of carbon-containing gases, for example, in steam reforming of hydrocarbons and Fischer-Tropsch synthesis. The carbon filament formation was detrimental for operation as they plugged reactors and deactivated catalysts.

Fiber type carbon nano materials, which are suitable as catalyst support can be classified into three types, namely, CNFs, CNTs, and SWNTs. CNFs have a number of special characteristics that make them materials of promise as catalyst supports [11] (Table 16.8).

TABLE 16.8   Classification of CNF Characteristics for Catalyst Support Application

CNFs special characteristics for catalyst support application
Chemical stability for corrosive attack in acidic and basic environments
Having inert nature and withstanding most organic solvents
Stability toward sintering and high-temperature gas reactions
Being conductive
Ability to apply in the field of electro catalysis such as fuel cell electrode

CNFs can be applied as catalyst supports in three ways:

§   Using small aggregates of entangled nanofibers loaded with the catalytic active phase. The typical application would be in slurry reactors with aggregate sizes on the order of 10 cm.

§   Application of larger aggregates (on the order of millimeters) of entangled CNF bodies to form a fixed bed. The fixed bed may be used as such in a single-phase operation or as a trickle bed for gas-liquid operation. Advantages of such a system would be its high porosity and low tortuosity

§   The CNFs can form layers on structured materials such as foams, monoliths, or felts; this helps to keep diffusion distances short. The structured materials of choice obviously will also determine the hydrodynamic behavior of the reactor [8].

It appears rather easy to grow and attach CNFs on structured materials, that is, monoliths, graphite felts, silica fibers, metal filters, and metal foams. In many cases, excessive formation of CNFs weakens the support material and therefore the conditions of CNF formation needs to be accurately controlled [176].

CNFs appear to be well attached to the supporting structures and a good explanation for this observation is still lacking. The layers of CNFs are indeed highly macroporous and should have low tortuosity, and the capability to allow fast mass and heat transfer has been demonstrated for hydrazine decomposition.

A comparable demonstration in liquid phase catalysis is lacking so far. In some cases, catalytic performance was claimed to be modified by the influence of the CNF-support material on either the metal particles or the mode of adsorption of reactants [12, 18].

Carbon nanofiber can be used as supporting agent for Pd catalysts. Pd catalysts have been applied to catalyze Heck reactions of various activated and non-activated aryl substrates. The activity increased exponentially with a decrease in Pd particle size. The high surface area, mesoporous structure of carbon nanofiber and highly dispersed palladium species on carbon nanofibers makes up one of the most active and reusable heterogeneous catalysts for Heck coupling reactions. Pd nanoparticles supported on platelet CNFs appear to be an excellent catalyst due to high activity, low sensitivity towards oxygen, almost no or low issues with leaching and high stability in multicycles [175, 177].

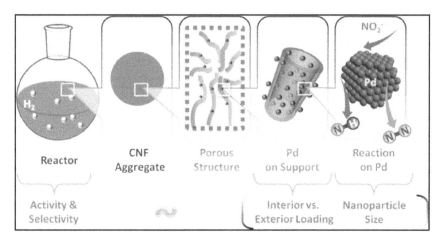

FIGURE 16.16   Carbon Nanofiber as Supporting Agent for Pd Catalysts.

16.16.6   CNF AS POLYMER REINFORCEMENT AGENTS

Early studies on electrospun based carbon nanofibers also included reinforcement of polymers. CNFs have an exceptional combination of mechanical and physical properties that make them ideal reinforcing materials for polymer composites. In order to properly incorporate CNFs into polymer composites, three major manufacturing challenges must be overcome:

§   dispersion of the CNFs in the matrix system;
§   uniform impregnation of the preform by the CNFs;
§   bonding and compatibility between the CNFs, matrix, and microsized reinforcement fibers [178].

As electrospun nanofiber mats have a large specific surface area and an irregular pore structure, mechanical interlocking among the nanofibers should occur. In general, the performance of a fibrous composite depends not only on the properties of the components, but also to a large degree on the coupling between the fiber and the matrix. In order to increase the internal laminar shear strength; numerous attempts have been made to improve bonding between the fiber and the matrix, consisting mostly of chemical and physical modification to the fiber surface [89].

Polymer/CNF nano composites can be prepared by different routes, including in situ polymerization, solution processing and melt mixing. The latter is the most common, given its simplicity and high yield, the compatibility with current industrial processes and the environmental advantage of a solvent-free procedure [179].

A novel approach in this field was the growing carbon nanofibers on the surface of conventional carbon fibers via carbon vapor deposition in the presence of a minuscule amount of metal catalyst [180]. The presence of the carbon nanostructures on the carbon fiber surface was found to enhance the surface area of perform from ~2 m²/g up to over 400 m²/g and consequently increased the interfacial bonding between the fiber and the matrix (Fig. 16.17).

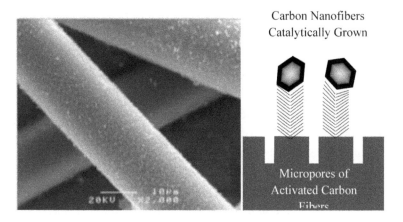

FIGURE 16.17   Carbon nanofiber on the surface of activated carbon fiber.

In order for a fiber to function successfully as reinforcement in high performance engineering materials, it must fulfill certain criteria:

1. A small diameter with respect to its grain size is needed since there will be a low probability of intrinsic imperfections in the material and experimental evidence shows that the strength of the fiber increases as its diameter decreases. In this regard, carbon nanofibers would appear to be superior to other types of carbon fibers since their diameters [147], which are controlled by the size of the catalyst particles responsible for producing them, can be as low as 2 nm [81].
2. A high aspect ratio is needed which ensures that a very large fraction of the applied load will be transferred through the matrix to the stiff and strong fiber. In general, carbon nanofibers possess extremely small diameters and high aspect ratios, typically [81, 89, 90].

3. A very high degree of flexibility is desirable for the complex series of op-
    erations involved in composite fabrication. When carbon nanofibers are
    produced in a helical conformation they were found to possess appreciable
    elastic properties.

FIGURE 16.18   Scanning electron micrograph showing the growth of carbon nanofibers on
the surface of a bundle of a carbon fiber.

So because of the exceptional properties exhibited by carbon nanofibers such
as their high tensile strength, modulus, and relatively low cost, such a material
have a tremendous potential for reinforcement applications in its own right.

Thermoplastics such as polypropylene, polycarbonate, nylon, and thermo
set, such as, epoxy, as well as thermoplastic elastomers such as butadiene–
styrene diblock copolymer, have been reinforced with carbon nanofibers, for
example. Carbon nanofibers with 0.5 wt.% loading were dry-mixed with polypropylene powder by mechanical means, and extruded into filaments by using a single screw extruder. Decomposition temperature and tensile modulus
and tensile strength have increased because of dispersion of CNF [121] (Fig.
16.19).

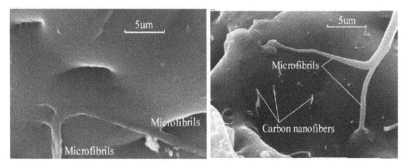

FIGURE 16.19   Fracture surface of neat and nanophased polypropylene (A: neat PP; B: CNF/PP) [181].

16.17   SUMMARY AND OUTLOOK

Carbon nanofibers have great advantages due to their high surface area to volume ratio. They have potential applications in the field of clean energy (solar cells, fuel cells and batteries), electronics, health (biomedical scaffolds, artificial organs), and environment (filter membranes). In summary, ACNF can be produced by different methods which stabilizing, carbonizing and activating the electrospun PAN nanofiber is prefer among them.

However, few studies have applied stretching to electrospun precursor nanofibers before stabilization. Additionally, although aligned nanofibers with various degrees of alignment have been obtained by using various modified collecting devices, tension is rarely applied to the nanofiber assembly during stabilization to prevent shrinkage of the fibers and to ensure molecular orientations along the fiber axis to a large extent. Therefore, there is still considerable room for further improvement of the microstructural, electrical and mechanical properties of the final CNFs. Considering the increased interest in electrospinning and the wide range of potential applications for electrospun CNFs, commercialized products can be expected in the future.

Current advances in electrospinning technology provide important evidence of the potential roles in energy conversion and storage as well as water, and air treatment applications. Though electrospinning has become an essential technique for generating 1D nanostructures, the research exploration is still young, but promising, in energy applications. One of the drawbacks of PAN Electrospinning is that, it has been difficult to obtain uniform nanofibers with diameters below 50 nm using electrospinning. Another drawback is the relatively low production rate. In the near future, it is likely that research efforts will be focused on engineering the electrospinning process, with the ultimate goal of producing carbon nanofibers with diameters below 50 nm, and at a faster rate. The average diameter of ACNF was approximately 250 nm, ranging from 200 nm to 400 nm. The specific surface area and micropore volume of the carbon nanofibers (CNs) were 853 m²/g, and 0.280 cm³/g, respectively [34].

There is still a long way to go and much work to be done, and it is time to begin the task of turning the positive results born of research into the viable solutions born of engineering. In spite of a little information on the chemistry of activated carbon nanofiber the structural study is expected to provide significant insight on the structure itself and the formation of pores contributing to the surface area of carbon materials.

KEYWORDS

- **activation**
- **adsorption**
- **carbon nanofiber**
- **polyacrylonitrile**
- **porosity**

REFERENCES

1. Zhang, Z. et al. (2009). Polyacrylonitrile and Carbon Nanofibers with Controllable Nanoporous Structures by Electrospinning, Macromolecules Material Engineering, *294*, 673–678.
2. Thavasi, V., Singh, G., & Ramakrishna, S. (2008). Electrospun Nanofibers in Energy and Environmental Applications, Energy & Environmental Science, *1*, 205–221.
3. Fowlkes, J. D. (2008). Size-Selectivity and Anomalous Subdiffusion of Nanoparticles through Carbon Nanofiber-Based Membranes, Nanotechnology, Nanotechnology, *19*, 415301–415313.
4. Rożek, Z. Potential Applications of Nanofiber Textile Covered by Carbon Coatings, Journal of Achievements in Materials and Manufacturing Engineering, 208, *27*, 1.
5. Torres, M. et al. (2007). Hydrogen Adsorption by Nanostructured Carbons synthesized by Chemical Activation, Microporous and Mesoporous Materials, *98*, 89–93.
6. Strobel, R. et al. (2006). Hydrogen Storage by Carbon Materials, Journal of Power Sources, *156*, 781–801.
7. Byung-Joo, K., Young-Seak, L., & Soo-Jin, P. (2008). A Study on the Hydrogen Storage Capacity of Ni-plated Porous Carbon Nanofibers, International Journal o f Hydrogen Energy, *33*, 4112–4115.
8. Keane, M. A., & Park, C. (2003). Use of Carbon Nanofibers as Novel Support Material: Phenol Hydrogenation over Palladium. Catalical Community, 7.

9. Sakoda, A., Nomura, T., & Suzuki, M. (1996). Activated Carbon Membrane for Water Treatments: Application to Decolorization of Coke Furnace Wastewater. Adsorption, 3, 93–98.

10. Jia-Zhia, W. (2008). Highly Sensitive Thin Film Sensor Based on Worm-like Carbon Nanofibers for Detection of Ammonia in Workplace. Chinese Journal of Chemistry, 26, 649–654.

11. Chinthaginjala, J., Seshan, K., & Lefferts, L. (2007). Preparation and Application of Carbon-Nanofiber Based Microstructured Materials as Catalyst Supports. Industrial & Engineering Chemistry Research, 46, 3968–3978.

12. Rodriguez, N. M., Kim, M. S., & Baker, R. T. K. (1994). Carbon Nanofibers: A Unique Catalyst Support Medium. Journal of Physical Chemistry, 98, 13108–13111.

13. Kim, C. & Lee, Y. H. (2003). EDLC Aapplication of Carbon Nanofibers Carbon Nanotubes Electrode Prepared by Electrospinning, in 203rd Meeting, Symposium Nanotubes, Nanoscale Materials, and Molecular Devices, The Electrochemical Society: Paris, France.

14. Kim, C., & Yang, K. (2003). Electrochemical Properties of Carbon Nanofiber Web as an Electrode for Supercapacitor prepared by Electrospinning, Applied Physics Letters, 83, 20–26.

15. Metz, K. et al. (2006). Ultrahigh Surface Area Metallic Electrodes by Templated Electroless Deposition on Functionalized Carbon Nanofiber Scaffolds, Chemical Materials, 18, 5398–5400.

16. McKnight, T. et al. (2003). Effects of Microfabrication Processing on the Electrochemistry of Carbon Nanofiber Electrodes, Journal of Physical Chemistry, 107, 10722–10728.

17. Huang, C. et al. (2007). Textural and Electrochemical Characterization of Porous Carbon Nanofibers as Electrodes for Supercapacitors, Journal of Power Sources, 172, 460–467.

18. Sun, J., Wu, G., & Wang, Q. (2004). Adsorption Properties of Polyacrylonitrile Based Activated Carbon Hollow Fiber, Applied Polymer Science, 93, 602–607.

19. Yu, Z. (2008). Pore Structure Analysis on Activated Carbon Fibers by Cluster and Watershed Transform Method, Applied Surface Science.

20. Oha, G. Y. et al. (2008). Adsorptions of Toluene on Carbon Nanofibers prepared by Electrospinning, Science of the Total Environment, 393, 341–347.

21. Song, X., Wangb, C., & Zhang, D. (2009). Surface Structure and Adsorption Properties of Ultrafine Porous Carbon Fibers, Applied Surface Science, 255, 4159–4163.

22. Quanming, L., & Wanxi, Z. (2009). Study on PAN-based Activated Carbon Fiber Prepared by KOH Activation Method. Carbon, 66, 70–75.

23. Reneker, D., & Chun, I. (1996). Nanometer Diameter Fibers of Polymer, produced by Electrospinning, Nanotechnology, 7, 216.

24. Ondarcuhu, T. C. J. (1998). Drawing a Single Nanofibre Over Hundreds of Microns, Euro-Physics Letters, 42, 21220–21225.

25. Martin, C. (1996). Membrane-based Synthesis of Nanomaterials. Chemistry of Materials, 8, 1739–1746.

26. Ma, P., & Zhang, R. (1999). Synthetic Nano-Scale Fibrous Extracellular Matrix, Journal of Biomedical Materials Research, 46, 60–72.

27. Whitesides, G. & Grzybowski, B. (2002). Self Assembly at all scales, Science, 295, 2418–2421.

28. Nataraj, S. K., Yang, K. S., & Aminabhavi, T. M. (2011). Polyacrylonitrile based Nanofibers: A state of the Art Review. Progress in Polymer Science, 37, 487–513.

29. Bhardwaj, N., & Kundu, S. C. (2010). Electrospinning: A Fascinating Fiber Fabrication Technique, Biotechnology Advances, 28, 325–347.

30. Ali, A., & El-Hamid, M. (2006). Electrospinning Optimization for Precursor Carbon Nanofibers, Composites: Part A, 37, 1681–1687.

31. Huang, Z. et al. (2003). A Review on Polymer Nanofibers by Electrospinning and their Applications in Nanocomposites, Composites Science and Technology, 63, 2223–2253.

32. Wang, C. et al. (2007). Electrospinning of Polyacrylonitrile Solutions at Elevated Temperatures, Macromolecules, *40*, 7973–7983.
33. Kim, J., Hong, I., & Lee, J. (1996). Preparation of PAN based Activated Carbon Fibers and its Application for Removal of Sox in Flue Gas.
34. Mitsubishi Rayon Co. L. (2001). Acrylonitrile based Precursor for Carbon Fiber and Method for Production thereof. .
35. Esrafilzadeh, D., Morshed, M., & Tavanai, H. (2009). An Investigation on the Stabilization of Special Polyacrylonitrile Nanofibers as Carbon or Activated Carbon Nanofiber Precursor, Synthetic Metals, *159*, 267–272.
36. Im, J. et al. (2008). The Study of Controlling Pore Size on Electrospun Carbon Nanofibers for Hydrogen Adsorption, Journal of Colloid and Interface Science, *318*, 42–49.
37. Kaixue, W. et al. (2009). Mesoporous Carbon Nanofibers for Supercapacitor Application, Journal of Physical Chemistry C, *113*, 1093–1097.
38. Lee, S., Kimb, T., & Kimb, A. (2007). Surface and Structure Modification of Carbon Nanofibers, Synthetic Metals, *157*, 644–650.
39. Li, C. et al. (2009). Porous Carbon Nanofibers Derived from Conducting Polymer: Synthesis and Application in Lithium-Ion Batteries with High-Rate Capability. Journal of Physical Chemistry C, *113*, 13438–13442.
40. Ji, L., & Zhang, X. (2009). Manganese Oxide Nanoparticle Loaded Porous Carbon Nanofibers as Anode Materials for High Performance Lithium Ion Batteries. Electrochemistry Communications, *11*, 795–798.
41. Liu, C. et al. (2009). Preparation of Carbon Nanofibers through Electrospinning and Thermal Treatment Society of Chemical Industry, *58*, 1341–1349.
42. Maciá-Agulló, J. A. et al. (2007). Influence of carbon fibres crystallinities on their chemical activation by KOH and NaOH. Microporous and Mesoporous Materials, *101*, 397–405.
43. Yusof, N. & Ismail, A. F. (2012). Post spinning and pyrolysis processes of polyacrylonitrile (PAN)-based carbon fiber and activated carbon fiber: A review. Journal of Analytical and Applied Pyrolysis, *93*, 1–13.
44. Kim, C. et al. (2008). Fabrications and Electrochemical Properties of Two Phase Activated Carbon Nanofibers from Electrospinning, Carbon, *42*, 34–39.
45. Suzuki, P. (1994). Activated Carbon Fiber, Fundamentals and Applications, Carbon, *32*, 577–586.
46. Bajaj, P., Streekumar, T., & Sen, K. (2002). Structure Development during Dry jet wet Spinning of Acrylonitrile Vinyl Acids and Acrylonitrile Methyl Acrylate Copolymers. Journal of Applied Polymer Science, *86*, 773–787.
47. Wilkinson, K. (2000). Process for the Preparation of Carbon Fiber.
48. Panels, J. et al. (2008). Synthesis and Characterization of Magnetically Active Carbon Nanofiber Iron Oxide Composites with Hierarchical Pore Structures, Nanotechnology, *19*, 455612–455619.
49. Wangxi, Z., Jie, L., & Gang, W. (2003). Evolution of Structure and Properties of PAN Precursors during their Conversion to Carbon Fibers, Carbon, *41*, 2805–2812.
50. Zhou, Z. et al. (2009). Development of Carbon Nanofibers from Aligned Electrospun Polyacrylonitrile Nanofiber Bundles and Characterization of their Microstructural, Electrical, and Mechanical Properties, Polymer, *50*, 2999–3006.
51. Jagannathan, S. (2008). Structure and Electrochemical Properties of Activated Polyacrylonitrile Based Carbon Fibers Containing Carbon Nanotubes. Journal of Power Sources, *185*, 676–684.
52. Aussawasathien, D. (2006). Electrospun Conducting Nanofiber Based Materials and their Characterization Affects of Fibers Characteristics on Properities and Applications, in the Graduate Faculty of the University of Akron, The University of Akron Ohio.

53. Fitzer, E. (1989). PAN-based Carbon-Fibers Present State and Trend of the Technology from the Viewpoint of Possibilities and Limits to Influence and to Control the Fiber Properties by the Process Parameters, Carbon, *27*, 621–45.
54. Eslami Farsani, R. et al. (2009). FT-IR Study of Stabilized PAN Fibers for Fabrication of Carbon Fibers, World Academy of Science, Engineering and Technology, *50*, 42–48.
55. Jimenez, V. et al. (2008). Chemical Activation of Fish Bone type Carbon Nanofiber.
56. Lozano-Castello, D. (2001). Preparation of Activated Carbons from Spanish Anthracite I, Activation by KOH, Carbon, *30*, 741–749.
57. Fu, R. et al. (2003). Studies on the Structure of Activated Carbon Fibers Activated by Phosphoric Acid, Journal of Applied Polymer Science, *87*, 2253–2261.
58. Huidobro, A., Pastor, A. C., & Rodrıguez-Reinoso, F. (2001). Preparation of Activated Carbon Cloth from Viscous Rayon, Part IV, Chemical Activation, Carbon, *30*, 389–398.
59. Purewal, J. J. et al. (2009). Pore Size Distribution and Supercritical Hydrogen Adsorption in Activated Carbon Fibers, Nanotechnology, *20*, 204012 (6pp).
60. Macia-Agullo, J. A. et al. (2004). Activation of Coal Tar Pitch Carbon Fibres: Physical Activation vs. Chemical Activation. Carbon, *42*, 1367–1370.
61. Guha, A. et al. (2001). Synthesis of Novel Platinum Carbon Nanofiber Electrodes for Polymer Electrolyte Membrane (PEM) Fuel Cells, Journal of Solid State Electrochemistry, *5*, 131–138.
62. Sun, J., & Wang, Q. (2005). Effects of the Oxidation Temperature on the Structure and Properties of Polyacrylonitrile-Based Activated Carbon Hollow Fiber. Journal of Applied Polymer Science, *98*, 203–207.
63. Nguyen, T., & Bhatia, S. (2005). Characterization of Activated Carbon Fibers using Argon Adsorption, Carbon, *43*, 775–785.
64. Mochida, I., & Kawano, S. (1991). Capture of Ammonia by Active Carbon Fibers Further Activated with Sulfuric Acid. Industrial & Engineering Chemistry Research, *30*, 2322–2327.
65. Song, Y. et al. (2007). Removal of Formaldehyde at Low Concentration Using Various Activated Carbon Fibers. Journal of Applied Polymer Science, *106*, 2151–2157.
66. Fitzer, E. (1986). Carbon Fibers and their Composites, Berlin: Springer-Verlang, 296 pages.
67. Casa-Lillo, M. et al. (2002). Hydrogen Storage in Activated Carbons and Activated Carbon Fibers, Journal of Physical Chemistry Part B, *106*, 10930–10934.
68. Purewal, J. J. (2009). Pore Size Distribution and Supercritical Hydrogen Adsorption in Activated Carbon Fibers, Nanotechnology, *20*, 204012–204018.
69. Candy Activated Carbon Fiber (ACF), 2002.
70. Singh, K. et al. (2002). Vapor-Phase Adsorption of Hexane and Benzene on Activated Carbon Fabric Cloth: Equilibria and Rate Studies. Industrial & Engineering Chemistry Research, *41*, 2480–2486.
71. Brasquet, C. & Cloirec, P. (1994). Adsorption onto Activated Carbon Fibers, Application to Water and Air Treatments, Carbon, *32*, 1307–1313.
72. Sun, J. et al. (2006). Effects of Activation Time on the Properties and Structure of Polyacrylonitrile-Based Activated Carbon Hollow Fiber. Journal of Applied Polymer Science, *99*, 2565–2569.
73. Kim, B., Lee, Y., & Park, S. (2007). A Study on Pore opening Behaviors of Graphite Nanofibers by a Chemical Activation Process, Journal of Colloid and Interface Science, *306*, 454–458.
74. Jiuling, C., Qinghai, C., & Yongdan, L. (2006). Characterization and Adsorption Properities of Porous Carbon Nanofiber Granules, China Particuology, *4*, 238–242.
75. Poole, C. & Owens, F. (2003). Introduction to Nanotechnology, Nalwa, H. S., ed. Hoboken, New Jersey, John Wiley & Sons, Inc. 396.

76. Dzenis, Y. (2004). Spinning Continuous Fibers for Nanotechnology, American Association for the Advancement of Science, *304*, 1917–1919.
77. Ramakrishna, S. et al. (2005). An Introduction to Electrospinning and Nanofibers, Singapore: World Scientific Publishing Co. Pte. Ltd.
78. Gu, S., Ren, J., & Vancso, G. (2005). Process Optimization and Empirical Modeling for Electrospun Polyacrylonitrile (PAN) Nanofiber Precursor of Carbon Nanofibers. European Polymer Journal, *41*, 2559–2568.
79. Kisin, E. R. et al. (2011). Genotoxicity of Carbon Nanofibers: Are they Potentially More or Less Dangerous than Carbon Nanotubes or Asbestos? Toxicology and Applied Pharmacology, *252*, 1–10.
80. Park, C., & Keane, M. (2001). Controlled Growth of Highly Ordered Carbon Nanofibers from Y Zeolite Supported Nickel Catalysts. Langmuir, *17*, 8386–8396.
81. Romero, A. et al. (2008). Synthesis and Structural Characteristics of Highly Graphitized Carbon Nanofibers Produced from the Catalytic Decomposition of Ethylene: Influence of the Active Metal (Co, Ni, Fe) and the Zeolite Type Support Microporous and Mesoporous Materials, *110*, 318–329.
82. Jong, K., & g. J. (2000). Carbon Nanofibers: Catalytic Sythesize and Application. Catalysis Reviews, Sciences and Engineering, *42*, 481–510.
83. Park, C., & Baker, R. (1998). Catalytic Behavior of Graphite Nanofiber Supported Nickel Particles. 2. The Influence of the Nanofiber Structure. Journal of Physical Chemistry, *102*, 5168–5177.
84. Kima, J. et al. (2008). Preparation of Polyacrylonitrile Nanofibers as a Precursor of Carbon Nanofibers by Supercritical Fluid Process, Journal of Supercritical Fluids, *47*, 103–107.
85. Wu, S. et al. (2008). Preparation of Pan-Based Carbon Nanofibers by Hot-Stretching, Composite Interfaces, *15*, 671–677.
86. Barranco, V. et al. (2010). Amorphous Carbon Nanofibers and their Activated Carbon Nanofibers as Supercapacitor Electrodes, Journal of Physical Chemistry C, *114*, 10302–10307.
87. Yoon, S. et al. (2004). Carbon Nano-Rod as a Structural Unit of Carbon Nanofibers, Carbon, *42*, 3087–3095.
88. Zheng, R. et al. (2006). Preparation, Characterization and Growth Mechanism of Platelet Carbon Nanofibers, Carbon, *44*, 742–746.
89. Rodriguez, N. M. (1993). A Review of Catalytically Grown Carbon Nanoflbers, Commentaries and Reviews, *8*, 12–19.
90. Kim, C. et al. (2004). Raman Spectroscopic Evaluation of Polyacrylonitrile-based Carbon Nanofibers Prepared by Electrospinning. Journal of Raman Spectroscopy, *35*, 928–933.
91. Lingaiah, S. et al. (2005). Polyacrylonitrile-based Carbon Nanofibers Prepared by Electrospinning.
92. Kurban, Z. et al. (2000). Graphitic Nanofibers from Electrospun Solutions of PAN in Dimethylsulphoxide.
93. Chun, I. et al. (1999). Carbon Nanofibers from Polyacrylonitrile and Mesophase Pitch, Journal of Advanced Materials, *31*, 36–41.
94. Wang, Y., Serrano, S., & Aviles, J. (2002). Conductivity Measurement of Electrospun PAN-Based Carbon Nanofiber, Journal of Materials Science Letters, *21*, 1055–1057.
95. Panapoy, M., Dankeaw, A., & Ksapabutr, B. (2008). Electrical Conductivity of PAN-Based Carbon Nanofibers Prepared by Electrospinning Method, Thammasat Int. J. Sc. Tech, *13*, 88–93.
96. Hou, H., & Reneker, D. (2004). Carbon Nanotubes on Carbon Nanofibers: A Novel Structure Based on Electrospun Polymer Nanofibers, Advanced Materials, *16*, 69–73.
97. Kim, S., & Lee, K. (2004). Carbon Nanofiber Composites for the Electrodes of Electrochemical Capacitors, Chemical Physics Letters, *400*, 253–257.

98. Jalili, R., Morshed, M., & Hosseini Ravandi, S. (2006). Fundamental Parameters Affecting Electrospinning of PAN Nanofibers as Uniaxially Aligned Fibers. Journal of Applied Polymer Science, *101*, 4350–4357.
99. Wang, Y., Serrano, S., & Santiago-Avile, J. (2003). Raman Characterization of Carbon Nanofibers Prepared using Electrospinning. Synthetic Metals, *138*, 423–427.
100. Kim, Ch. et al. (2004). Raman Spectroscopic Evaluation of Polyacrylonitrile-based Carbon Nanofibers Prepared by Electrospinning, Journal of Raman Spectroscopy, *35*, 928–933.
101. Jalili, R., Morshed, M., & Ravandi, S. H. (2006). Fundamental Parameters Affecting Electrospinning of PAN Nanofibers as Uniaxially Aligned Fibers, Journal of Applied Polymer Science, *101*, 4350–4357.
102. Wang, Ch. et al. (2007). Electrospinning of Polyacrylonitrile Solutions at Elevated Temperatures, Macromolecules, *40*, 7973–7983.
103. Zussman, E. et al. (2005). Mechanical and Structural Characterization of Electrospun PAN-Derived Carbon Nanofibers, *Carbon, 43*, 2175–2185.
104. Ci, L. et al. (2001). Carbon Nanofibers and Single walled Carbon Nanotubes Prepared by the Floating Catalyst Method, *Carbon, 39*, 329–335.
105. Kalayci, V. E. (2005). Charge Consequences in Electrospun Polyacrylonitrile (PAN) Nanofiber, Polymer, *46*, 7191–7200.
106. Blackman, J. et al. (2006). Activation of Carbon Nanofibres for Hydrogen Storage, Carbon, *44*, 44–48.
107. Yördem, O., Papila, M., & Menceloglu, Y. (2001). Prediction of Electrospinning Parameters for targeted Nanofiber Diameter.
108. Vicky, V., Katerina, T., & Niko, S. C. (2006). Carbon Nanofiber Based Glucose Biosensor, Analytical Chemistry, *78*, 5538–5542.
109. Deitzel, J. M. et al. (2001). The Effects of Processing Variables on the Morphology of Electrospun Nanofibers and Textiles, Polymer, *42*, 261–272.
110. Boland, E. D. (2001). Tailoring Tissue Engineering Scaffold Using Electrostatic Processing Techniques: A Study of Poly (Glycolic Acid) Electrospinning. Journal of Macromolecular Science, Pure and Applied Chemistry, *12*, 1231–1243.
111. Ryu, Y. J. (2003). Transport Properties of Electrospun Nylon *6* Nonwoven Mats. European Polymer Journal, *39*, 1883–1889.
112. Katti, D. S. et al. (2004). Bioresorbable Nanofiber-Based Systems for Wound Healing and Drug Delivery, Optimization of Fabrication Parameters, Journal of Biomed Mater Res B Appl Biomater, *70*, 286–296.
113. Reneker, D., & Chun, I. (1996). Nanometer Diameter Fibers of Polymer, produced by Electrospinning, Nanotechnology, *7*, 216.
114. Demir, M. M. (2003). Electrospining of Polyurethane Fibers, Polymer, *43*, 3303–3309.
115. Sukigara, S. et al. (2004). Regeneration of Bombyx Mori Silk by Electrospinning, Part 2 Process Optimization and Empirical Modeling using Response Surface Methodology, Polymer, *45*, 3701–3708.
116. He, J., Wan, Y., & Yu, J. (2008). Effect of Concentration on Electrospun Polyacrylonitrile (PAN) Nanofibers, Fibers and Polymers, *9*, 140–142.
117. Agend, F., Naderi, N., & Alamdari, R. (2007). Fabrication and Electrical Characterization of Electrospun Polyacrylonitrile Derived Carbon Nanofibers Farima. Journal of Applied Polymer Science, *106*, 255–259.
118. Jalili, R., Hosseini, A., & Morshed, M. (2005). The Effects of Operating Parameters on the Morphology of Electrospun Polyacrilonitrile Nanofibers, *Iranian Polymer Journal, 14*, 1074–1081.
119. Ma, G., Yanga, D., & Nie, J. (2009). Preparation of Porous Ultrafine Polyacrylonitrile (PAN) Fibers by Electrospinning, Polymer Advanced Technology, *20*, 147–150.

120. Baker, S. et al. (2006). Functionalized Vertically Aligned Carbon Nanofibers as Scaffolds for Immobilization and Electrochemical Detection of Redox Active Proteins. Chemistry of Materials, *18*, 4415–4422.
121. Hasan, M., Zhou, Y., & Jeelani, S. (2006). Thermal and Tensile Properties of Aligned Carbon Nanofiber Reinforced Polypropylene. Materials Letters, *61*, 1134–1136.
122. Chenga, J. et al. (2004). Long Bundles of Aligned Carbon Nanofibers obtained by Vertical Floating Catalyst Method, Materials Chemistry and Physics, *87*, 241–245.
123. Weisenbergera, M. et al. (2009). The Effect of Graphitization Temperature on the Structure of Helical-ribbon Carbon Nanofibers. Carbon, *47*, 2211–2218.
124. Zhang, L., & Hsieh, Y. (2009). Carbon Nanofibers with Nanoporosity and Hollow Channels from Binary Polyacrylonitrile Systems, European Polymer Journal, *45*, 47–56.
125. Wang, Y., & Santiago, J. (2002). Early Stages on the Graphitization of Electrostatically Generated PAN Nanofibers, in Conference on Nanotechnology (IEEE-NANO 2002): Pensylvania, USA, 29–32.
126. Wang, Y., Serrano, S., & Santiago-Avile, J. (2003). Raman Characterization of Carbon Nanofibers Prepared Using Electrospinning. Synthetic Metals, *138*, 423–427.
127. Ko, T., Kuo, W., & Hu, C. (2001). Raman Spectroscopic Study of Effect of Steam and Carbon Dioxide Activation on Microstructure of Polyacrylonitrile- Based Activated Carbon Fabrics. Journal of Applied Polymer Science, *81*, 1090–1099.
128. Yamashita, Y. et al. (2008). Carbonization Conditions for Electrospun Nanofiber of Polyacrylonitrile Copolymer, Indian Journal of Fiber and Textile Research, *33*, 345–353.
129. Su, C. et al. (2012). PAN-based Carbon Nanofiber Absorbents Prepared Using Electrospinning. Fibers and Polymers, *13*, 436–442.
130. Shimada, I., & Takahagi, T. (1986). FT-IR Study of the Stabilization Reaction of Polyacrylonitrile in the Production of Carbon Fibers, Journal of Polymer Science, Part A: Polymer Chemistry, *24*, 1989–1995.
131. Coleman, M., & Petcavich, R. (1987). Fourier Transform Infrared studies on the Thermal Degradation of Polyacrylonitrile, Journal of Polymer Science, Part A: Polymer Chemistry, *16*, 821–832.
132. Zhang, S. et al. (2008). Structure Evolution and Optimization in the Fabrication of PVA-based Activated Carbon Fibers. Journal of Colloid and Interface Science, *321*, 96–102.
133. Wang, P., Hong, K., & Zhu, Q. (1996). Surface Analyses of Polyacrylonitrile-Based Activated Carbon Fibers by X-ray Photoelectron Spectroscopy Journal of Applied Polymer Science, *62*, 1987–1991.
134. Yoon, S. et al. (2004). KOH Activation of Carbon Nanofibers, Carbon, *42*, 1723–1729.
135. Jim´enez, V. et al. (2009). Microporosity Development of Herringbone Carbon Nanofibers by RbOH Chemical Activation, Research Letters in Nanotechnology, *5*, 52–56.
136. Tavanai, H., Jalili, R., & Morshed, M. (2009). Effects of Fiber Diameter and CO_2 Activation Temperature on the Pore Characteristics of Polyacrylonitrile Based Activated Carbon Nanofibers. Surface and Interface Analysis, *41*, 814–819.
137. Seo, M., & Park, S. (2009). Electrochemical Characteristics of Activated Carbon Nanofiber Electrodes for Supercapacitors, Materials Science and Engineering, *164*, 106–111.
138. Im, J., Jang, J., & Lee, Y. (2009). Synthesis and Characterization of Mesoporous Electrospun Carbon Fibers Derived From Silica Template. Journal of Industrial and Engineering Chemistry, *15*, 914–918.
139. Lillo-Ro´denas, M., Cazorla-Amoro´s, D., & Linares-Solano, A. (2003). Understanding Chemical Reactions Between Carbons and NaOH and KOH An Insight into the Chemical Activation Mechanism. Carbon, *41*, 267–275.
140. Chuang, C. et al. (2008). Temperature and Substrate Dependence of Structure and Growth Mechanism of Carbon Nanofiber, Applied Surface Science, *254*, 4681–4687.

141. Ji, L., & Zhang, X. (2009). Electrospun Carbon Nanofibers containing Silicon Particles as an Energy-Storage Medium, Carbon, 47, 3219–3226.
142. Kim, C. et al. (2006). Fabrication of Electrospinning-Derived Carbon Nanofiber Webs for the Anode Material of Lithium-Ion Secondary Batteries, Advanced Functional Materials, 16, 2393–2397.
143. Oh, G. et al. (2008). Preparation of the Novel Manganese Embedded PAN-Based Activated Carbon Nanofibers by Electrospinning and Their Toluene Adsorption. Journal of Analytical applied pyrolysis, 81, 211–217.
144. Vasiliev, L. et al. (2007). Hydrogen Storage System Based on Novel Carbon Materials and Heat Pipe Heat Exchanger. International Journal of Thermal Sciences, 46, 914–925.
145. Ahn, C. et al. (1998). Hydrogen Desorption and Adsorption Measurements on Graphite Nanofibers, Applied Physics Letters, 73, 77–81.
146. Tibbetts, G., Meisner, G., & Olk, C. (2001). Hydrogen Storage Capacity of Carbon Nanotubes, Filaments, and Vapor-Grown Fibers, Carbon, 39, 2291–2301.
147. Sharon, M. et al. (2004). Synthesis of Carbon Nano-Fiber from Ethanol and its Hydrogen Adsorption Capacity, Carbon, 61, 21–26.
148. Chahine, R., & Bénard, P. (2001). Assessment of Hydrogen Storage on different Carbons, in Metal Hydrides and Carbon for Hydrogen Storage, Richard Chahine, Canada.
149. Browning, D. et al. (2002). Studies into the Storage of Hydrogen in Carbon Nanofibers: Proposal of a Possible Reaction Mechanism. Nano letters, 2, 201–205.
150. Rzepka, M. et al. (2005). Hydrogen Storage Capacity of Catalytically Grown Carbon Nanofibers, Journal of Physical Chemistry C, 109, 14979–14989.
151. Byung-Joo, K., Young-Seak, L., & Soo-Jin, P. (2008). Preparation of Platinum-Decorated Porous Graphite Nanofibers and their Hydrogen Storage Behaviors, Journal of Colloid and Interface Science, 318, 530–533.
152. Cook, B. (2001). An Introduction to Fuel Cells and Hydrogen Technology, Vancouver.
153. Zeches, R. (2002). Carbon Nanofibers As A Hydrogen storage Medium for Fuel Cell Applications in the Transportation Sector.
154. Kim, D. et al. (2005). Electrospun Polyacrylonitrile-Based Carbon Nanofibers and their Hydrogen Storages. Macromolecular Research, 13, 521–528.
155. Ghasemi, M. et al. (2011). Activated Carbon Nanofibers as an Alternative Cathode Catalyst to Platinum in a Two-Chamber Microbial Fuel Cell, International Journal of Hydrogen Energy, 36, 13746–13752.
156. Lee, Y., & Sun, J. I. M. (2010). Preparation of Functionalized Nanofibers and their Applications in Nanofibers, Kumar, A., Editor.
157. Tran, P., Zhang, L., & Webster, T. (2009). Carbon Nanofibers and Carbon Nanotubes in Regenerative Medicine, Advanced Drug Delivery Reviews, 61, 1097–1114.
158. Seidlits, S. K., Lee, J. Y., & Schmidt, C. E. (2008). Nanostructured Scaffolds for Neural Applications, Nanomedicine, 3, 183–199.
159. Tavangarian, F., & Li, Y. (2012). Carbon Nanostructures as Nerve Scaffolds for Repairing Large Gaps in Severed Nerves, Ceramics International, 38(8), 6075–6090.
160. Ma, C. et al. (2012). Phenolic-based Carbon Nanofiber Webs Prepared by Electrospinning for Supercapacitors, Materials Letters, 76, 211–214.
161. Lipka, S. (1998). Carbon Nanofibers and their Applications for Energy Storage in Battery Conference on Applications and Advances I. X. Library, Editor: Long Beach, C. A., USA 373–374.
162. Ji, L. et al. (2009). Porous Carbon Nanofibers from Electrospun Polyacrylonitrile SiO2 Composites as an Energy Storage Material, Carbon, 47, 3346–3354.
163. Jian, F. et al. (2008). Applications of Electrospun Nanofibers, Chinese Science Bulletin, 53, 2265–2286.

164. Mukherjee, R. et al. (2012). Nanostructured Electrodes for High-Power Lithium Ion Batteries Nano Energy, *1*, 518–533.
165. Donough, J. et al. (2009). Carbon Nanofiber Supercapacitors with Large Areal Capacitances Applied Physics Letters, *95*, 243109–243111.
166. Su, Y. et al. (2009). Activation of Ultra-Thin Activated Carbon Fibers as Electrodes for High Performance Electrochemical Double Layer Capacitors. Journal of Applied Polymer Science, *111*, 1615–1623.
167. Tao, X. et al. (2006). Synthesis of Multi-Branched Porous Carbon Nanofibers and their Application in Electrochemical Double-Layer Capacitors, Carbon, *44*, 1425–1428.
168. Bae, S. D. et al. (2007). Preparation, Characterization, and Application of Activated Carbon Membrane with Carbon Whiskers, Desalination, *202*, 247–252.
169. Tahaikt, M. et al. (2007). Fluoride Removal from Groundwater by Nanofiltration, Desalination, *212*, 46–53.
170. Mostafavia, S., Mehrnia, M., & Rashidi, A. (2009). Preparation of Nanofilter from Carbon Nanotubes for Application in Virus Removal from Water, Desalination, *238*, 88–94.
171. Park, C. et al. (2000). Use of Carbon Nanofibers in the Removal of Organic Solvents from Water, Langmuir, *16*, 8050–8056.
172. Cuervo, M. et al. (2008). Effect of Carbon Nanofiber Functionalization on the Adsorption Properties of Volatile Organic Compounds, Journal of Chromatography, 1188, 264–273.
173. Guo, Y. et al. (2009). Adsorption of DDT by activated Carbon Fiber Electrode, in International Conference on Energy and Environment Technology.
174. Lee, S. (2010). Application of Activated Carbon Fiber (ACF) for Arsenic Removal in Aqueous Solution, Korean Journal of Chemical Engineering, *27*, 110–115.
175. Shuai, D. et al. (2012). Enhanced Activity and Selectivity of Carbon Nanofiber Supported Pd Catalysts for Nitrite Reduction. Environmental Science and Technology, *46*, 2847–2855.
176. Coelho, N. et al. (2008). Carbon Nanofibers: a Versatile Catalytic Support. Materials Research Society, *11*, 353–357.
177. Zhu, J. et al. (2009). Carbon Nanofiber-supported Palladium Nanoparticles as Potential Recyclable Catalysts for the Heck Reaction Applied Catalysis A: General, *352*, 243–250.
178. Rodriguez, A. et al. (2011). Mechanical Properties of Carbon Nanofiber Fiber Reinforced Hierarchical Polymer Composites Manufactured with Multiscale Reinforcement Fabrics Carbon, *49*, 937–948.
179. Novais, R., Covas, J., & Paiva, M. (2012). The Effect of Flow Type and Chemical Functionalization on the Dispersion of Carbon Nanofiber Agglomerates In Polypropylene Composites: Part A, *43*, 833–841.
180. Lim, S. et al. (2004). Surface Control of Activated Carbon Fiber by growth of Carbon Nanofiber, Langmuir, 5559–5563.
181. Hasan, M. M., Zhou, Y., & Jeelani, S. (2006). Thermal and Tensile Properties of Aligned Carbon Nanofiber Reinforced Polypropylene. Materials Letters, *61*, 1134–1136.

INDEX

A

Absorption spectrum, 145
  AgNPs, 145
  Co/AgNPs, 145
Acetone vapor, 20, 21, 54, 55
Acrylonitrile unit, 365
Activated carbon (AC), 17, 18, 20–22, 49, 54–56, 369
Activated carbon fibers, 366, 380
Activated carbon hollow fibers, 366
Adsorption process, 26, 27, 74, 75
Aerosol, 28, 76
Aggregates, 7, 26, 28, 352, 405
  clusters, 74
Aging time of luminescence, 89
Air purification, 320, 368
Alignment speed, 380
Alkaline single-cell gel electrophoresis assay, 146
American Type Culture Collection, 141
Ames mutagenicity test, 151
Amorphous-crystalline polymers, 18, 50
Anderson model, 3
Anderson periodic model, 2
ANOVA technique, 69
Anticancer immunity, 140
Applied Mechanics Institute, 118
Argon plasma, 76, 77
Aromatic structure, 362
Arsenic, 403
Aspect ratio, 247
Asphalt mixtures, 94, 99
ASTM standards, 63
Atomic force microscope, 173, 391
Atomistic approach, 181
Automobile industries, 45
Avogadro number, 207
Axial direction, 172, 248, 255, 299
Axial/coaxial electrospinning, 172, 269–274, 333

B

Badminton rackets, 365
Barrett–Joiner–Halenda method, 382
Bead-rod model,
Bead-spring chain model, 23
  see, Rouse chain
Bead-spring model, 237
Beckman Institute (USA), 118
Benzene vapor, 19, 54
Bessel equation, 326
Bhatnagar-Gross-Krook, 265
Bioimaging, 141
Biological applications, 14
  cancer treatment, 14
  drug delivery, 14
  magnetic resonance imaging enhancement, 141
Biomedicine, 141
Bitumen, 9
  see, modified aged bitumen
Bleomycin drug, 142, 147, 148
Blood lymphocytes, 142
Boltzmann equation, 3, 5, 265, 266
Boltzmann's constant, 198
Bottom-up process, 108
Boubaker Polynomials Expansion Scheme (BPES), 273
Boundary element method (BEM), 261, 263
Boundary integral method, 251
Brabender (Germany), 29, 77
Bragg peaks, 32–34
Brownian dynamics, 236
Brownian motion force, 240
Brunner–Emmett–Teller method, 382
Butadiene-acrylonitrile elastomer, 40

C

Calcium hydroxide, 69
Calcium-Silicate-Hydrate (C-S-H), 60

California University (Berkeley), 129
Cancer therapy, 140
Cancer treatment, 141
Capillary forces, 28
Capillary number, 247
Carbocyclic carbon skeleton, 21
Carbon nanofiber, 410
Carbon nanostructures, 2
Carbon nanotube, 2, 3, 60, 197, 198
Carbonization, 381
Cellulose, 16–18, 22, 48
Cement matrix, 60
Cement paste, 61
Ceramic boat, 371
Chebysheva, 35, 82
Chemical activation, 389
Chemical methods, 177, 179
Chemical vapor deposition method, 373
Chromatid type breaks, 150
Chromosomal level, 140
Civil engineering infrastructure, 94
Classical theory, 262
Cloisite 15A montmorillonite, 40
Coaxial electrospun nanofibers, 314
Cold plasma, 178
Comet assay, 146
Comet tail, 142
Composite films, 29–36, 77–84, 92
Compressive strength, 63
Computational fluid dynamics (CFD), 207
Computer-aided exchange procedure, 64
Concrete Mixture, 67
Conductivity, 12, 188, 189, 209, 248, 260, 273, 286, 293, 324, 374
Conglomerate, 134
Continuum Mechanics theory, 267
Cool light, 319
Copper/carbon nanocomposite, 158
Core-shell nozzle, 259
Core–shell nanoparticles, 76
Coulomb interaction, 3
Coulomb repulsion, 3
Coulomb's law, 226
Crystal lattice, 3
Crystallinity characteristics, 83
Curing agents, 160
Cytotoxic effect, 151

D

D-optimal method, 64
Dasri model, 251
Deborah number, 247
Design of Experiment analysis, 60
Dichlorodiphenyltrichloroethane, 403
Dielectric constant ratio, 247
Dielectric Effect, 189
Diffractograms, 81
Diffusion coefficient, 2, 3, 6, 9–12
Diffusion process, 2
Dilithiation process, 316
Direct Monte Carlo Simulation (DSMC) Multi-scale approach, 181
Discrete node model, 219
DNA, 140
Double strand breaks (DSB), 147
Droplet formation model, 249
Drug delivery, 141
Drug release modeling, 321
Dry-wet spinning, 362
Drying temperature, 90
Drying time, 90
Dumbbell model, 237, 239
Dynamics interactions, 108, 136

E

Eagle's minimal essential medium (EMEM), 141
Elasticity modulus, 354, 357, 358
Electric field, 2
Electric Peclet number, 247
Electric vehicles, 298
Electrochemical double-layer capacitors, 400
Electrohydrodynamic model, 248
Electroluminescent composites, 75
Electromagnetic model, 250
Electron dispersion law, 9
Electron energy loss spectroscopy, 381
Electrospinning dilation, 221
Electrospinning process, 186, 195
Electrospinning simulation, 251
Electrospinning technology, 183
Electrospraying, 312
Electrospun fibers, 371

Electrostatic force, 125, 247
Energetic characteristics, 156, 164, 166, 168
Environmental problems, 360
Environmental safety, 75
Epitaxial growth, 178
Epoxy resins, 159
Euclidean space, 354
Euler number, 247, 312
Extracellular signal-regulated kinases
  (ERKs), 151

F

Fair alignment, 379
Faradaic process, 400
Fatigue test results, 10
    Nottingham experiment, 103
Feed Rate, 189
FENE dumbbell model, 197
Fermi annihilation, 4
Fiber morphology, 376
Fiber reinforced concrete, 59, 60, 71
Fibrous network, 202
Fick's law, 220
Fine dispersed suspension, 158, 160, 166
Fine particle homogeneous scatter, 97
Fine powders, 49
Finite element method, 251, 257, 260
Finite volume method, 257
Fischer-Tropsch synthesis, 404
Fishing rods, 365
Flexural strength, 63, 71
Fluorescent microscope, 143
Fluorescent spectroscopy, 90
Formaldehyde, 402
Fourier law, 220
Fourier series, 6, 278
Fourier transform infrared spectroscopy,
  381
Fourier transformation, 4
Fourth-level organization, 200
Fractal clusters, 76
Fractal space, 35
    fractal lattice, 353
Froude number, 246
FTIR spectra, 28, 76
Fuel cells, 20
    alkaline, 20

direct methanol, 20
protonexchange mat (PEM), 20
solid oxide, 209

G

Gas cycle, 77
Gas jets, 76
Gas phase method, 17
    chemical vapor condensation, 17
    chemical vapor reaction, 177
    heating heat pipe gas reaction, 177
    laser induced chemical vapor reaction,
    177
    plasma enhanced chemical vapor reac-
    tion, 17
    gas-phase evaporation method, 177
    electric heating evaporation, 177
    electron beam heating, 177
    high frequency induction heating, 177
    laser heating, 177
    plasma heating, 177
    resistance heating, 177
    vacuum deposition, 17
    sputtering, 177
Gas–vapor mixture, 76
Gauss law, 277, 285
Gaussian distribution, 111
Gaussian electrostatic system, 240
Genotoxicity results, 142
Genotoxicity test (comet assay), 142
Genotoxicity, 150
Gibbs's model, 275
Giesekus equation, 232
Granulometric structure, 77
Graphene bilayer graphene, 3
Graphene nanoribbons, 3
Graphite carbon, 316
GraphPad Prism 5.0 software, 144
Gravimat-4303, 7
    automated vacuum adsorption apparatus,
    76
Green function method, 4
Guinier diffractometer, 30, 78

H

Hamiltonian Anderson model, 2

Hardwood sawdust, 19
Hartree–Fock methods, 175
Heat-treatment temperature, 363
Heck coupling reactions, 405
Helium gas, 178
Heptanes vapor, 21
Herringbone structure, 372
Herringbone-type carbon nanofibers, 373
Heterogeneous systems, 18
High temperature shearing (HTS), 16, 48
High-density polyethylene, 352
High-strength polyethylene films, 74
High-tech bicycles, 365
High-temperature shift crushing, 22, 56
Homogeneous scatter, 97, 104
Hooke's law, 231, 233
Hookean springs, 237
Horvath–Kawazoe method, 382
Hot plasma, 178
Human Epidermoid cancer cells, 141
Hydrocarbons, 16
Hydrogen atoms, 2
Hydrogen bonds, 112
Hyper thermic techniques, 140
Hyperthermia, 140
Hypocrystalline phases, 110
Hypoploidy, 148

I

I-methyl tetra hydro phthalic anhydrate, 160
In vitro viability assay, 145
Incubation, 140
Industrial revolution, 173
Instability analysis, 292
Institute of Standards, 78
Interfacial transition zone (ITZ), 61, 70, 7
    aggregate, 7
    cement, 71
International System for human Cytogenetic
    Nomenclature (ISCN), 143
Inviscid model, 306, 307
IR spectra, 384
IR spectroscopy, 160, 168, 381
Iranian nano clay, 94
Iranian Pavement Regulations, 99
ISIS software, 14
    ISIS imaging system, 143

Iso methyl tetra hydro phthalic anhydrate,
    158
Isolation technique, 360
Isotropic distribution, 109
ITZ, 61, 71

J

Joint Supercomputer Center, 135

K

Kinetics of Luminescence System, 90
Kutta-Merson method, 251, 257

L

Lagrangian axial strain, 240
Langevin differential equations, 11
Laplace equation, 327
Large electrical relaxation time limit, 310
Laryngeal cancer cells, 151
Larynx, 141
Laser based methods, 180
Laser optics, 91
Laser-induced decomposition, 74
Lasers, 175
Lattice Boltzmann method, 251, 26
    Thermal Lattice Boltzmann method, 265
LCD systems, 89
Light control tests, 147
Light emitting diode (LED), 140, 142, 175
Liquid crystal template, 91
Liquid crystals, 8
    lyotropic, 88
Liquid phase method, 17
    emulsion, 17
    hydrolysis, 17
    oxidation reduction, 17
    precipitation, 17
    radiation chemical synthesis, 17
    sol-gel processing, 17
    solvent evaporation pyrolysis, 17
    solvent thermal method, 17
    spray, 178
Lithium ion batteries, 314, 316, 398
Lorentz force law, 235
Low density polyethylene (LDPE), 16, 29,
    48

Low-frequency arc discharge, 76
Luminescence, 7
    Spectrum, 90
Luminescent characteristics, 89
Lymphocytes layer, 142
Lyotropic systems, 91

M

Macroscopic approach, 18
    hydrodynamic models, 181
Magnetic resonance imaging enhancement,
    141
Maleine anhydride, 352
Maltenes, 9
    oily liquid matrix, 97
Marshal experiments, 99
Marshal samples, 94, 96
Marshal stability, 99
Mathematical modeling methods, 110
Maxwell's equation, 235, 274
Mechanical mixing method, 54, 55, 104
Mechanical resonance method, 375
Medical applications, 396
Medical tablet, 55
Mesoscopic model, 201
Metal/carbon nanocomposites, 152, 156,
    168
Methyl acrylate, 365
Methyl methacrylate, 365
Microbial fuel cells, 395
Microcracks, 70
Microencapsulation/nanoencapsulation
    process, 274, 324, 333
Micronucleus test, 151
Microscopic models, 236
Miller's index, 34, 82
Missiles, 365
Mitotic index, 148
Modified aged bitumen, 100
Modulus experiment, 10
    Resilient Modulus Results, 102
Modulus test results, 102
Molecular dynamics (MD), 181, 236
Molecular electronics, 91
Molecular mechanics, 108, 110, 136
Momentum balance, 223
Monochromatic blue light, 140

Monosilane, 74
Monte Carlo method, 236
Montmorillonite (MMT), 40
Montmorillonite, 40, 42, 46, 352
Morphology, 184, 188, 190, 204, 211, 358
Morse and Lennard-Johns interaction, 117
Morse potential, 126
Mother Nature, 212
Motor oil, 45
Multicomponent systems, 92
Multiplexed bioassays, 141
Multiscale method, 20
    bridge method, 20
    coarse-grain method, 20
    dissipative particle dynamics method, 20
    quasi-continuum, 201
Multiwall carbon nanotubes, 370
Mutagenicity, 151, 152
MYLAR (Chemplex Industries Inc.), 30

N

N,N-Dimethylformamide, 374
Nano-carbon, 56
Nano-electro-mechanical systems (NEMs),
    191
Nano-scale materials, 61
Nanocarbon (NC), 16, 48
Nanocoatings, 193
Nanocrystal surface, 74
Nanocrystalline silicon, 26–29, 37, 74, 84
Nanocrystals, 26, 173
Nanodisk, 173
Nanoelements, 10
    nanofibers, 10
    nanoparticles, 10
    nanotubes, 108
Nanofiber technology, 182
Nanofibers production methods, 18
    drawing, 18
    electrospinning, 185
    see, taylor conephase separation method,
    184
    platelet-like structure, 18
    self-assembly, 18
    template synthesis, 184
Nanofiltration, 401
Nanoparticles connection types, 12

coupling, 12
merging, 129
Nanoparticles interaction, 12
electrostatic forces, 12
Hall-Petch, 12
interaction force, 12
packing type, 123, 12
potential energy plot, 12
Van-der-Waals forces, 12
vibration, 125
Nanoplatelet, 173
Nanostructural elements, 109, 136
Nanostructure materials, 17
0D nanoparticles, 175
lasers, 175
light emitting diodes, 175
single-electron transistors, 175
solar cells, 17
1D nanoparticles, 176
hierarchical nanostructures, 176
nanobelts, 176
nanofibers, 176
nanoribbons, 176
nanorods, 176
nanotubes, 176
nanowires, 17
2D nanoparticles, 176
branched structures, 176
continuous islands, 176
nanodisks, 176
nanoplates, 176
nanoprisms, 176
nanosheets, 176
nanowalls, 17
3D nanoparticles, 176
nanoballs, 177
nanocoils, 177
nanocones, 177
nanoflowers, 177
nanopillers, 177
Nanostructures, 173
Nanotechnology, 94, 173
National and international safety laws, 392
National Institute of Laser Enhanced Sciences (NILES), 141
National Science Foundation (NSC), 182
Navier-Stokes fluid, 267

New Jersey University, 129
Newton's law, 220, 221, 223
Nitrile-butadiene rubber, 40, 45
Non-Faradaic process, 400
Nonlinear method, 35
Nottingham device test (NDT), 96
Nottingham experiment, 103
Nozzle regime, 215
Nyquist plots, 388

O

Ohnesorge number, 247
One-dimensional nanomaterials, 172
One-way ANOVA Turkey's Multiple Comparison Test, 144
Ordinary differential equations (ODE), 253
Ordinary Portland cement, 62
Organomontmorillonite, 42
Orientation mesophases switches, 91
Ostwald–de Waele power law, 231, 333

P

Paramagnetic properties, 165
Partial differential equations (PDE), 256
Penicillin, 142
Peptizers, 160
Peripheral blood lymphocytes, 148
Petrol, 45
Petroleum industries, 45
Petroleum, 16
Phenyl trimethoxysilane (PhTEOS), 88
Philips NED microscope, 28
Philips NED microscope, 76
Phosphate Buffer Saline (PBS), 142
Photo thermal therapy, 140, 152
Photoluminescent composites, 75
Physical methods, 177
Physicomechanical properties, 29
Phytohaemagglutinin, 142
Pipette tip, 187
Pitch, 385
Plasma based methods, 17
cold plasma, 17
hot plasma, 178
Plasma chemical deposition, 75
Plasma evaporator, 77

Plasmochemical method, 74
Plasmon resonance, 141
Platelet carbon nanofibers, 372
Platelet structure, 372
Plug-in hybrid electric vehicles, 398
Plumbum zirconate-titanate (PZT), 129
Poisson equation, 263
Poisson probability distribution, 202
Polar media, 165
Polarization-optical microscopy (POM), 88
Polarized light, 90
Poly(methyl methacrylate), 391
Polyacrylonitrile, 364
Polycrystal nucleus, 135
Polyethylene of low density (LDPE), 77
Polyethylene polyamine, 160
Polyethylene, 16, 26, 352
    films, 26
Polymer electrolyte membrane, 394
Polymer matrix, 26, 74
Polymer melting, 29
Polymer nanocomposites, 16, 37, 352
Polymer–clay nanocomposites (PCN), 94
Polymeric compositions solvents, 160
Polymeric matrix, 166
Polypropylene (PP), 40, 60
Polypropylene fiber, 65
Polyvinylidene chloride, 362
Porosity, 410
Porousfibrous structures, 191
Pressure vessels, 365
Pure crystal silicon, 80
Pyrolysis, 178–180, 333

Q

Quantum chemistry, 2
Quantum dot formation, 175, 178

R

Radio frequency, 75
Raman spectra, 386
Raman spectral probing, 140
Raman spectroscopy, 381, 391
Raw materials, 18
Rayon (viscose), 362
Redox reaction, 40
    see, Faradaic process
Refractometry, 168
Relaxation method, 251, 264
Relaxation process, 116, 120
Reneker model, 249
Representative volume element (RVE), 200
Representative volume, 263
Research Program of the Ural Branch, 135
Response Surface Methodology (RSM), 64
Responsive nanoparticle (RNP) applications, 192
Reynolds number, 246, 308
RF plasma method, 178
Rouse chain (Rouse-Zimm chain), 23
    see, Hookean springs
RSM plots, 64, 69
Rubber, 43
Runge–Kutta method, 116
Russian Foundation for Basic Research, 2

S

SBS polymer, 94
Scaling law, 290
Scanning electron microscope, 65, 94, 96, 391
Scientific and Pedagogical Shots of Innovative Russia, 84
Self-assembly, 172
Self-organizing process, 109
Self-replication, 172
SEM results, 10
    bitumen, 10
    Iran University of Science and Technology, 10
    mechanical mixing method, 10
    thin layer laboratory, 103
Semi-inverse method, 251, 267
Semiconductor technology, 173
Semiconductors, 177
Semicrystalline polymer, 354, 355, 358
Shapes, 11
    asymmetric nanoparticles, 11
    globe-like, 11
    spherical centered, 11
    spherical eccentric, 11
    spherical icosahedral nanoparticles, 119
Shear-lag model, 201

Shimadzu Lab XRD-6000 diffractometer, 28, 76
Shimifaraiand Company, 62
Silica fume, 70
Silica gel, 360
Silicate matrix, 88, 89, 92
Silicon nanoclusters, 27, 75
Silicon oxide/oxynitride shell 76
Silicon powder, 75
Silver nanoparticles, 141, 14
    catheters, 14
    wound dressings, 149
Single strand breaks (SSB), 147
Single wall carbon nanotubes, 370
Single-electron transistors, 175
Slattery, 276
Sleeve film, 77
Slender body, 248
Small electrical relaxation time limit, 309
Small platforms, 130
Sodium p- sulfophenyl methallyl ether, 365
Sodium p-styrene sulfonate, 365
Sol temperature, 89
Sol-gel synthesis, 89
Sol-gel technology, 89
Solar cells, 175
Solid electrolyte interphase, 318
Solid matrices, 7
    α-SiO2, 75
Solid nanoparticles, 74
Solid phase method, 17
    milling, 17
    solid-state reaction, 17
    spark discharge, 17
    stripping, 17
    thermal decomposition, 178
Solid propellant rocket motors, 365
Sorbents, 56
Space applications, 362
Spatial distribution, 130
Spatial orientation, 115
Spectra, 29
Spectrophotometric measurement, 142
Spectroscopy, 37
Spin-electronics, 174
Spintronics, 174
Spivak and Dzenis model, 248

Splaying, 216
Splitting, 216
Stability theory, 249
Static creep tests, 100
Statistical methods, 64
Steam drawing, 362
Strain, 94
Streptomycin, 142
Stress-strain relation, 231
Styrene-co-acrylonitrile, 316
Sun-protection films, 27, 75
Super molecules, 109, 156, 164, 168
Super small quantities, 156, 158, 160, 168
Surface tension, 189
Surface-to-volume ratio, 174
Systematic study, 107, 136

T

Taylor cone, 185, 292
TEM micrographs, 144
Tennis rackets, 365
Tetrahydrofuran, 385
The Ural Branch of the Russian Academy of Sciences, 118
Theoretical Biophysics Group, 118
Thermo set, 40
    epoxy, 408
Thermobaric pressing, 18
Thermolysis and pyrolysis, 179
Thermoplastic elastomers, 40
    butadiene–styrene diblock copolymer, 408
Thermoplastic, 40, 40
    nitrile-butadiene rubber, 4
    nylon, 40
    polycarbonate, 40
    polypropylene, 4
    polypropylene, 408
2D Timoshenko beam-network, 202, 205
Thin film oven test (TFOT), 96
Thinning jet, 21
    jet stability, 214
Time derivative, 23
    Oldroyd derivative, 232
Toluene, 90
Topeka & Binder percentage, 96
Topeka and Binder gradation, 104

Translational motion, 115
Transmission Electron Microscope, 28, 76,
    141, 381
    see, Philips NED microscope
Transmittance spectra, 79
Transport coefficients,
    conductivity coefficient,
    diffusion coefficient, 3
Trial mixtures, 62
Trypan blue exclusion test, 147
Trypan blue stain, 142
Tubular structure, 372
Tumor cell, 151
Tyrene–butadiene–styrene block copolymer
    (SBS), 94

U

Ultrasensitive bio detection, 141
Ultrasonic machining, 159
Uniform particles arrays, 17
    see, quantum dots
University of Illinois, 118
Up-down process, 108
UV protectors, 28, 7
    environmental safety, 75
UV radiation, 32, 74
UV-protective film, 37, 84

V

Valence orbitals,
    p-type, 3
Van der Waals contacts, 112
Van-der-Waals forces, 125
Vaporization method, 65
Verlet scheme, 116
Vinyl acetate, 365
Vinyl trimethoxysilane (VMOS), 88
Viscoelastic elements, 172
Viscosity, 188, 247
Volatile organic compounds, 402
Volgograd State University, 2

Voltage, 189
Vulcanization, 46

W

Wan-Guo-Pan model, 250
Wastewater treatment, 368
Water absorption, 60, 63
Water penetration, 69
Water purification, 368
Water/cement percentage (W/C), 62
Weber number, 246, 312
Weibull statistical distribution, 375
Whipping instability, 216
Whipping jet, 216, 217, 220, 333
    jet instability, 216
WinX^{POW} program, 30, 36, 78, 83,

X

X-ray analysis, 74
X-ray diffraction analysis, 30, 35, 37, 78,
    83,
X-ray diffraction patterns, 74
X-ray diffractometry, 391
X-ray photoelectron spectroscopy, 156, 168
X-ray powder diffraction analysis, 7
    see, Shimadzu Lab XRD-6000 diffrac-
    tometer
X-ray, 2
    analysis, 2
    patterns, 2
    powder diffraction, 2
    scattering diffractograms, 32

Y

Young's nanotube modulus, 201, 364, 409

Z

Zigzag nanotubes, 2–9, 12
    see, carbon nanotubes

Milton Keynes UK
Ingram Content Group UK Ltd.
UKHW031139141024
449569UK00024B/1224